ENZYMOLOGY

LABFAX

The LABFAX series

Series Editors:

B.D. HAMES Department of Biochemistry and Molecular Biology, University of Leeds, Leeds LS2 9JT, UK

D. RICKWOOD Department of Biology, University of Essex, Wivenhoe Park, Colchester CO4 3SQ, UK

MOLECULAR BIOLOGY LABFAX

CELL BIOLOGY LABFAX

CELL CULTURE LABFAX

BIOCHEMISTRY LABFAX

VIROLOGY LABFAX

PLANT MOLECULAR BIOLOGY LABFAX

IMMUNOCHEMISTRY LABFAX

CELLULAR IMMUNOLOGY LABFAX

ENZYMOLOGY LABFAX

PROTEINS LABFAX

Forthcoming titles

NEUROSCIENCE LABFAX

MOLECULAR BIOLOGY: GENE CLONING LABFAX

MOLECULAR BIOLOGY: GENE ANALYSIS LABFAX

ENZYMOLOGY

EDITED BY
P.C. ENGEL
Department of Biochemistry,
University College Dublin,
Belfield, Dublin 4,
Eire

βIOS
SCIENTIFIC
PUBLISHERS

ACADEMIC PRESS

©BIOS Scientific Publishers Limited, 1996

Published jointly by Academic Press, Inc., 525 B Street, San Diego, CA 92101-4495, USA, and BIOS Scientific Publishers Limited, 9 Newtec Place, Magdalen Road, Oxford OX4 1RE, UK.

A CIP catalogue record for this book is available from the British Library.

ISBN 0 12 238840 2

Distributed exclusively throughout the world by Academic Press, Inc., 525 B Street, San Diego, CA 92101-4495, USA pursuant to agreement with BIOS Scientific Publishers Limited.

Typeset by Marksbury Typesetting Ltd, Midsomer Norton, Bath, UK.
Printed by Biddles Ltd, Guildford, UK.

CONTENTS

CONTENTS   **vii**

CONTENTS   **ix**

CONTENTS   **xi**

8. EPR SPECTROSCOPY IN ENZYMOLOGY   249

9. SOLUTIONS USED IN ENZYMOLOGY   269

CONTRIBUTORS

A. BERRY

Department of Biochemistry and Molecular Biology, University of Leeds, Leeds LS2 9JT, UK

K. BROCKLEHURST

Laboratory of Structural and Mechanistic Enzymology, Department of Biochemistry, Queen Mary and Westfield College, University of London, Mile End Road, London E1 4NS, UK

R. CAMMACK

Centre for the Study of Metals in Biology and Medicine, Kings College, University of London, Campden Hill Road, Kensington, London W8 7AH, UK

A.R. CLARKE

Department of Biochemistry, School of Medical Sciences, University of Bristol, University Walk, Bristol BS8 1TD, UK

A. CORNISH-BOWDEN

Laboratoire de Chimie Bactérienne, Centre National de la Recherche Scientifique, 31 Chemin Joseph-Aiguier, B.P. 71, 13402 Marseille Cedex 20, France

F.M. DICKINSON

Department of Applied Biology, University of Hull, Cottingham Road, Hull HU6 7RX, UK

P.C. ENGEL

Department of Biochemistry, University College Dublin, Belfield, Dublin 4, Eire

S.E. HARDING

Department of Applied Biochemistry and Food Science, University of Nottingham, Sutton Bonington LE12 5RD, UK

C.R. LOWE

Institute of Biotechnology, University of Cambridge, Tennis Court Road, Cambridge CB2 1QT, UK

D. PATEL

Department of Biology, University of Essex, Wivenhoe Park, Colchester CO4 3SQ, UK

N.C. PRICE

Department of Biological and Molecular Sciences, University of Stirling, Stirling FK9 4LA, UK

D. RICKWOOD

Department of Biology, University of Essex, Wivenhoe Park, Colchester CO4 3SQ, UK

A.J. ROWE

National Centre for Macromolecular Hydrodynamics, Department of Biochemistry, University of Leicester, Leicester LE1 7RH, UK

J.K. SHERGILL

Centre for the Study of Metals in Biology and Medicine, Kings College, University of London, Campden Hill Road, Kensington, London W8 7AH, UK

L. STEVENS

Department of Biological and Molecular Sciences, University of Stirling, Stirling FK9 4LA, UK

J.A. THOMAS

Institute of Biotechnology, University of Cambridge, Tennis Court Road, Cambridge CB2 1QT, UK

K.F. TIPTON

Department of Biochemistry, Trinity College, Dublin 2, Eire

A. WALMSLEY

Department of Molecular Biology and Biotechnology, University of Sheffield, PO Box 594, Firth Court, Western Bank, Sheffield S3 7HF, UK

ABBREVIATIONS

ADP	adenosine diphosphate
AMP	adenosine monophosphate
APS	ammonium persulfate
ATP	adenosine triphosphate
AUC	analytical ultracentrifuge
BAPNA	benzoyl arginine p-nitroanilide
BCA	bicinchoninic acid
BPTI	bovine pancreatic trypsin inhibitor
BSA	bovine serum albumin
CAT	chloramphenicol acetyltransferase
CDP	cytidine diphosphate
CE	capillary electrophoresis
CL	cross-linked
CMC	critical micellar concentration
CMP	cytidine monophosphate
CoA	coenzyme A
CP	ceruloplasmin
CTP	cytidine triphosphate
DMPO	5, 5'-dimethyl-1-pyrroline-N-oxide
DMSO	dimethylsulfoxide
DNA	deoxyribonucleic acid
d.p.m.	disintegrations per minute
DTNB	5, 5'-dithiobis(2-nitrobenzoic acid)
DTT	dithiothreitol
EDTA	ethylenediaminetetraacetic acid
EGTA	ethylene glycol-bis(β-aminoethyl ether) N, N, N', N'-tetraacetic acid
ELISA	enzyme-linked immunosorbent assay
EMR	electronic magnetic resonance
ENDOR	electron–nuclear double resonance
EPR	electron paramagnetic resonance
ESEEM	electron spin-echo envelope modulation
ESMS	electrospray ionization mass spectrometry
ESR	electron spin resonance
ETF	electron transfering flavoprotein
EtOH	ethanol
FAD	flavin adenine dinucleotide
FCS	fetal calf serum
FDH	formate dehydrogenase
FMN	flavin mononucleotide
FPLC	fast protein liquid chromatography
FSCE	free-solution capillary electrophoresis
GdnHCl	guanidine hydrochloride
GTP	guanosine triphosphate
3HF	3-hydroxyfluoran
HLB	hydrophile–lipophile balance
HMG	high-mobility group proteins

HPLC	high performance liquid chromatography
HSA	human serum albumin
IEF	isoelectric focusing
IMP	inosine monophosphate
IPG	immobilized pH gradients
IR	infra-red
LC	liquid chromatography
LDH	lactate dehydrogenase
LD-TOF	laser desorption-time of flight
LGT	low gelling temperature
MALDMS	matrix-assisted laser desorption mass spectrometry
MAO	monoamine oxidase
MCR	methyl coenzyme-M reductase
MeOH	methanol
MNP	2-methyl-2-nitrosopropane
MS	mass spectrometry
MWC	Monod–Wyman–Changeux
NAD(H)	nicotinamide adenine dinucleotide (reduced)
NADP(H)	nicotinamide adenine dinucleotide phosphate (reduced)
NCS	newborn calf serum
NEPHGE	nonequilibrium pH gradient electrophoresis
NMR	nuclear magnetic resonance
NP-40	Nonidet P-40
NPGB	4-nitrophenyl 4′-guanidinobenzoate
OPA	o-phthalaldehyde
PAGE	polyacrylamide gel electrophoresis
PBN	N-tert-butyl-α-phenylnitrone
PCA	perchloric acid
PDMS	plasma-desorption mass spectrometry
Pi	inorganic phosphate
PPi	inorganic pyrophosphate
PPO	2,5-diphenyloxazole
PVDF	polyvinyldifluoride
RES	resonance energy transfer
RF	radiofrequency
RNA	ribonucleic acid
r.p.m.	revolutions per minute
SDS	sodium dodecyl sulfate
SR	sarcoplasmic reticulum
TCA	trichloroacetic acid
TEMED	N, N, N', N'-tetramethylethylenediamine
TOPA	6-hydroxydopa quinone
t-PA	tissue plasminogen activator
TPP	thiamine pyrophosphate
Tris	tris(hydroxymethyl)amino methane
tRNA	transfer RNA
TTP	thymidine triphosphate
UV	ultraviolet
v/v	volume/volume
w/v	weight/volume
X-gal	5-bromo-4-chloro-3-indoyl-β-D-galactopyranoside
XO	xanthine oxidase

PREFACE

When the project of putting together an *Enzymology Labfax* volume was first broached, my initial concern in defining its scope was the overlap with the existing *Biochemistry Labfax*, which of course covers enzymology. In general, the present volume covers its narrower area in much greater depth, but a few sections have merely been updated from the earlier book. When this book was already underway, concerns about potential new areas of overlap arose with the proposal for a separate volume on proteins. Nick Price (the editor of *Proteins Labfax*) and I have endeavored to minimize overlap. Nevertheless each volume needs to be able to stand alone; hence some sections initially destined for *Enzymology Labfax* now appear in *Proteins Labfax* and a small amount of material appears in both.

In assembling my contributors I looked for a relatively large number of experts, giving each a correspondingly circumscribed brief. In putting their contributions together we have accepted the distinctive flavor of each. Industrial contributors have their own preferences with regard to terminology, units, symbols, etc. The discerning reader may accordingly spot 'inconsistencies' in style and presentation that would certainly not be there if the whole were the work of one author. I have taken the view that, so long as everything is clearly defined, this variety does not matter. Indeed for students it is perhaps as well to realize that a variety of symbols can be used to describe the same thing! There are also some areas of slight overlap which have been left in order to leave the completeness of each section undisturbed.

There are some missing topics that I had originally hoped to include but which do not appear owing to either lack of space or a shortage of willing contributors able to tackle the subject. It would certainly be helpful if readers could let me know of any particularly glaring or irritating gaps that might be plugged in any future revised edition.

I wish to express my thanks to all my contributors and to David Rickwood for his very substantial help. My colleague Dr Suren Aghajanian gave invaluable help with his diagrams. Thanks are also due to Mrs Mena Lincoln who patiently typed the successive versions of my own contributions and my wife Sue for patient help with the index.

I hope that the finished volume will prove useful.

Paul C. Engel

CHAPTER 1
CLASSIFICATION AND MEASUREMENT OF ENZYME ACTIVITY

A. Cornish-Bowden

A. CLASSIFICATION OF ENZYMES

1. ORGANIZATION

New enzymes are classified by the Nomenclature Committee of IUBMB and published in successive editions of *Enzyme Nomenclature* (most recently in 1992 (1)) with addenda published periodically in the *European Journal of Biochemistry* (most recently in 1995 (2)). Suggestions for new entries or corrections to existing ones should be sent to Professor K.F. Tipton, Biochemistry Department, Trinity College, Dublin 2, Ireland.

2. CHARACTERISTICS

Enzyme Nomenclature classifies catalytic activities, not protein structures. Thus the name superoxide dismutase refers to at least two entirely different protein structures that happen to catalyse the same reaction; conversely, the same protein can have two entirely different names, such as homoserine dehydrogenase and aspartate kinase if, as is the case in *Escherichia coli*, the two activities are found in the same protein molecule.

Although many enzymes are named according to the direction of reaction considered more important physiologically, this is not always appropriate, either because both directions are physiologically important (e.g. alcohol dehydrogenase is named as a detoxification enzyme, but the reaction proceeds in the opposite direction in fermentation), or because naming in the physiological direction would obscure the relationship to other enzymes in the same class (e.g. pyruvate kinase).

If reclassification results in deletion of an entry, the number is never reassigned to another entry. For example, the present list contains no entry for EC 2.7.1.9, and no future list will contain this entry unless acetylaminodeoxyglucose kinase is reinstated.

A typical entry in *Enzyme Nomenclature* contains information under seven headings, as illustrated in *Table 1* for phosphofructokinase.

3. SYSTEM OF CLASSIFICATION

Enzymes are placed in six classes, as listed in *Table 2*, with numerous sub-classes and sub-sub-classes, of which only the most important are listed in *Table 2*. The sub-class and sub-sub-class normally define the nature of the reaction more precisely, or specify a particular set of donors or acceptors. Numbers in the range 99, 98, 97, etc., are used for miscellaneous or 'other' entries: sub-class 1.97 is for 'other oxidoreductases'; sub-sub-class 1.1.99 is for oxidoreductases acting on CH–OH groups with 'other acceptors'. These classifications should be regarded as provisional and are likely to be changed in future editions of *Enzyme Nomenclature* (e.g. sub-class 1.99 has been replaced in the current edition by sub-classes 1.13 and 1.14).

Table 1. Presentation of an entry in *Enzyme Nomenclature* (1)

Rubric	Entry	Explanation or comments
EC number	2.7.1.11	Class 2: transferases; sub-class 2.7, transferring phosphorus-containing groups; sub-sub-class 2.7.1, with an alcohol group as acceptor; enzyme 2.7.1.11, unique identifier for 6-phosphofructokinase
Recommended name	6-Phosphofructokinase	The recommended name is often slightly different from the name commonly used; the more explicit name is necessary in contexts where 6-phosphofructokinase must be distinguished from 1-phosphofructokinase (EC 2.7.1.56), but otherwise 'phosphofructokinase' is acceptable
Reaction	ATP + D-fructose 6-phosphate = ADP + D-fructose 1,6-bisphosphate	This identifies as precisely as possible the catalytic activity to which the entry refers
Other name(s)	Phosphohexokinase; phosphofructokinase 1	Alternative (often older) names that are no longer given in the list, but the more important ones appear in the index followed by (Obs.) or (Mis.)
Systematic name	ATP-D-fructose-6-phosphate 1-phosphotransferase	This type of name is too cumbersome for everyday use, but defines the reaction much more precisely than does the recommended name
Comments	D-Tagatose 6-phosphate and sedoheptulose 7-phosphate can act as acceptors. UTP, CTP and ITP can act as donors. Not identical to EC 2.7.1.105	Additional information useful to the reader. Frequently indicates the specificity more fully than the systematic name can do, or indicates any cofactors that the enzyme may contain
References	212, 3075, 3654, 3781, 3963, 4643, 5194	References to additional sources of information. For some enzymes these references are very old or out-of-date: the compilers of *Enzyme Nomenclature* welcome information about such cases, especially if references to current reviews exist

Table 2. Classification system used in *Enzyme Nomenclature* (1)

Class	Reaction type	Style of systematic name (example)	Principal sub-classes
1. Oxidoreductases[a]	Oxidation/reduction	Donor:acceptor oxidoreductase (EC 1.1.1.1, alcohol:NAD$^+$ oxidoreductase)	1.1 Acting on CH–OH group 1.2 Acting on aldehyde or oxo group 1.3 Acting on CH–CH group 1.4 Acting on CH–NH$_2$ group 1.n.1 NAD(P) as acceptor 1.n.2 Cytochrome as acceptor 1.n.3 Oxygen as acceptor
2. Transferases	Group-transfer reactions[b]	Donor:acceptor group-transferase (EC 2.1.3.2, carbamoyl-phosphate: L-aspartate carbamoyltransferase)	2.1 Transferring one-carbon groups 2.3 Acyltransferases 2.4 Glycosyltransferases 2.6.1 Transaminases 2.7 Transferring phosphorus-containing groups 2.7.1 Alcohol group as acceptor
3. Hydrolases	Hydrolysis reactions[c]	Substrate hydrolase[d] (EC 3.1.1.7, acetylcholine acetylhydrolase)	3.1 Esterases 3.2 Glycosidases 3.4 Peptidases
4. Lyases	Elimination reactions	Substrate group-lyase[e] (EC 4.1.1.1, 2-oxo-acid carboxy-lyase)	4.1 Carbon–carbon lyases 4.2 Carbon–oxygen lyases 4.3 Carbon–nitrogen lyases
5. Isomerases	Isomerization reactions	Substrate reactionase[f] (EC 5.3.1.1, D-glyceraldehyde-3-phosphate ketol-isomerase)	5.1 Racemases and epimerases 5.3 Intramolecular oxidoreductases 5.4 Intramolecular transferases (mutases)
6. Ligases[g]	Bond synthesis coupled to hydrolysis (e.g. of ATP)	X–Y ligase (product-forming) (EC 6.1.1.1, tyrosine-tRNA^{Tyr} ligase (AMP-forming))	6.1 Forming carbon–oxygen bonds 6.2 Forming carbon–sulfur bonds 6.3 Forming carbon–nitrogen bonds

[a]Recommended names contain the word 'dehydrogenase' when possible; 'oxidase' is used only when O_2 is the acceptor.
[b]Other than enzymes in classes 1 and 3, or intramolecular group-transfer reactions (sub-class 5.4).

Table 2. Continued

^cOther than hydrolysis reactions coupled to bond-forming reactions (class 6).

^dMany recommended names are formed by adding the suffix '-ase' to the name of the substrate. For proteolytic enzymes (sub-class 3.4) many trivial names survive as recommended names (pepsin, trypsin, etc.).

^eThe hyphen is an essential part of the name: e.g. hydroxy-lyase should not be written hydroxylyase, with risk of confusion with hydroxylase. Recommended names often include the words 'decarboxylase', 'aldolase' or 'dehydratase' if the elimination reaction is the physiologically important one, or 'synthase' if the reverse reaction is to be emphasized.

^fWhere 'reactionase' stands for racemase, epimerase, mutase, etc., as appropriate.

^gIn earlier editions of *Enzyme Nomenclature* ligases were called synthetases and recommended names were constructed in the style substrate synthetase (e.g. glutamine synthetase) rather than X–Y ligase (glutamate–ammonia ligase). The distinction between synthetases (class 6 only) and synthases (used in classes 1–5 for enzymes that it was desirable to name according to their physiological products) is now no longer used in the list as it was found confusing in practice. However, enzymes in class 6 may still be named as synthetases by authors who prefer this form, and any enzyme may be named as a synthase when appropriate.

4 ENZYMOLOGY LABFAX

4. ISOENZYMES

Enzyme Nomenclature does not at present deal with isoenzymes in any systematic way. In most cases their existence is either ignored or mentioned only under 'Comments'. In a few cases, an isoenzyme has sometimes been considered a separate enzyme and misclassified under a name and EC number that properly belongs to a different enzyme. For example, liver hexokinase is often called glucokinase, and *Enzyme Nomenclature* recognizes this usage; however, it should not be classified under EC 2.7.1.2, which belongs to true glucokinases.

5. PROBLEM AREAS

Certain enzymes acting on polymeric substrates have proved difficult to name and classify according to the principles generally adopted in *Enzyme Nomenclature*. Enzymes with low specificity, most notably the proteolytic enzymes in class 3.4, have been studied for many years, and many are still known by trivial names, such as trypsin, that convey no information. Despite continuing efforts to arrange these enzymes rationally, it is likely that their lack of specificity will continue to defeat these efforts. At the other extreme the site-specific deoxyribonucleosidases (restriction nucleases) and other enzymes that act on nucleic acids are highly specific, recognizing particular base sequences, but are used more as tools than studied as enzymes, and experts consulted by the Nomenclature Committee of IUBMB considered that classifying them in the same way as other enzymes would serve no useful purpose.

6. CLASSIFICATION OF COMMON ENZYMES

Table 3 lists the EC numbers and recommended names of some well-known enzymes.

Table 3. Common enzymes

Acetylcholinesterase, 3.1.1.8
Adenylate kinase, 2.7.4.3
Alcohol dehydrogenase, 1.1.1.1
Aldehyde dehydrogenase, 1.2.1.3
Alkaline phosphatase, 3.1.3.1
Amino acid–tRNA ligase,
  6.1.1-22
Aminoacyl-tRNA synthetase: see
  amino acid–tRNA ligase
α-Amylase, 3.2.1.1
β-Amylase, 3.2.1.2
Aspartate carbamoyltransferase,
  2.1.3.2
Aspartate transaminase, 2.6.1.1
Carboxypeptidase A, 3.4.17.1
Catalase, 1.11.1.6
Chymotrypsin, 3.4.21.1
β-Fructofuranosidase, 3.2.1.26
Fructose-bisphosphatase, 3.1.3.11
Fructose-bisphosphate aldolase,
  4.1.2.13
Fumarase: see fumarate hydratase
Fumarate hydratase, 4.2.1.2
β-Galactosidase, 3.2.1.23

Glucose oxidase, 1.1.3.4
Glucose 6-phosphate
  dehydrogenase, 1.1.1.49
Glutamate–ammonia ligase,
  6.3.1.2
Glutamate dehydrogenase, 1.4.1.2
Glutamine synthetase: see
  glutamate–ammonia ligase
Glyceraldehyde 3-phosphate
  dehydrogenase (phosphorylating),
  1.2.1.12
Hexokinase, 2.7.1.1
Homoserine dehydrogenase,
  1.1.1.3
Invertase: see β-fructofuranosidase
Isocitrate dehydrogenase (NAD^+),
  1.1.1.41
D-Lactate dehydrogenase, 1.1.1.28
L-Lactate dehydrogenase, 1.1.1.27
Lysozyme, 3.2.1.17
Malate dehydrogenase, 1.1.1.37
Nitrogenase, 1.18.6.1
Pancreatic ribonuclease, 3.1.27.5
Papain, 3.4.22.2

Pepsin A, 3.4.23.1
Peroxidase, 1.11.1.7
Phosphofructokinase, 2.7.1.11
Phosphoglucomutase, 5.4.2.2
Phosphorylase, 2.4.1.1
Pyruvate carboxylase, 6.4.1.1
Pyruvate decarboxylase: see
  pyruvate dehydrogenase
  (lipoamide)
Pyruvate dehydrogenase, 1.2.1.51
Pyruvate dehydrogenase
  (lipoamide), 1.2.4.1
Pyruvate kinase, 2.7.1.40
Subtilisin, 3.4.21.62
Succinate dehydrogenase, 1.3.5.1
Superoxide dismutase, 1.15.1.1
Thrombin, 3.4.21.5
Triose-phosphate isomerase,
  5.3.1.1
Trypsin, 3.4.21.4
Tryptophan synthase, 4.2.1.20
Urease, 3.5.1.5
Xanthine oxidase, 1.1.3.22

B. UNITS OF ENZYME ACTIVITY

Regardless of kinetic details, most steady-state rate equations for enzymes can be written in the form $v = e_0 f(a, i, \ldots)$, where e_0 is the total enzyme concentration and f is a function of the concentrations of substrates, inhibitors, pH, temperature and all other conditions *except* the enzyme concentration. If this is true (there are exceptions for enzymes that associate or dissociate in the concentration range of the measurements), measurement of v provides a way of measuring e_0: the value of $f(a, i, \ldots)$ does not have to be known, but it does have to be fixed, which means that all of the conditions that affect it must be standardized.

Such measurements give the value of e_0 in *relative* units, not molarity (unless the conversion factor is independently known). Usually v is measured in units of concentration per unit time (e.g. mM min^{-1}), equivalent to amount of substance per unit volume per unit time (mmol l^{-1} min^{-1}). However, the volume is, in principle, irrelevant, because it cancels when one determines the amount of enzyme (i.e. if $v = e_0 f(a, i, \ldots)$ is valid then multiplying v by the volume gives the same amount of enzyme regardless of the assay volume). Thus assays are carried out in whatever volume is convenient, but for expressing the amount of enzyme activity the rate should be converted to units of amount of substance per unit time (i.e. it should be converted to a *rate of conversion* (3)).

The traditional 'enzyme unit' was normally defined as $1 \text{ U} = 1$ μmol min^{-1}, and this unit is still widely used. The definition should always be given explicitly, because other definitions have also been used. The SI (Système International) unit of amount of substance per unit time is 1 mol sec^{-1}, and in 1972 the IUPAC–IUB Commission on Biochemical Nomenclature (4) suggested the name katal (symbol kat) for the SI unit of enzyme activity (i.e. 1 kat = 1 mol sec^{-1}). This name and symbol have not been widely adopted, however, and the current recommendations of the Nomenclature Committee of IUBMB (5) no longer actively promote them, though they remain valid for biochemists who wish to use them.

Regardless of what unit is used for enzyme activity, the exact conditions of the assay must always be specified, as the theory above assumes that $f(a, i, \ldots)$ is fixed.

7. REFERENCES

1. Nomenclature Committee of IUBMB (1992) *Enzyme Nomenclature*. Academic Press, San Diego.

2. Nomenclature Committee of IUBMB (1995) *Eur. J. Biochem.*, **232**, 1.

3. Nomenclature Committee of IUB (1982) *Eur. J. Biochem.*, **128**, 281; *Arch. Biochem. Biophys.*, **224**, 732; *Biochem. J.*, **213**, 561.

4. IUPAC–IUB Commission on Biochemical Nomenclature (1973) *Enzyme Nomenclature*. Elsevier, Amsterdam, p. 26.

5. International Union of Biochemistry (1992) *Biochemical Nomenclature and Related Documents* (2nd edn). Portland Press, London.

CHAPTER 2
PURIFICATION AND ANALYSIS OF ENZYME PREPARATIONS

A. DYE–LIGAND CHROMATOGRAPHY IN ENZYME PURIFICATION

C.R. Lowe and J.A. Thomas

1. INTRODUCTION

Affinity chromatography is a highly efficient technique which separates biological molecules based on the specific interaction of a covalently bound ligand with its complementary biological counterpart. The types of immobilized ligand used can be separated into two classes: biospecific ligands such as receptors, enzyme substrates, antibodies, nucleotides and coenzymes and pseudo-biospecific ligands such as dyes, hydrophobic compounds and metal ions. From this list, dyes are rapidly becoming the most frequently tested for use as affinity ligands in the purification of a wide range of enzymes and other proteins.

There are several advantages to using dyes as pseudo-biospecific ligands compared to natural biological ligands, including their stability to chemical and enzymatic degradation, ease of immobilization, relatively low cost, availability of a large number of chromophores with a wide range of specificities, ease of chemical modification to alter specificity and the generation of adsorbents with very high protein-binding capacities. Enzymes react differently with each immobilized dye and optimization of a purification protocol therefore requires screening a range of media and eluents. Screening immobilized dye matrices for affinity often enables purification protocols to be developed with very little prior knowledge of the structural or physical properties of the target enzyme.

2. GENERAL USE OF IMMOBILIZED DYE MATRICES

Many of the commercially available dyes have undisclosed structures and this, together with a lack of detailed information regarding most enzymes, precludes a truly rational approach to the design of a dye-based affinity purification protocol. Therefore, the development of a purification scheme may require a screen of a number of different immobilized dyes.

2.1. Protein adsorption
The enzyme to be purified first needs to be adsorbed to the ligand immobilized on the matrix.

A screen can be established to find a suitable affinity matrix for purification using conditions that maintain the integrity of the enzyme to be purified. The immobilized dye matrix is pre-equilibrated by washing with 10 column volumes of an appropriate buffer. Up to 10 mg of the crude protein solution (dialysed and filtered) is incubated with a library of immobilized dye matrices (0.5 ml settled volume), at a range of pH values, each in a small tube. Alternatively, the protein solution can be applied to a series of small columns (for example, using Pasteur pipettes plugged with glass wool) containing 0.5 ml settled-bed volume of the immobilized dye matrix. Ready prepared screening kits are also currently available for screening a range of dyes (Affinity Chromatography Ltd, Amicon and Sigma). A matrix displaying affinity for the enzyme will bind and remove it from solution. Where several dyes are able to bind the enzyme, the appropriate choice would be the matrix which binds the least contaminants. If no binding is observed, then the ionic strength and pH of the sample should be reduced or metal ions added.

The mixed ionic/hydrophobic character of many dye molecules facilitates further optimization of an enzyme purification by altering one or more of several variables, including the pH, ionic strength, flow rate, temperature, and presence or absence of metal ions, ethylenediamine-tetraacetic acid (EDTA), coenzymes and redox reagents. The screens above can be used to optimize the conditions for enzyme binding by varying any of these parameters in a systematic fashion. *Table 1* summarizes common adsorption conditions.

2.2. Elution of the target enzyme

Once the contaminating solutes are washed away, the enzyme can be eluted nonspecifically, either by a change in pH (1), usually by increasing pH, or by increasing the salt concentration stepwise to 1–1.5 M NaCl or by using NaSCN (2). The purification may be improved by eluting the enzyme selectively with a competing substrate (3) or nucleotide (4–6); however, this approach is often not favored, partly owing to cost and partly because it may require the subsequent removal of the eluting ligand. These problems can be outweighed, however, by achieving spectacular purifications; for example, the purification of bovine lens aldose reductase resulted in a 12 600-fold purification on a Procion red H-3B agarose column (7). *Table 2* summarizes elution conditions.

Buffers used for enzyme purification usually have concentrations between 5 and 100 mM and pH values between 5 and 9. Operating temperatures can affect the affinity by increasing or decreasing binding, depending on the interaction; temperatures have been used between 0 and 50°C. Room temperature is usually used, although, if the enzyme is labile, a temperature of 4°C will often help to stabilize it.

3. DYE COLUMN REGENERATION AND STORAGE

Dye–ligand columns can be used hundreds of times provided they are regularly regenerated to remove any accumulation of materials such as denatured proteins, lipids and lipoproteins.

Table 1. Common conditions used for enzyme adsorption to immobilized dye affinity matrices

Buffer	Concentration: 5–150 mM
	Acetate, pK_a 4.76; citrate, pK_a 4.76 and 6.4; Hepes, pK_a 7.48;
	    Mes, pK_a 6.10; Mops, pK_a 7.20; phosphate, pK_a 2.15 and 7.2;
	    succinate, pK_a 4.21 and 5.64; tricine, pK_a 8.05; triethanolamine,
	    pK_a 7.76; Tris, pK_a 8.06

pH	5–9
Temperature	4–50°C
Lincar flow rate	10–30 cm h^{-1}
Additives	Salts to increase hydrophobic interactions: $MgCl_2$, $MgSO_4$, $Mg(CH_3COO)_2$, KCl, NaCl, NH_4Cl, $(NH_4)_2SO_4$ Reducing agents: dithiothreitol, β-mercaptoethanol, L-cysteine Detergents
Buffer polarity	10–50% (v/v) sucrose, glycerol or ethylene glycol
Metal ions	Na^+, K^+, Mg^{2+}, Ca^{2+}, Sr^{2+}, Ba^{2+}, Zn^{2+}, Cu^{2+}, Co^{2+}, Ni^{2+}, Mn^{2+}, Fe^{2+}, Al^{3+}, Fe^{3+}, Cr^{3+}

Table 2. Common parameters used for elution of enzymes from the dye-immobilized matrix

Salts	0.025–6 M, NaCl, KCl, CaCl$_2$, NH$_4$Cl, (NH$_4$)$_2$SO$_4$, gradient not usually used
Chelating agents	EDTA
Detergents	0.1% (w/v) SDS, Triton X-100
pH	Usually increased, pH gradient can improve resolution
Substrates	
Inhibitors	
Specific ligands	
Polyols	To reduce buffer polarity, add 10–50% (v/v) glycerol or ethylene glycol. This method has limited use
Chaotropes	Urea, NaSCN, KSCN. These act by weakening hydrogen bonds and hence denature labile enzymes and should be used with care

Regeneration of columns is usually achieved by washing in 1 M NaOH, but other regimes use solutions such as 8 M urea in 0.5 M NaOH (this causes reversible color changes in some dyes) or guanidine-HCl, detergents (1% (w/v) sodium dodecylsulfate or Triton X-100), ethylene glycol (<50%(v/v)) or thiocyanate (1–4 M) (8, 9). If there is lipid fouling, a chloroform/methanol solution can be used.

Immobilized dye columns may be stored wet or moist at 4°C, or at ambient temperature in the presence of a suitable bacteriocide.

4. STRUCTURE AND CHEMISTRY OF DYES

The most common dyes used for affinity chromatography are based around a 1,3,5 trichloro-*s*-triazine ring (*Figure 1*). These dyes originate from the 1950s when they were developed for the textile industry by ICI (the 'Procion' range) and Ciba-Geigy (the 'Cibacron' range). Each chlorine is susceptible to nucleophilic displacement with alkyl or arylamines or hydroxyls (10). As each chlorine is displaced, the remaining chlorines become subsequently less reactive. Dyes are formed when a chromophoric base is used as a substituent, while the second substituent can be any moiety which may confer further affinity, specificity or solubility, and the third chlorine is left unreacted ready for subsequent immobilization to an activated support. Examples of chromophores which have been used include anthraquinone moieties giving blue dyes, phthalocyanine moieties to give green dyes, or azo-containing chromophores to give red, orange or yellow dyes. Metal complexes of the azo dyes provide further variations of color. A selection of dye structures is shown in *Figure 1*. Many dyes are commercially available for subsequent immobilization (*Table 3*).

Recent advances have shown how dye structures can be manipulated to improve the affinity of the dye for a particular protein (15). Analogs of the terminal ring of Cibacron Blue F3G-A were made for the purification of alkaline phosphatase (16). A range of modifications were made to Cibacron Blue F3G-A for the purification of horse liver alcohol dehydrogenase (17, 18). Recently, ligands based on the dye chemistry have been designed and synthesized *de novo* for the purification of kallikrein (19, 20).

Figure 1. The structures of some reactive dyes: (a) Procion Blue MX-R; (b) Procion Blue H-B; (c) Procion Red H-3B; (d) Procion Scarlet MX-G; and (e) Procion Yellow MX-R.

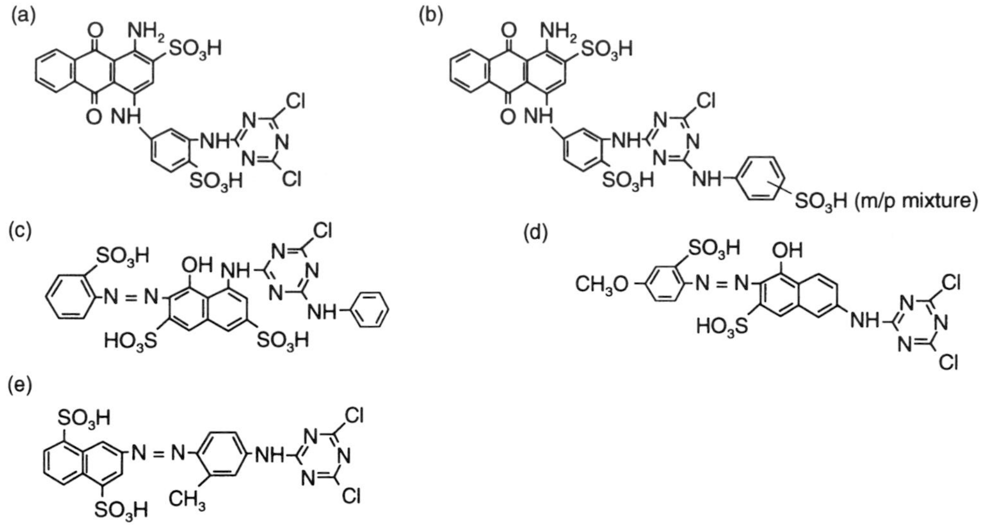

Table 3. Examples of requirements for metal ions for proteins to bind to the dye affinity matrices (11–14)

Immobilized dye	Metal	Concentration (mM)	Protein
Red H-8BN	Zn^{2+}	2	Carboxypeptidase G_2
Yellow H-A	Zn^{2+}	2	Alkaline phosphatase
Blue F3G-A	Zn^{2+}	2	Alkaline phosphatase
Green H-4G	Mg^{2+}	10	Hexokinase
Blue HE-RD	Al^{3+}	2	Ovalbumin
	Mg^{2+}	10	Hexokinase
	Zn^{2+}	2	Tyrosinase

5. DYE IMMOBILIZATION

There is a wide range of ready prepared immobilized dye matrices available for purchase, some of which are listed in *Table 4*. However, an effective purification protocol can occasionally require a customized immobilized dye matrix. Customization of the affinity matrix by making one's own immobilized dye matrix allows optimization by changing the orientation of the dye on the support (this method improves the purification of alcohol dehydrogenase by immobilizing the dye via the anthraquinone group rather than the triazine ring (21), or the use of spacer arms (purification of kallikrein using immobilized ligand only occurred in the presence of a spacer arm (15, 19, 22)) and increasing or decreasing the ligand concentration.

5.1. Choice of the support matrix

There is a range of activated supports that can be used to immobilize dye molecules, and these vary in their activation chemistry and/or their support matrix (see ref. 23 for detailed methods

Table 4. Properties of some commercial immobilized affinity matrices

Name	Matrix	Activation	Spacer	Ligand concentration	Capacity	Proteins bound
Affinity Chromatography Ltd						
Mimetic Red 1	CL6% agarose	?	Yes			
Mimetic Red 2	CL6% agarose	?	Yes			
Mimetic Red 3	CL6% agarose	?	Yes			
Mimetic Orange 1	CL6% agarose	?	Yes			
Mimetic Orange 2	CL6% agarose	?	Yes			
Mimetic Orange 3	CL6% agarose	?	Yes	2.2–2.8 μmol ml^{-1}	$\leq$45 mg protein ml^{-1}	
Mimetic Yellow 1	CL6% agarose	?	Yes			
Mimetic Yellow 2	CL6% agarose	?	Yes			
Mimetic Green 1	CL6% agarose	?	Yes			
Mimetic Blue 1	CL6% agarose	?	Yes			
Mimetic Blue 2	CL6% agarose	?	Yes			
Bio-Rad						
Cibacron Blue 3G-A	CL agarose	?	?	6.7–7.9 μmol ml^{-1}	11 mg albumin ml^{-1}	
ICN flow						
Cibacron Blue F3G-A	Agarose			2–5 μmol ml^{-1}		$NAD(P)^+$-dependent oxidoreductases, interferon, steroid receptors, and 1,25-dihydroxyvitamin D_3-receptor
CM-Cibacron Blue						
F3G-A	Agarose			2–5 μmol ml^{-1}		Antibody from serum, tissue culture supernatant and ascites
DEAE-Cibacron Blue						
F3G-A	Agarose			2–5 μmol ml^{-1}		
Reactive Green	Agarose			2–5 μmol ml^{-1}	3–5 mg HSA ml^{-1}	Nucleotide enzymes, complement factors, HSA

CL, cross-linked; HSA, human serum albumin.

of immobilizing ligands on supports which are not ready activated). Due to the perceived problem of dye leakage from the column, it is important to use a stable ether linkage between the dye and the matrix and to use a matrix which is acid stable, as leakage has been attributed to matrix hydrolysis (24). To detect dye leakage, an enzyme-linked immunosorbent assay (ELISA) has been developed (25). Activated supports can be purchased from a wide range of companies, including Alltech, Affinity Chromatography Ltd, Americon Qualex, Amicon, Beckman, Bio-Probe, Bio-Rad, Bodman, Boehringer Mannheim, Calbiochem, Chromatochem, IBF Biotech, ICN, Pharmacia Biotech, Pierce, Polyscience, Schleicher and Schüell, Sigma, Sterogene, Tosohaas and Waters. Supports are often agarose based and usually cross-linked to varying extents in order to maintain rigidity and thereby allow fast flow rates. There is an increasing range of new polymeric support matrices coming to the marketplace, which confer various advantages, especially the facilitation of fast flow rates. However, the nature of affinity chromatography limits the usefulness of flow rate improvements since the maximum flow that can be used is determined by the rate of enzyme binding to the ligand. For the ideal support a number of considerations should be taken into account. Ideally the support will:

(i)   have very low nonspecific adsorption.
(ii)   Have good mechanical properties to resist shear forces and high pressure and flow rates.
(iii)   Have good chemical stability to extremes of pH, detergents, dissociating agents, organic solvents and high temperatures.
(iv)   Be homogeneous.
(v)   Be uncharged to reduce nonspecific ion-exchange properties.
(vi)   Have a hydrophilic surface with an abundance of functional groups suitable for coupling the affinity ligand.
(vii)   Have good flow properties for the particular application. There is a compromise to be made between good flow rates and capacity, since larger beads with fast flow have smaller surface areas for ligand immobilization.
(viii) Not swell or shrink when the solvent or pH is altered.
(ix)   Be inexpensive.

5.2. Choosing the activation chemistry
Reactive dyes are ideally and most commonly immobilized via their chlorotriazine function directly to hydroxylic matrices at pH values greater than or equal to 8.5. Epoxy activation can also be used to immobilize dyes via amino and hydroxyl groups, and is also ideal since the stability of the resulting ether linkage is excellent.

There is a variety of alternative activation chemistries available for amino-spacer-modified dyes, including cyanogen bromide, cyanuric chloride, bis-epoxide, carbonyldiimidazole, N-hydroxy-succinimide and sulfonylchlorides (*Table 5*). However, consideration should be given to whether the link formed will be resistant to leakage and hydrolysis, how efficient the coupling chemistry is and the ease of coupling. The amine group on the dye is frequently involved with protein binding and hence dyes coupled using this group often show low affinity. Activated supports with spacer arms are available and are used to allow small ligands to interact with deep binding sites of proteins where there would be steric hindrance if the dye were coupled directly to the matrix. However, if the arm is too long this will often lead to nonspecific binding and reduce the specificity of the medium.

5.3. Immobilizing the dye
Immobilizing 2 µmol of dye per ml of gel is usually sufficient for affinity media, although if the affinity for the enzyme is low, this concentration can be increased to improve binding (26). Optimum efficiency is generally achieved when the total reaction volume is between 1.5 and 4.5 ml ml^{-1} of gel.

Table 5. Summary of the properties of various activated gels

Activation chemistry	Coupling group on ligand	Optimal pH for coupling	Reactivity	Linkage type	Stability of linkage
CNBr	$-NH_2$	8–10	Good	Isourea	Moderate
Triazine	$-NH_2/-OH$	8–10 for dyes	Excellent	Aminotriazine/ alkoxytriazine	Good
Epoxy	$-NH_2/-OH/-SH$	10–12	Moderate	Secondary amine ether/thioether	Excellent
N-hydroxysuccinimide	$-NH_2$	Neutral	Excellent	Amide	Moderate

The activated gel may be dry or pre-swollen when purchased. If dry, the gel can be swollen in water for approximately 30 min. The swollen gel must then be washed thoroughly with at least 20 volumes of buffer (washing the gel is fast under reduced pressure, when the washing solution is pulled through a suitable sized sintered funnel, provided that the gel is not allowed to become dry) and then transferred to a tube for ligand coupling.

The ligand is added to 10 mg ml^{-1} (can be altered to change the degree of substitution) and the slurry agitated (not by magnetic stirrers, which can cause bead damage). Base is added to obtain pH 11–12 and the addition of sodium chloride to 2% (w/v) can improve the extent and rate of reaction. Immobilization may take several hours or even days at room temperature, but this time can be considerably reduced by raising the temperature to 40–80°C. Dye–ligand coupling can usually be monitored by measuring the absorbance of the residual dye in the supernatant at intervals using an appropriate wavelength (*Table 6*).

The gel should be washed extensively with any desired buffer, or by alternating high- and low-pH buffers, until all the free ligand is removed, and then washed with the purification protocol eluting buffer to ensure no further ligand is eluted. Washing with distilled water is not recommended as this process can remove the counter ions from the immobilized dye, leaving the bound dye present as an acidic species capable of causing matrix degradation (28). Before use, the gel should be washed with 0.01 M NaOH or 2 M NaCl or 6 M urea, followed by the equilibration buffer.

The concentration of the immobilized dye can be determined spectrophotometrically. A known quantity of the gel is acid hydrolysed in 5 M HCl, incubating until the gel is dissolved, followed by returning the solution to the appropriate pH by addition of buffer. The absorbance is measured and the concentration calculated using the extinction coefficient for the dye (*Table 6*).

If a spacer arm is required, activated gels (epoxy or CNBr) containing spacer arms are available for immobilizing the dye. Alternatively, an aminoalkane spacer can be substituted on one of the chlorines and immobilized via the amine group using epoxy- or CNBr-activated matrices (29).

Table 6. Absorbance properties of some reactive dyes (27)

Dye	λ max	Molar extinction coefficient (1 mol^{-1} cm^{-1})
Cibacron Blue F3G-A	610	12 300–13 600
Procion Blue MX-R	600	4100
Procion Green H-4G	675	57 400
Procion Green HE-4BD	630	20 800
Procion Red HE-3B	522	30 000
Procion Red H-8BN	546	21 300
Procion Brown MX-5BR	530	15 000
Procion Yellow H-A	385	8900

6. EXAMPLES USING DYE CHROMATOGRAPHY

Immobilized dyes have been used to purify an extensive range of proteins, including dehydrogenases, phosphotransferases, nucleases, aminoacyl-tRNA-synthetases, nucleic acid synthetases, phosphodiesterases, phosphatases, restriction endonucleases, albumin, coagulation factors, complement proteins, growth factors, glycoproteins, lipases, kinases, collagenases, penicillin-binding proteins, enterotoxins and interferons (8, 27, 28, 30–35). Cibacron Blue F3G-A was the first dye discovered to have specificity towards proteins and is still the most frequently used, in particular for the purification of nucleotide-binding proteins, since it is known to compete for the coenzyme binding sites of many enzymes (27). This led to the hypothesis that the dye acts as a nucleotide analog. However, Cibacron Blue F3G-A also binds albumin (36) and plasma-mediated OXA-2-β-lactamase (37), which do not have nucleotide-binding sites, and appears to compete with ligand-binding sites; for example, in monkey liver serine hydroxymethyl transferase it binds at the tetrahydrofolate binding site. Since the discovery of Cibacron Blue F3G-A, a wide range of polysulfonated aromatic triazine dyes has been used successfully as pseudo-affinity ligands (*Table 7*).

Procion Red HE-3B has been shown to have some preference for NADP(H) coenzyme binding enzymes (39), although this is not always strictly the case (40). A number of papers have been published comparing the relative affinities of groups of dyes (41–45). However, these types of studies often do not fully characterize the matrices and it should be borne in mind that the concentration of dye on the matrix considerably affects the protein binding.

Table 7. Examples of dyes that mimic substrates, used for the purification of a range of enzymes (38)

Natural biological ligand	Biomimetic dye	Typical enzyme purified
NAD$^+$(NADH)	Cibacron Blue F3G-A	Lactate dehydrogenase
NAD$^+$(NADH)	Procion Blue MX-R	Alcohol dehydrogenase
NAD$^+$	Procion Red HE-3B	Glucose 6-phosphate dehydrogenase
IMP	Procion Blue H-B	IMP dehydrogenase
GTP	Procion Blue H-B	Adenylosuccinate synthetase
HMG-CoA	Cibacron Blue F3G-A	HMG-CoA reductase
Folate	Procion Red H-8BN	Carboxypeptidase G-2
Dihydrofolate	Cibacron Blue F3G-A	Dihydrofolate reductase
Phosphate esters	CI Reactive Yellow 13	Alkaline phosphatase
Phosphate esters	Cibacron Blue F3G-A phosphate analog	Alkaline phosphatase
Peptides	Cationic dyes	Trypsin, kallikrein, urokinase

HMG, 3-hydroxy-3-methyl glutamyl; IMP, inosine 5′-monophosphate.

7. LIGAND DESIGN

The screening of textile dyes for the affinity separation of enzymes can often produce good purifications in high yield. However, some enzymes still require further purification or cannot be purified using standard dyes. Recent work has been carried out to improve the standard dyes to make them more selective for specific targets and some are currently at the stage of being designed *de novo* (15). This work often utilizes the known crystal structure of the enzymes to be purified and exploits molecular graphics to bind theoretically and modify the dye ligands to improve specificity and affinity. The ligand is designed to fit into a known binding pocket to make Van der Waals contacts with the hydrophobic regions and appropriate electrostatic interactions with the hydrophilic regions of the enzyme. It is important to limit the conformational flexibility of the ligand so that it presents itself to the enzyme in a suitable conformation for binding, while at the same time allowing some flexibility so that it can adopt the optimum shape in the binding site. Ease of ligand synthesis, solubility and immobilization also have to be considered when designing the ligand. Once the novel biomimetic ligand has been designed, it is synthesized, immobilized and tested for affinity with the enzyme. Another cycle of design improvement may then be undergone to improve further the ligand binding. Examples where this approach has been successful include the design of ligands for the purification of alcohol dehydrogenase (15, 38) and the protease kallikrein (19, 22). The ligands designed to purify kallikrein are particularly novel since they are cationic moieties, unlike the polysulfonated dyes used in conventional affinity purifications.

B. CRITERIA OF ENZYME PURITY

A. Berry

The purification of an enzyme is usually only the starting point for a more detailed study of the protein in question. The nature of these further studies will determine the quantity of purified protein, whether the protein is required in an active form, the time and cost of the purification and the level of purity required. If the protein is to be used in research, the amounts needed may be small but the purity is of the utmost importance (usually $\geq 95\%$) and the removal of interfering contaminants is essential. However, if the enzyme is needed for industrial applications, large quantities are required and the purity of the sample (typically 80–90%) may have to be compromised. For therapeutic uses, all contaminants must be removed, and the criteria used to check this should include assays for self-aggregation, contaminating proteins, DNA, lipid, carbohydrate, endotoxins and any additive used during the preparation. Specific steps may be needed to remove such contaminants. In general, the research enzymologist should attempt to obtain a sample of the protein of the highest possible purity and catalytic activity. This section deals with the methods available to assess these features (*Table 8*).

8. DETECTION OF NONPROTEIN CONTAMINANTS

Since most procedures for isolating enzymes are highly successful at removing noncovalent, nonprotein contaminants, this section will concentrate on a discussion of the techniques which can be used to detect other proteins contaminating the enzyme of interest. However, the measurement of the amount and nature of nonprotein material in a sample may be critical in some cases. The determination of the amount of nucleic acid in a protein sample can be easily achieved by the spectrophotometric method of Warburg and Christian (46) (*Figure 2*) and this method may also be used to estimate the protein concentration of the sample. The absorbance of a suitably diluted aliquot of protein solution is measured at both 260 nm and 280 nm and the A_{280}/A_{260} ratio is calculated. *Figure 2* relates the values of A_{280}/A_{260} ratio with the % nucleic acid in the sample and a factor, F, for calculating the protein concentration.

The values of F in *Figure 2* were calculated from the absorption coefficients of solutions (each 1 mg ml^{-1}) of pure crystalline yeast enolase ($\varepsilon_{280} = 2.06$ cm^2 mg^{-1}; $\varepsilon_{260} = 1.18$ cm^2 mg^{-1}) and yeast nucleic acid ($\varepsilon_{280} = 24.8$ cm^2 mg^{-1}; $\varepsilon_{260} = 50.8$ cm^2 mg^{-1}) and the method is therefore liable to some small error as other proteins and nucleic acids will have different extinction coefficients. Despite this, it is a generally useful method for estimating the amount of nucleic acid in a protein sample.

Detailed discussion of the methods of detecting and analysing carbohydrate and lipid components in a preparation is beyond the scope of this section and the reader is referred to references 47–49 for descriptions of these methods.

Table 8. Methods for assessing the purity of enzyme preparations

Method	Quantity required	Comments
Specific activity measurement	Varies	Essential measurement for enzymes. Purification to a constant specific activity should be attained. Useful as a comparison with previously reported values
Electrophoresis (nondenaturing)	ng–µg	Samples should be electrophoresed at several pH values as two proteins might run together at a single pH
Electrophoresis in the presence of SDS	ng–µg	Very useful for detecting impurities that differ in subunit molecular weight. Useful for detecting proteolysis of samples. Problems with enzymes composed of nonidentical subunits
Isoelectric focusing/ chromatofocusing	ng–µg	Very sensitive method for detecting impurites. Very small differences in isoelectric point (about 0.01 pH units) can be detected. Some problems caused by presence of ampholytes after technique has been applied
Capillary electrophoresis	ng–pg	Rapidly becoming popular. Variety of modes appropriate for proteins. Very small amounts of sample required (pg by mass, nl by volume). Rapid separations (1–30 min)
Chromatography (HPLC/ FPLC/conventional)		Analytical scale separations possible based on:
gel filtration	µg	molecular size
ion exchange	µg	charge
affinity chromatography	µg	specific binding
Ultracentrifugation		Difficult to detect low levels of impurities (e.g. < 5%)
velocity	µg	Some problems with associating–dissociating systems
equilibrium	µg	
Mass spectrometry	ng–µg	Accurate measurement of mass possible (±0.01%). Can resolve species of very similar mass (0.1%). May detect minor species, such as proteolytic fragments and post-translational modifications. Small amounts of protein required. Rapid analyses (15–30 min)
Amino acid analysis/ N-terminal analysis/ N-terminal sequencing		Compare with amino acid composition inferred from DNA sequence or previously purified enzyme. N-terminal analysis or sequencing should show the expected number of polypeptide chains, with the expected sequences. Problems with blocked N-termini
Active site titrations	Varies	Useful to determine whether an apparently homogeneous preparation contains inactive forms of the enzyme

FPLC, fast protein liquid chromatography; HPLC, high performance liquid chromatography.

Figure 2. The quantification of nucleic acid contaminants in protein samples. The absorbance of a suitably diluted aliquot of protein solution is measured at both 260 nm and 280 nm and the A_{280}/A_{260} ratio is calculated. From the value of the A_{280}/A_{260} ratio, the percentage nucleic acid in the sample can be found (– – –), and the factor, F (——), can be used to calculate the protein concentration according to the equation:

protein concentration (mg ml^{-1}) = $A_{280} \times F \times 1/d$,

where d is the path length in centimeters.

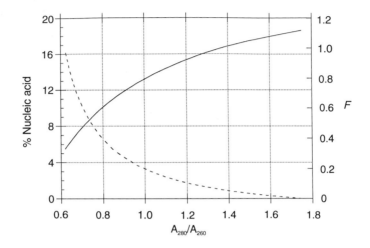

9. DETECTION OF PROTEIN CONTAMINANTS

Many of the preparative methods used during enzyme purification may be used on an analytical scale to detect contaminating proteins (*Table 8*). Each method is based on a different physical property of the protein. The choice between the methods will depend on the following parameters:

(i)   the quantity of protein available.
(ii) The nature of the suspected impurity.
(iii) The accuracy of the test needed.
(iv) The sensitivity of the test needed.
(v)   Specific factors depending on the particular enzyme.

Each technique can only demonstrate the presence of an impurity rather than prove its absence, and a single test will not, therefore, be sufficient to prove that the preparation is homogeneous. However, since each method will establish a certain degree of probability that the preparation is pure (dependent on the sensitivity of the method), a combination of two or more methods should demonstrate whether the preparation is homogeneous to a very high degree and hence that the protein can be accepted as pure.

Many of the tests of purity are associated with fractionation of the sample, with the criterion of purity being the presence of a single component. Similarly, the determination of some of the molecular properties of the purified protein and the comparison with those expected is important in ensuring that the protein is in the required, active form. Practical details of many of these methods are given elsewhere in this book and are not repeated here. Instead, this section will attempt to provide guidance in the selection of appropriate methods and will discuss some of the problems and pitfalls associated with each method.

10. SPECIFIC ACTIVITY MEASUREMENT

In addition to assessment of purity based on physical properties, other tests based on the activity of the enzyme are fundamentally important. From measurements of protein concentration and activity, the specific activity of the enzyme may be calculated and used as a very sensitive assay of the efficiency of any purification procedure and the quality of the product. Given that the level of contamination may be very low ($< 1\%$), it may be useful to measure specifically the catalytic activity of suspected interfering enzymes, and such tests are often carried out and reported by commercial suppliers of enzymes. For example, commercially available lactate dehydrogenase purified from rabbit muscle (Sigma Chemical Company Ltd) contains $<0.01\%$ of the potential interfering contaminants, pyruvate kinase, myokinase, malate dehydrogenase, glutamate–oxaloacetic transaminase and α-glycerophosphate dehydrogenase.

The purification of an enzyme should be followed at every step, if at all possible, by measurement of the activity of the preparation and of the amount of protein present. These measurements can then be combined to provide information of the specific activity of the enzyme of interest and of the degree of purity achieved at each step in the purification:

$$\text{specific activity} = \frac{\text{activity of enzyme of interest (units)}}{\text{total protein (mg)}}$$

$$\text{degree of purification} = \frac{\text{specific activity at step } (n+1)}{\text{specific activity at step } (n)}$$

The specific activity of the protein should increase during the purification procedure until a constant specific activity is obtained for the pure protein. The calculation of the degree of purification obtained at each step can be used to develop purification procedures to an optimum yield and degree of purification by attempting to modify or replace steps which yield only low degrees of purification. The specific activity measurement is an indirect measure of the purity of the enzyme since it must be compared with that *expected* or *previously reported* for the 'pure' enzyme.

Methods for estimating the total protein present in a sample are given in Part C of this chapter. It is clearly beyond the scope of this section to discuss in detail assay methods for specific enzymes; however, some of the general principles are set out below. Assays of enzymes involve measurement of the rate of formation of product or of disappearance of substrate by a variety of techniques, the simplest of which is direct observation by spectroscopy (absorption or fluorescence), or by coupling to other enzyme reactions. Other methods include sampling the reaction at different times and analysing the levels of product formed. Only if direct methods cannot be found should a discontinuous sampling method be used, as this will almost certainly be more laborious than a continuous assay and will involve sampling errors and potentially considerable work in separating and estimating the products.

11. ELECTROPHORETIC METHODS

Useful electrophoretic methods (50) for assessing the purity of enzymes include nondenaturing gel electrophoresis, denaturing electrophoresis in the presence of SDS and isoelectric focusing (for practical details of these methods see Section D of this chapter and references 51–53). Any of these methods can be used independently to detect contaminating proteins, and one of these methods is likely to be first choice when assessing the purity of a protein because of the simplicity, cost and sensitivity. *Table 9* gives an outline guide to the choice of technique,

Table 9. The choice of electrophoretic method for the determination of protein purity

Difference between protein of interest and expected contaminants	Amount of protein needed	Method of choice
Difference in molecular weight	Low (ng)	SDS–PAGE plus silver stain *or* SDS–protein molecular weight analysis by capillary electrophoresis
	High (ng–µg)	SDS gel electrophoresis
Different amino acid composition	Low (pg–ng)	Free-solution capillary electrophoresis (FSCE) in coated or uncoated capillaries
	High (ng–µg)	Native gel electrophoresis
Small differences in pI	Low (pg–ng)	Capillary electrophoresis; isoelectric focusing
	High (ng–µg)	Isoelectric focusing

SDS, sodium dodecyl sulfate; PAGE, polyacrylamide gel electrophoresis.

depending on the nature of the expected contaminants and the amount of protein available for analysis. The method of sample preparation, electrophoresis and visualization of the protein in each of these methods is described in Part D. The protein of interest is electrophoresed according to the appropriate method and the protein is judged as pure if no other protein band is detected after electrophoresis. One important point to consider, when using these techniques to look for protein impurities, is the amount of sample to be used in the experiment. *Table 9* gives guidelines for the minimum amount of protein required for each method. However, since the contaminating protein will constitute a fixed proportion of the sample, the use of larger amounts of sample will always give a better chance of detecting the contaminant. For the same reason, the most sensitive staining techniques (if required) should be utilized. However, these two factors may introduce some problems with the technique. Large sample volumes may cause band broadening and, therefore, poor resolution (54), while sensitive staining techniques do not permit the recovery of sample.

Denaturing gel electrophoresis is the most commonly used technique for assessing an enzyme's purity, because of its ease. Since one cannot know the expected molecular weights or electrophoretic mobility of every possible contaminant, the choice of gel concentration to be used in the experiment is an important decision and the use of gradient gels may be appropriate. Isoelectric focusing (50, 55, 56) may also be used to separate the required protein from any contaminant. This technique is particularly sensitive. Ampholytes can be used to generate pH gradients spanning the range 2–10 but the sensitivity of the technique may be further enhanced by the use of narrow pH ranges, and it is then possible to separate proteins differing in isoelectric point by only 0.01 pH units. Isoelectric focusing may also be combined with other electrophoretic techniques in a second dimension to further enhance its resolution.

Electrophoretic techniques are usually the first choice for any assessment of enzyme purity. However, care must be taken in the interpretation of the results of the experiments. *Table 10*

Table 10. Some common problems encountered in electrophoresis and their remedies

Erroneous conclusion	Cause	Remedy
False positive (i.e. pure sample appears contaminated)	Nonuniform gel	Clean gel plates and repour gel. Ensure good mixing of gel components and stable temperature for polymerization. Ensure persulfate does not exceed recommended quantity
	Residual oxidant in gel	Allow gel to set for 24 h. Pre-run gel. Include thioglycollate (2 mM) in upper reservoir
	Thermal effects; gel run too fast	Run at lower current
False negative (i.e. contaminated sample appears pure)	Contaminant comigrates with protein of interest	Try different pH gel
	Contaminant too big to enter gel	Stain and examine whole gel, including stacking gel, for staining material. Use lower percentage gel
	Contaminant protein does not stain	Stain with more than one stain, e.g. Coomassie brilliant blue followed by silver stain
	Net charge on contaminant is zero or opposite in charge to that of interest	Important in native gels. Run at more than one pH

lists some of the causes of such mistakes and the steps which should be taken to ensure correct interpretation of the results (see also Part D of this chapter).

12. CAPILLARY ELECTROPHORESIS

Capillary electrophoresis (CE) is a recently introduced technique which is gaining widespread use in the characterization of proteins, and will soon be a commonly used method in the analysis of protein purity. Although some of the basic principles underlying capillary electrophoresis are the same as those which are well understood in conventional gel electrophoresis, free-solution electrophoresis (i.e. without gels) may seem more complex. It is interesting to note that the first electrophoretic separations were in a free-solution mode (57) and the gels now commonly used were introduced to control the convective currents which were generated and which resulted in poor resolution. Capillary electrophoresis overcomes the problems of convection by the use of narrow bore (<150 µm internal diameter) capillaries which dissipate heat very efficiently, obviating the need for a gel matrix. Nowadays, however, gel matrices are used in some applications of CE to affect the mode of separation. *Table 11* summarizes some of the possible modes of use of CE, and each is described in *Figure 3*. CE is now finding a much increased role in purity assessment because of the ease, speed and low amounts of material needed for the analysis.

Table 11. The uses of capillary electrophoresis

Method	Direction of flow	Basis of separation	Sample types
Free-solution capillary electrophoresis in uncoated capillaries (FSCE) or micellar CE	Towards cathode	Net charge Field strength Frictional coefficient Endosmotic flow	Inorganic ions Small charged molecules Proteins Peptides Carbohydrates
Free-solution capillary electrophoresis in coated capillaries (FSCE)	Generally towards anode (depends on coating)	Net charge Field strength Frictional coefficient Endosmotic flow	Inorganic ions Small charged molecules Proteins Peptides Carbohydrates
Capillary electrophoresis isoelectric focusing	Towards cathode	pI	Proteins Peptides
SDS–protein molecular weight by CE	Towards anode	Size	Large peptides Proteins

In all cases, impurities are detected in the sample by the presence of extra peaks in the electrophoretogram. In free-solution capillary electrophoresis (FSCE) in uncoated capillaries, (*Figure 3a*) substances migrate in the electric field as a result of their charge, frictional coefficient, the strength of the electric field and the endosmotic flow (the flow of liquid through the capillary due to the electric field). FSCE can be carried out without gel in uncoated capillaries from pH 2 to pH 12. Although charged compounds migrate towards their oppositely charged electrode, endosmotic flow (μeo) is greater than their mobility and carries everything towards the cathode. The magnitude of the endosmotic flow is dependent on the charge on the capillary wall which is, in turn, dependent on the pH of the buffer. Low pH buffers generate low endosmotic flow.

FSCE is sensitive enough to detect even small changes in the protein such as those caused by different degrees of glycosylation. Some problems occur in FSCE when buffers with pH values below the pI of the protein are used. Under these conditions, the protein will have a positive charge and interacts with the negatively charged wall of the capillary. To counteract this tendency, coated capillaries or buffer additives can be used. In FSCE in coated capillaries (*Figure 3b*) the surface wall of the capillary is coated with positively charged molecules, which minimizes the absorption of the positively charged proteins on to the wall over a wide range of pHs. In this case, the polarity must be reversed because endosmosis (μeo) moves towards the anode. As in the use of uncoated capillaries, the endosmosis carries the proteins past the detector.

Further refinements of CE techniques have involved the introduction of capillary electrophoresis isoelectric focusing (CE-IEF), described in *Figure 3c*, and the introduction into the capillary of matrices of linear molecules that form an entangled network in order to carry out SDS-protein molecular weight analysis by CE (*Figure 3d*). The sieving matrix separates proteins using the same principle as SDS–polyacrylamide gel electrophoresis (PAGE). The accuracy and resolution are comparable with SDS–PAGE but the amount of sample required and the time of a single separation are both improved.

Figure 3. Capillary electrophoresis can be used in a variety of modes. The principles underlying these are illustrated diagramatically: (a) free-solution capillary electrophoresis in uncoated capillaries. Charged samples are attracted to the oppositely charged electrode. Endosmotic flow is greater than the migration towards either electrode, and all samples generally move towards the cathode. (b) Free-solution capillary electrophoresis in coated capillaries. The wall of the capillary is coated with material to alter the charge properties of the wall (e.g. to reverse the charge). This minimizes the adsorption of positively charged protein samples to the wall and allows the use of a wide range of buffer pH values. (c) Capillary electrophoresis isoelectric focusing. (d) SDS–protein molecular weight analysis by capillary electrophoresis. The gel matrix separates proteins using the same principles as SDS–PAGE. Small molecules migrate fastest through the gel matrix.

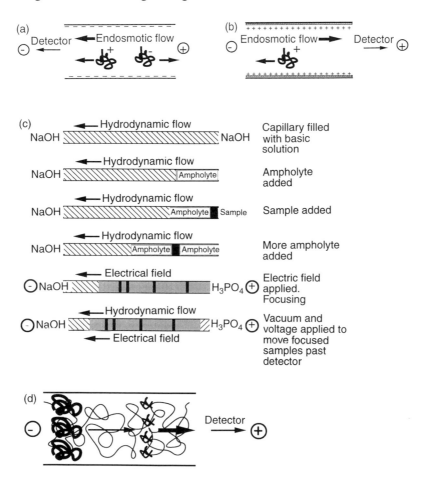

13. CHROMATOGRAPHIC TECHNIQUES

Many chromatographic techniques can be used to detect contaminating proteins. These methods include gel filtration, ion-exchange and affinity chromatography, and all may be carried out either in a conventional manner, by high performance liquid chromatography (HPLC) or by fast protein liquid chromatography (FPLC). These form some of the simplest methods for assessing the levels of impurities and are often used. In general they are less sensitive than the electrophoretic methods described above; however, they have the advantage that they are usually nondestructive and the sample can be recovered. The amount of material required varies on the type of separation being effected and the type of detection being employed, but since the sample will be diluted during passage through the column, the starting concentration must be well above the minimum detectable.

13.1. Gel filtration

Molecules are separated on the basis of molecular size (58, 59). Detection methods, such as continual measurement of the absorbance of eluate (e.g. at 280 or 220 nm) are suitable for detecting eluting proteins. A further refinement can be incorporated by specific assay of the fractions from the separation. The elution profile of a pure protein should be Gaussian, but often a slight skew towards the trailing edge is observed. Impurities are detected either as separate peaks in the chromatogram, as a shoulder on the peak of interest, as a strongly skewed peak or by the observation that the profiles monitored by absorbance and specific assay do not overlap. The use of a calibrated column for this analysis is strongly recommended in order that the apparent molecular weight of contaminating proteins may be determined. Such information may be useful in further purification steps. One of the major drawbacks of gel filtration is that proteins may associate with each other or with the column matrix, and this will produce results similar to sample heterogeneity. Rechromatography or chromatography under different conditions of ionic strength or pH should allow one to distinguish between these two conditions.

13.2. Ion exchange

Ion-exchange chromatography depends on ionic interaction between the protein and the gel matrix of the ion-exchange resin. Ion-exchange chromatography finds major use in the purification of proteins because of its high capacity and high resolution at reasonable cost (50, 60, 61). Ion-exchange resins may also serve on an analytical scale to assess the purity of a protein. In this respect HPLC or FPLC ion-exchange chromatography have the best resolution and speed of analysis. The presence of a single peak of protein in the eluate of an ion-exchange chromatography run is evidence for the purity of the sample, *if* it can be shown that all of the applied protein has been eluted from the column. This final proviso is important, since a contaminated protein might appear as a single peak in the eluate if conditions have not been used to ensure that all the protein has eluted.

13.3. Affinity chromatography

Affinity chromatography (61–65), taking advantage of the specific interaction of an enzyme with its substrate or competitive inhibitor, is a very powerful method for purifying proteins. In principle, the method could be used to evaluate the purity of a protein, although it is not often used for this purpose. Contaminants which do not bind to the immobilized ligand will elute in the void volume of the column and can be detected by monitoring the absorbance of the eluate (e.g. at 280 nm). Proteins other than the one of interest which also bind to the immobilized ligand may be eluted under different conditions and appear as separate peaks in the chromatographic profile. As with ion-exchange chromatography, it is fundamentally important when assessing protein purity by this method to ensure that all the protein elutes from the column under the conditions chosen.

13.4. Chromatofocusing

In chromatofocusing (66), proteins are absorbed on to a poly(ethyleneimine) agarose ion-exchange resin by electrostatic interactions. A pH gradient is then generated by the addition to the column of the acid form of a mixture of ampholyte-type buffers of high buffering capacity.

This results in titration of the groups on the ion-exchange resin which generates a decreasing pH gradient. As the pH falls the proteins are eluted in the order of their isoelectric points. The method has high resolving power, but some of the materials are costly. Chromatofocusing has found use therefore in the later stages of protein purifications. It can be used to detect impurities in the sample, which show up as separate peaks in the elution profile.

14. CENTRIFUGATION METHODS

Ultracentrifuges are capable of producing intense centrifugal fields. In such fields, macromolecules (whose density is usually greater than that of the solution) tend to sediment (67–69). This property can be used in either of two ways in order to determine the molecular weight of the macromolecule and is often used to assess the state of aggregation or association of a protein sample. These two methods, sedimentation velocity and sedimentation equilibrium, may also be used to detect the presence of other components in the sample (70). Neither method is ideally suited to detecting small amounts of impurities ($<5\%$) and use of the ultracentrifuge has to some extent become less common in assessment of purity of proteins, being surpassed by other techniques which are less demanding in equipment, amount of material required and expertise in physical chemistry. However, the ultracentrifuge still plays an important role in the study of protein structure and provides the method of choice for proteins that show reversible association and dissociation.

14.1. Sedimentation velocity

This is a relatively simple, nondestructive method for assessing the purity of protein preparations. In very high centrifugal fields the forces are great enough to sediment the protein. The rate of sedimentation of the enzyme can be monitored by suitable optical means and the sedimentation coefficient calculated. If the diffusion coefficient of the molecule (D), its partial specific volume (v) and the density of the solution (ρ) are known, we can calculate the molecular weight of the protein according to *Equation 1* in *Table 12*. Impurities in the protein preparation are detected as a second sedimenting species. A major limitation of the method is its insensitivity to small differences between molecules. The methodology is described in Section F.

Table 12. Equations used in molecular weight determinations by ultracentrifugation

Equation 1. Sedimentation velocity

$$M_r = \frac{RTs}{D(1 - \bar{v}\rho)}$$

where R is the gas constant, T is the temperature, s is the sedimentation coefficient, D is the diffusion coefficient of the macromolecule, v is the partial specific volume of the macromolecule and ρ is the density of the solution.

Equation 2. Sedimentation equilibrium

$$M_r = \frac{2RT}{(1 - \bar{v}\rho)\omega^2} \frac{\mathrm{d}\ln c}{\mathrm{d}r^2}$$

where R is the gas constant, T is the temperature, ω is the angular velocity in rad sec^{-1}, v is the partial specific volume of the macromolecule and ρ is the density of the solution. The value of (d ln c/r^2) is obtained from the slope of the plot of ln c against r^2.

14.2. Sedimentation equilibrium

This is another nondestructive method for detecting impurities based on their molecular weight. Under conditions of low centrifugal field, the tendency of the protein to sediment will be balanced by its tendency to diffuse from the region of high concentration at the base of the tube. At this equilibrium position, measurement of the concentration distribution of the protein (c) as a function of the distance along the tube (r) can be made, and from this the molecular weight of the protein can be calculated (*Equation 2, Table 12*). This method has several advantages. Since the measurements are made at equilibrium there is no dependence on the shape of the molecule or on the viscosity of the solution, and in suitably small tubes equilibrium may be reached in a matter of a few hours. Sedimentation equilibrium may be used to detect impurities in a sample and to detect heterogeneity in a sample which undergoes self-association. If the sample is pure, the concentration dependence of the apparent molecular weight (determined from the local slopes of a graph of $\ln c$ versus $r^2/2$) should be independent of rotor speed. The sample can be demonstrated to be pure if constant molecular weights versus concentration are obtained at several rotor speeds. If the molecular weight curves increase at higher concentrations, either heterogeneity or mass action equilibrium are indicated. If the curves are systematically displaced to lower apparent molecular weight at higher rotor speeds, the solution contains a mixture of higher and lower molecular weight components. If, on the other hand, the curves of apparent molecular weight as a function of c are superimposable, a mass action equilibrium is indicated. Similarly, for samples at different protein-loading concentrations run at a single speed, heterogeneity results in a lower apparent molecular weight being calculated for the most concentrated solution, whereas self-association results in superimposable curves. The acquisition and analysis of these data has been automated and nonlinear fitting techniques (71) are also available to speed up and simplify the analysis of sedimentation equilibrium data.

15. MASS SPECTROMETRY

Until recently, mass spectrometry would not have merited inclusion in a chapter on criteria of enzyme purity. However, the recent development of methods for producing intact molecular ions such as electrospray ionization mass spectrometry (ESIMS) (72) and matrix-assisted laser desorption-time of flight (MALDI-TOF) (73) have revolutionized the use of mass spectrometry in biochemistry to the point at which it is now becoming a standard technique. The new techniques are very sensitive, requiring only pmol quantities of protein, and have very high resolving power (typically the molecular mass of a protein can be measured to $\pm 0.01\%$ over a range of molecular masses from 40 Da to 300 kDa (74, 75)). This gives mass spectrometry an unrivalled place in the assessment of protein purity.

ESIMS employs a very mild form of ionization since it occurs at temperatures and pressures close to ambient. This method of ionization produces multiply-charged intact protein ions with up to about 100 charges on a large molecule arising through the protonation or deprotonation of side chains. The large number of charges means that most proteins produce ions with mass-to-charge ratios in the range of 100–4000 Da and this range can be detected on a relatively low-cost quadrupole instrument. For details see *Figure 4a*.

The majority of ions produced by matrix-assisted laser desorption mass spectrometry (MALDI-MS) carry only a single charge, and the mass : charge spectrum therefore gives a straightforward read-out of the singly-protonated mass ($M + H^+$). Some use has also been made of an earlier method, plasma-desorption mass spectrometry (PDMS), to produce molecular ions of large peptides (<25 kDa). However, its application to proteins has been very limited. The different properties of the two main forms of mass spectrometry are given in *Table 13*. Some problems may be encountered in the use of mass spectrometry to analyse proteins. First, not all

Figure 4. The basis of electrospray and matrix-assisted laser desorption mass spectrometry. In electrospray mass spectrometry (a) ionization is achieved by pumping the solution of protein at 1–5 µl min^{-1} through a needle at high potential. The liquid is broken into small droplets (about 1 µm) and evaporation is then effected by a current of warm nitrogen. As the charge intensity on the surface of the droplet increases it becomes unstable and further breaks up. Eventually the charge transfers to the protein molecule, itself now in the vapor phase. The multiply-charged ions pass into the evacuated portion of the spectrometer, the mass analyser. Diagramatic representations of typical recorded and transformed spectra are shown. In matrix-assisted laser desorption mass spectrometry (b) samples are prepared by adding the protein (1 pmol or less) to a concentrated solution of matrix material. This is a UV-absorbing, low mass molecule such as cinnamic or sinapinic acid. The sample is dried down and inserted into the spectrometer where it is pulsed with UV-radiation from a laser. Some of the matrix and the embedded protein are volatilized by this treatment. The ions generated are accelerated by focusing electrodes to a flight tube which is electric field free. The time taken from the pulse of the laser to the arrival of the ions at the detector is proportional to the mass.

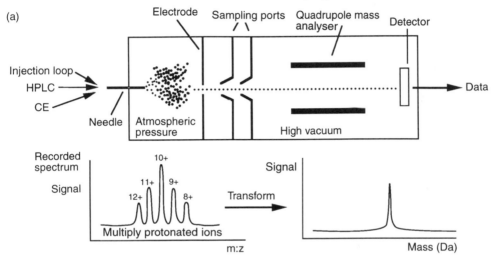

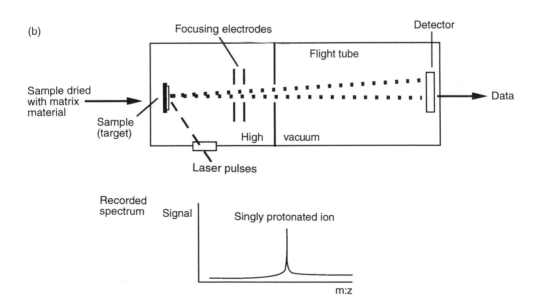

Table 13. The properties of electrospray and matrix-assisted laser desorption mass spectrometry

Property	Electrospray MS (ESIMS)	Matrix-assisted laser desorption MS (MALDIMS)
Mass range	Up to 150 kDa	Up to 300 kDa
Mass accuracy	1 in 20 000	up to 1 in 10 000
Mass resolution	1 in 1000	1 in 400
Mass sensitivity	1 pmol	< 1 pmol
Interference by salts	Yes	No
Ability to:		
analyse complex mixtures	No	Yes
couple to HPLC/CE	Yes	No

components will produce ions ('fly') in the spectrometer in every case, and secondly, ESIMS may be sensitive to the presence of salt in the sample. Despite these drawbacks, mass spectrometry has already made a great impact on the assessment of sample purity.

Many natural and recombinant proteins have now been studied by ESIMS or MALDIMS. The 'soft' ionization produced by the two methods leaves covalent bonds (76) and even relatively labile bonds, such as those formed in acyl enzyme intermediates, intact. With most natural proteins above about 70 kDa, heterogeneity seems to be the norm. Mass spectrometry has often revealed closely related variants of the protein which would be difficult or impossible to distinguish by any other method. One of the most common forms of heterogeneity occurs with glycoproteins because different glycoforms of the protein may vary by only one or two sugar groups. Such glycoforms may easily be detected by MS and the nature of the change in carbohydrate may be identified from a knowledge of the masses of various standard sugars. MS has also been utilized to investigate the purity of recombinant proteins. In this case the measured molecular mass may be compared with that inferred from the nucleic acid sequence. Observed differences may correspond either to errors in the determined amino acid sequence (recently shown for bovine serum albumin (77)), or to post-translational modifications of the protein, such as the formation of disulfide bridges, processing of the N- or C-terminus, removal of the initiation methionine, N-terminal modifications, glycation of protein (78), lipid attachment (79), acetylation of lysines (80), glycosylation (81) or phosphorylation (82, 83) (*Table 14*). Another important use of MS is in the detection of C-terminally truncated products present at low abundance, as produced, for example, by proteolytic degradation of the protein.

16. AMINO ACID ANALYSIS AND SEQUENCE ANALYSIS

16.1. Amino acid composition
Analysis of the amino acid composition of a protein is an indirect method of determining its purity. In general, the amino acid composition is compared with that expected for the protein sequence (e.g. inferred from the DNA sequence). However, since the technique is automated and amino acid analysis is often readily available, it is often used as a guide to purity. In general, the observed values of the number of each amino acid in the protein should be within 5–10% of that expected . Values outside that range may indicate the presence of impurities. Some problems encountered with amino acid analyses can given rise to observed values outside this range. These are detailed in *Table 15* along with corrective measures which can be taken.

Table 14. Some common post-translational modifications and the associated change in molecular mass

Modification	Mass change (Da)
Formation of N-pyrrolidone carboxyl	−17.00
Disulfide bridge formation	−2.02
C-terminal amide	−0.98
Deamidation of Asn or Gln	0.98
N-methylation	14.00
Oxidation of Met or Trp	16.00
Proteolysis of peptide bond	18.02 (gives two peptides)
N-formylation	28.00
Acetylation	42.04
Phosphorylation	79.98
Glycation	162.14
N-myristoylation	210.00
Addition of N-linked glycan 'core'	892.80

Table 15. Problems encountered in amino acid analysis

Amino acid	Problem	Solution
Hydroxyamino acids (Ser, Thr)	Partially destroyed in acid hydrolysis (5–10% in 24 h)	Perform hydrolyses for different times, e.g. 24, 48 and 72 h and extrapolate to zero time
(Tyr)	Chlorination	Include phenol in hydrolysis
Hindered amide bonds e.g. between branched side-chains	Some amide bonds (e.g. Val–Val) resistant to hydrolysis	Extended periods of hydrolysis necessary (e.g. 120 h at 105°C)
Amide amino acids	Hydrolysed to Asp and Glu	Glu and Asp values include values for Gln and Asn. Can be distinguished by sequencing
Sulfur-containing amino acids (Cys, Met and cystine)	Subject to oxidation	Determine as derivatives after oxidation with performic acid prior to hydrolysis
Tryptophan	Destroyed in acid hydrolysis	Determine from alkaline hydrolysis or by spectrophotometry

16.2. N-terminal analysis and amino acid sequence analysis

Identification of the N-terminal amino acid of a polypeptide chain can be used to determine the number of different polypeptide chains present and hence can be used as a guide to enzyme purity. The method consists simply of labeling the N-terminal amino acid in the intact protein, followed by hydrolysis and identification of the amino acid derivative formed. Several reagents are available for N-terminal labeling and these are summarized in *Table 16*. The method is most easily applied for the detection of impurities when the protein of interest is known to be a single

Table 16. Some reagents for N-terminal labeling of proteins

Reagent	Abbreviation	Structure	Notes/reference
Fluorodinitrobenzene			Sanger's reagent (84)
1-Dimethylaminophthalene-5-sulfonyl chloride	DNS-Cl or dansyl chloride		85
Phenylisothiocyanate	PITC		Edman's reagent (86, 87)
4,N,N,-Dimethylaminoazobenzene-4'-isothiocyanate	DABITC or dabsyl isothiocyanate		88, 89

polypeptide. In this case contaminating proteins *with different N-terminal amino acids* will be detected by the presence of a second labeled amino acid derivative. Problems are encountered when the contaminants have the same N-terminus as the protein of interest, or if the N-terminus of a protein does not react with the reagent used.

Further complications arise when the protein of interest is composed of multiple (heterologous) subunits, since the pure form of such a protein would give rise to multiple N-terminal residues. Problems associated with identifying impurities having the same N-terminus can be overcome by N-terminal sequence analysis. This technique will allow impurities to be detected in most cases, although it will not distinguish between similar proteins having different C-termini, arising, for example, by proteolysis of the protein. C-terminal amino acid sequencing is at present in its infancy and C-terminal analysis is more complicated than N-terminal analysis because suitable reagents have yet to be found. The normal method used is to carry out carefully controlled digestion of the protein with carboxypeptidase and identify the order in which amino acids are released.

17. ACTIVE-SITE TITRATION (for a more detailed discussion see Part E of this chapter)

Preparations of proteins which have been demonstrated to be 'pure' by the techniques discussed above, may still contain a certain proportion of the enzyme in an inactive form. In some cases it is possible to use active-site titration (90) to determine the amount of active enzyme present. This method may be used when an enzyme-bound intermediate accumulates during a reaction. Under conditions where the intermediate breaks down slowly, if at all, to regenerate the enzyme, the measurement of the rapid release of a product of the enzyme gives a measure of the active enzyme, since inactive enzyme does not produce product. *Figure 5* shows the general scheme for active-site titration of an enzyme.

Figure 5. Active-site titration. If k_2, the rate of breakdown of an enzyme intermediate (EI) is zero, the concentration of the first product, P_1, gives the concentration of active enzyme. If $k_2 < k_1$, but not zero, the overall release of products is linear with time after an initial 'burst' phase. The linear portion of the curve may be extrapolated back to zero time to give the magnitude of the burst, equivalent to the amount of active enzyme.

$$E + S \xrightarrow{k_1} \underset{+ P_1}{EI} \xrightarrow{k_2} E + P_2$$

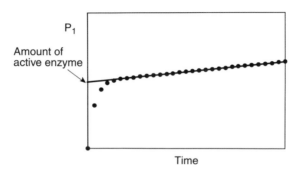

C. THE DETERMINATION OF PROTEIN CONCENTRATION
N.C. Price

18. INTRODUCTION

The purpose of this section is to help the reader decide which method of determining protein concentration may be the most appropriate in any particular situation. Section 19 outlines some of the reasons why a knowledge of protein concentration is important. The principal methods are set out in a summary form (*Table 17*), and a brief discussion of each method with a note of its advantages and disadvantages is given in Section 20.

19. WHY DETERMINE PROTEIN CONCENTRATION?

(i)   The determination of total protein at each stage in a purification procedure is an essential part of monitoring the course of the purification and establishing the success (or otherwise) of that particular step.

(ii) A knowledge of the amount of protein present in a purified sample is essential in order to define the intrinsic molecular properties of that protein such as the specific activity (in units of μmol min^{-1} mg^{-1} or katal kg^{-1}; or expressed as turnover numbers, sec^{-1}), the ligand binding capacity, the stoichiometry of modification in chemical modification experiments, or the molar ellipticity in far UV circular dichroism studies (which can be used to give estimates of the secondary structure content of the protein). These intrinsic properties can then be used to define the protein and make comparisons between batches of the protein.

The methods listed in *Table 17* and discussed in Section 20 yield values for the total proteinaceous material present in a sample. In order to establish how much of the material is active, it is necessary to perform assays of activity; for example, using active enzyme titration (see Section 17 above, Part E and ref. 91) or measurements of ligand binding capacity (Chapter 6) or other tests of biological activity.

20. THE CHOICE OF METHOD

The principal factors involved in deciding which method is the most appropriate are:

(i)   what is the sensitivity of the method – how much protein is required?
(ii)   How complex is the method?
(iii) Is recovery of the sample possible?
(iv)   Is any special equipment required?
(v)   Is the method capable of giving an absolute (or near absolute) value or only a value relative to a certain standard protein (e.g. bovine serum albumin)?
(vi)   What substances interfere with the method, and can the effect of these be overcome or taken into account?

The principal methods are listed in *Table 17*. Any of the methods can, in principle, be used to determine either the total protein in a complex mixture or a single purified protein. The gravimetric, UV absorbance, amino acid analysis and reaction with *o*-phthalaldehyde (OPA) are, however, most commonly used to determine the amount of a single, purified, protein. Detailed descriptions of many of the methods can be found in the review articles published elsewhere (92–94).

Table 17. Principal methods for the determination of protein concentration

Method	Amount of protein required (μg)	Recovery of protein	Complexity of method[a]	Response of identical masses of different proteins	Reference protein used	Major sources of interference[b]	Equipment needed
Gravimetric	5000–20 000	No	4	Identical	No	Other solutes	Microbalance
Biuret	500–5000	No	2	Very similar	Yes	Tris, NH_4^+, sucrose, glycerol	Spectrophotometer
UV absorbance 280 nm	100–1000	Yes	1	Variable	No	Nucleic acids, chromophores such as heme	Spectrophotometer
far UV	5–50	Yes	1	Very similar	No	Many buffers and other compounds absorb strongly in the far UV	Spectrophotometer
Amino acid analysis	10–200	No	4	Variable	No	Other contaminating proteins	Amino acid analyser
Lowry	5–100	No	3	Variable	Yes	Some amino acids, NH_4^+, zwitterionic buffers, nonionic detergents, thiol compounds, sucrose	Spectrophotometer
Bicinchoninic acid (BCA)	5–100 (1–10 in microprotocol)	No	3	Variable	Yes	Glucose, NH_4^+, EDTA	Spectrophotometer
Coomassie blue binding	5–50 (1–10 in microprotocol)	No	2	Variable	Yes	Triton, SDS	Spectrophotometer
Reaction with o-phthalaldehyde	0.1–2	No	3	Variable	Yes	Tris, glycine, NH_4^+	Spectrofluorimeter

[a] Graded on a 1–4 scale. 1 involves pipetting of the sample only; 2 involves mixing the sample with one reagent solution; 3 involves mixing the sample with more than one reagent solution; 4 involves lengthy manipulation of the sample.
[b] In some cases (see Section 20) the interference can be minimized or overcome.

20.1. Gravimetric determination (95)

Procedure. The protein solution (at about 5 mg ml^{-1}) should be dialysed against buffered 1 mM EDTA to remove metal ions, and then extensively against distilled water or a volatile buffer (e.g. ammonium acetate). A solution containing in the range 5–20 mg protein is added to a preweighed drying vessel. The solution is frozen, then thawed gently, evaporated at 30°C, and finally heated at 120–140°C *in vacuo* (10^{-5} mmHg). After cooling, air is admitted, the vessel stoppered and reweighed. A further round of heating *in vacuo* is carried out to ensure that drying is complete. The accuracy is ±0.1 mg (±2% for 5 mg protein).

Advantages. Can give the absolute weight of protein.

Disadvantages. A large amount of sample is required; manipulations are lengthy; interference occurs from traces of inorganic material or other macromolecules.

20.2. Biuret method (93, 96)

This method depends on the formation of a purple complex between Cu^{2+} ions and adjacent peptide bonds under alkaline conditions. Because the reaction is specific for amide bonds, which occur at a nearly constant proportion in proteins (one amide bond per 110 mass units), there is relatively little variation in the response between different proteins. Ammonium ions interfere with the reaction, so the method cannot be used for the determination of protein in samples where ammonium sulfate precipitates have been redissolved.

Procedure.

(i)   Dissolve 1.5 g CuSO$_4$.5H$_2$O and 6.0 g sodium potassium tartrate in 500 ml water.
(ii) Add 300 ml 10% (w/v) NaOH, and make up to 1 l with water. Addition of 1 g KI allows the solution to be kept almost indefinitely in a plastic container.
(iii) To 0.5 ml of a sample containing up to about 5 mg protein, add 2.5 ml reagent.
(iv) Allow to stand for 20–30 min before reading the A$_{540}$ relative to a blank containing no protein.
(v)   Known amounts of bovine serum albumin are used to construct a standard curve.

Advantages. The method is simple; the reagent stable; there is relatively little variation between proteins so near-absolute determination of protein is possible.

Disadvantage. Relatively large amounts of protein are required.

20.3. UV absorbance

Near UV (280 nm) (92, 94)
Absorption of radiation in the near UV depends on the tyrosine and tryptophan content of a protein (and to a very small extent on the content of disulfide bonds). The actual A$_{280}$ for a 1 mg ml^{-1} solution varies widely between different proteins (e.g. from zero for some parvalbumins to 2.6 for hen egg white lysozyme and even up to 4 for some tyrosine-rich wool proteins), although most values are in the range 0.5–1.5 (97, 98). The actual value for a given protein must be determined by some absolute method (e.g. gravimetric or amino acid analysis), or can be calculated from the amino acid composition (99):

$$A_{280} \ (1 \text{ mg ml}^{-1}) = (5690n_W + 1280n_Y + 120n_C)/M_r$$

where n_W, n_Y and n_C are the numbers of Trp, Tyr and Cys residues in the polypeptide of mass M_r.

Strictly speaking, this value applies to the protein in 6 M guanidine hydrochloride (GdnHCl), but the value in buffer is generally within 10% of this value and the relative absorbances in GdnHCl and buffer can be easily determined by parallel dilutions from a stock solution.

If the solution is turbid, the apparent A_{280} will be increased by light scattering. Filtration (through a 0.2 μm Millipore filter) or centrifugation can be carried out. For turbid solutions a convenient approximate correction can be applied by subtracting the A_{310} (proteins do not absorb at this wavelength) from the A_{280}.

The presence of nonprotein chromophores (e.g. heme, pyridoxal) can increase A_{280}. If nucleotides or nucleic acids are present, one of the following formulae can be applied to give estimates of protein concentration:

protein (mg ml^{-1}) = 1.55 A_{280} − 0.76 A_{260} (100)
protein (mg ml^{-1}) = 0.183 A_{230} − 0.075 A_{260} (101)
protein (mg ml^{-1}) = (A_{235} − A_{280})/2.51 (102).

Advantages. The method is simple and the sample recoverable.

Disadvantages. There can be interference from other chromophores; the specific absorption value for a given protein must be determined.

Far UV absorbance (92, 94, 103)
The peptide bond absorbs strongly in the far UV with a maximum at about 190 nm. Because of difficulties caused by absorption by oxygen and the low output of conventional spectro-photometers at this wavelength, measurements are more conveniently made at 205 nm, where the absorbance is about half that at 190 nm. Various side chains, including Trp, Phe, Tyr and His, and to a lesser extent Arg, Met and Cys, make small contributions to the A_{205}. Scopes (103) proposed an empirical formula to calculate the A_{205} for a 1 mg ml^{-1} solution of protein:

A_{205} (1 mg ml^{-1}) = 27 + 120 (A_{280}/A_{205}).

An alternative formula, based on measurements at longer wavelengths, was suggested by Waddell (104):

Protein (mg ml^{-1}) = 0.144 (A_{215} − A_{225}).

Procedure. The sample should be diluted in 0.1 M K_2SO_4 containing 5 mM potassium phosphate buffer adjusted to pH 7.0 (103). At low concentrations, protein can be lost from solution by adsorption on the cuvette; the high ionic strength helps to prevent this. Inclusion of a nonionic detergent (0.01% Brij 35) in the buffer may also help to prevent these losses (94).

Advantages. The method is simple and sensitive; the sample is recoverable; there is little variation in response between different proteins, permitting near-absolute determination of protein.

Disadvantages. Accurate calibration of the spectrophotometer is necessary in the far UV. Many buffers and other components absorb strongly in this region.

20.4. Amino acid analysis (105, 106)
The amount of protein in a sample can be evaluated by determining the amounts of the various amino acids present after hydrolysis. Since some amino acids (e.g. Trp) are destroyed by acid hydrolysis and there can be significant losses of others (e.g. Ser and Thr) a more reliable method *for proteins whose amino acid composition is known* is to relate the amount of protein to the amounts of various stable and reasonably abundant amino acids (e.g. Ala, Arg, His, Lys and Tyr) present in the sample.

Procedure. The weight of protein typically used for analysis is in the range 10–200 μg; 0.2–1 μmol taurine or norleucine is added to serve as an internal standard for the analysis. Hydrolysis is carried out using standard conditions (24 h, 6 M HCl, 110°C *in vacuo*). Analysis usually involves derivatization with OPA (see Section 20.8) prior to separation by reverse-phase HPLC.

The relative amounts of the amino acids (Ala, Arg, His, Lys and Tyr) should correspond with their known relative abundances in the protein. If they do not, contamination by another protein or proteins should be suspected. The method can, in principle, be used with considerably (at least 100 times) less protein, but problems with trace impurities from buffers, etc., can be considerable. The amount of protein which is desirable reflects the amount needed for the manipulations (such as hydrolysis) prior to analysis; 0.1 μg is easily sufficient for the analysis itself.

Advantages. Can give absolute amounts of protein, provided that the sample is not contaminated by other proteins; can be used if other components which do not react with the OPA reagent (e.g. nonionic detergents) are present.

Disadvantages. Lengthy manipulations are involved; an amino acid analyser is required.

20.5. Lowry method (92–94)
The method is based on reduction of the phosphomolybdic–tungstic mixed acid chromogen present in the Folin–Ciocalteu reagent by a protein to give a blue species (absorption maximum at 750 nm). There is a rapid reaction of the chromogen with the side chains of Tyr, Trp and possibly also His and Cys, and a slower reaction with the copper chelates of the peptide chain and/or polar side chains.

Procedure. The reagent solutions are:
(A) 1% (w/v) $CuSO_4.5H_2O$ in water.
(B) 1% (w/v) sodium potassium tartrate.$4H_2O$ in water.
(C) 2% (w/v) Na_2CO_3 in 0.1 M NaOH.
These solutions are stable for several months at room temperature. Working solutions (WS1 and WS2) are prepared freshly each day:
(WS1) (A) and (B) are mixed in a 1 vol.:1 vol. ratio and 98 vols of (C) are added.
(WS2) Commercial Folin–Ciocalteu reagent is diluted 1 vol.:1 vol. with distilled water.

(i)   To 0.2 ml of sample containing 5–100 μg protein is added 2.1 ml of WS1 (above) and the mixture allowed to stand for 10 min.
(ii) 0.2 ml of WS2 (above) is then added *with immediate mixing.*
(iii) After standing for 30 min at room temperature, the A_{750} is read against a blank containing no protein.
(iv) A standard curve is constructed using bovine serum albumin.

Modifications. The basic procedure can be modified to take account of interfering substances. Soluble interfering substances can be removed by precipitating the protein with sodium deoxycholate and trichloroacetic acid. After washing, the pellet can be resuspended in water or buffer.

Advantages. A sensitive, widely used method with variations developed to deal with particular problems (nonlinearity; presence of detergents, etc.).

Disadvantages. Nonlinear response; large number of interfering substances; significant variation in the responses of different proteins; working solutions must be prepared fresh daily.

20.6. Bicinchoninic acid (BCA) reagent (94, 107, 108)

This method relies on the ability of BCA to combine specifically with the Cu^+ produced by reduction of Cu^{2+} by a protein under alkaline conditions.

Procedure. The reagent solutions are:
(A) 1% (w/v) BCA-Na$_2$, 2% (w/v) Na$_2$CO$_3$.H$_2$O, 0.16% (w/v) disodium tartrate, 0.4% (w/v) NaOH and 0.95% (w/v) NaHCO$_3$. If necessary, NaOH or NaHCO$_3$ is added to the solution to bring the pH to 11.25.
(B) 4% (w/v) CuSO$_4$.5H$_2$O in water.

These reagents are stable indefinitely at room temperature. The working solution (WS) is prepared by mixing (A) and (B) in a 50 vol. : 1 vol. ratio. It is stable for up to a week.

(i) 0.1 ml sample containing 5–100 µg protein is mixed with 2 ml of WS (above). The mixture is incubated at 37°C for 30 min, before being allowed to cool to room temperature. It is important to use a standard set of conditions, since the color development depends to some extent on the temperature and duration of the incubation.
(ii) The A_{562} is read against a blank containing no protein.
(iii) Known amounts of bovine serum albumin are used to construct a standard curve.

Smith *et al.* (107) also describe a microprotocol for the determination of 1–10 µg protein.

Advantages. Compared with the Lowry method: less interference by nonionic detergents and simple buffer salts; a more stable working solution; more nearly linear response.

Disadvantage. The response shows a marked dependence on the time and temperature of the incubation, requiring carefully controlled conditions to be employed.

20.7. Coomassie blue binding (93, 94, 109–111)

The method relies on the change in absorption spectrum of the dye when bound to proteins. The protonated form of the dye is a pale orange-red color, whereas the unprotonated form is blue. On binding to proteins in acid solution, the protonation of the dye is suppressed by the positive charges on amino acid side chains (principally Arg) and a blue color results. In addition to these electrostatic effects, hydrophobic interactions between dye and protein are also involved.

It has been reported that the Serva blue G preparation has a greater dye content than the Coomassie blue G250 (color index (CI) 42655) material and the procedure outlined below uses the former preparation. Coomassie blue R250, used for staining gels, is unsuitable.

Procedure. The reagent solution is prepared by dissolving 100 mg Serva blue G in a mixture of 100 ml 85% phosphoric acid and 50 ml 95% methanol. After the dye has dissolved, water is added to bring the final volume to 1 l. (If necessary, the solution can be filtered through Whatman no. 1 paper to remove any insoluble material). This solution is stable for several weeks when stored (in the dark) at room temperature.

(i) To a 0.05 ml sample containing 5–50 µg protein is added 2.5 ml of the reagent solution. The mixture is allowed to stand for 5 min at room temperature before the A_{595} is read against a blank containing no protein.
(ii) A calibration curve is constructed using known amounts of bovine serum albumin.

Bradford (109) also describes a microprotocol for determination of 1–10 µg protein.

The dye–protein complex tends to precipitate over periods longer than an hour. This is more of a problem with quartz cuvettes, so glass or plastic cuvettes should be used. In the procedure described by Sedmak and Grossberg (110), the dye solution contains perchloric acid instead of phosphoric acid and the ratio of A_{620} to A_{465} is measured; this gives a linear response to the amount of protein.

Advantages. A sensitive, simple method involving addition of one (stable) reagent with relatively little interference by other components.

Disadvantages. Some variation in response between different proteins, correlated to some extent with the content of basic side chains; nonlinear response (Bradford procedure); interference by Triton and SDS at the 1% but not at the 0.1% level.

20.8. Reaction with OPA (92, 112)

OPA reacts with primary amines under alkaline conditions to form a fluorescent product (excitation and emission maxima of 340 and 450 nm, respectively). The reaction is with the ε-amino groups of Lys side chains and the α-amino group of the polypeptide chain, and thus the fluorescence yield depends on the Lys content of the protein. The method can be adapted to measure protein after total hydrolysis (6 M HCl, 24 hours, 110°C) in which case the response is much more uniform between proteins and the sensitivity is improved by at least tenfold.

Procedure. The OPA reagent solution is prepared by dissolving 120 mg OPA in 1.5 ml ethanol and added to 100 ml of 1 M potassium borate buffer, pH 10.4, followed by 0.6 ml 30% (v/v) Brij 35. This solution is stable in the dark at room temperature for at least 3 weeks. The WS is made up on the day of use by adding 3 µl 2-mercaptoethanol per 1 ml stock OPA solution. (This should be done at least 30 min prior to use.)

(i)   100 µl WS (above) is added to the sample containing 0.1–2 µg protein in 10 µl. After mixing and incubation for 15 min, 2 ml 0.5 M NaOH is added and the fluorescence is measured (excitation and emission wavelengths 340 nm and 450 nm, respectively). The fluorescence reading should be taken only once, since prolonged irradiation of the sample decreases the fluorescence.
(ii) A standard curve is constructed by using known amounts of BSA in the 0.1–2 ug range.

Advantage. Very sensitive (subnanogram amounts of hydrolysed protein can be detected, although 0.1–2 µg of intact protein is more generally employed).

Disadvantages. Requires a spectrofluorimeter; primary amines (e.g Tris, glycine, ammonia, etc.) interfere; interference from impurities in water, buffers and other components; considerable variation in response between different proteins.

20.9. Overall conclusions

It is recommended that the absolute amount of protein in a sample is determined, if possible, using amino acid analysis. This result can then be used to calibrate a convenient routine procedure (e.g. based on UV absorbance, dye binding, BCA or Lowry). If amino acid analysis is not possible, use of a method based on peptide bond content [e.g. biuret, or far UV measurements (more sensitive)] will give a good calibration. Measurements of A_{280} are convenient and, for most pure proteins, can be used to determine the amount of protein reasonably precisely, provided that the A_{280} for a 1 mg ml^{-1} solution can be calculated from the amino acid composition.

21. APPENDIX

21.1. Bovine serum albumin as a protein standard

BSA solutions can be made up by weight, but concentrations should be checked by A_{280} measurements (1 mg ml^{-1} has A_{280} of 0.66).

Stock solutions should be kept frozen at 1 mg ml^{-1} or 10 mg ml^{-1}. More dilute solutions can be prepared freshly by dilution as required.

21.2. Spectrophotometry

Quartz cuvettes should be cleaned with 5 M nitric acid (care – strong oxidizing agent, corrosive, dangerous fumes) to remove traces of adsorbed protein, and then rinsed thoroughly with distilled or deionized water before drying.

Absorbances should be measured if possible in the 0.05–1.0 range (corresponding to absorption of 10% and 90%, respectively, of the incident radiation). The greatest accuracy is achieved at absorbances of about 0.3 (50% absorption). Absorbance readings above 2.0 (99% absorption) should not be used routinely. The sample could be diluted by a known factor to bring the absorbance into the desired range.

D. FRACTIONATION AND ANALYSIS OF ENZYMES BY GEL ELECTROPHORESIS

D. Patel and D. Rickwood

22. INTRODUCTION

This section does not seek to duplicate the detailed methods for fractionating and analysing enzymic proteins offered in many manuals; the reader should consult these manuals for such information. Rather, it aims to provide key information required to analyse enzymes by gel electrophoresis. Due to limitations on the length of this section, the following tables are not intended to be exhaustive and there are recipes and referenced techniques other than those listed.

Many chemicals commonly used for gel electrophoresis are toxic, while the status of others is unknown. Readers must acquaint themselves with the precautions required for handling the chemicals mentioned in this section. Acrylamide and bisacrylamide are both toxic. Acrylamide is a known potent neurotoxin. Great care must be taken when handling such reagents. All reagents must be of the highest purity available. It is recommended that the reagents are purchased from suppliers who have quality-tested the products for use in electrophoretic techniques.

23. RECIPES FOR GELS AND BUFFERS

23.1. Separation of enzymes by nondenaturing gel electrophoresis

Electrophoresis under native conditions is used in circumstances where the enzymatic activity is to be maintained. Native electrophoresis techniques can only be applied to enzyme samples which are soluble and which will not precipitate or aggregate during electrophoresis. It is essential to select the correct buffer to ensure optimal separation of the enzyme of interest and the correct direction of migration. It is advantageous to use a discontinuous system to obtain stacking of the proteins and thus maximize resolution of the protein bands. Buffers and recipes for gels for the nondenaturing discontinuous system are given in *Tables 18* and *19*, respectively.

23.2. Determination of the molecular weights of enzymes by gel electrophoresis on SDS-denaturing gels

In SDS separations, migration is determined by the molecular weight. SDS is bound by the enzyme, and in doing so confers a net negative charge to the enzyme. The enzymes then have a

Table 18. Buffers for nondenaturing discontinuous systems

Low pH discontinuous (115)	Stacks at pH 5.0, separates at pH 3.8
Stacking gel buffer	Acetic acid–KOH pH 6.8: 2.9 ml glacial acetic acid and 48.0 ml 1 M KOH mixed. Adjusted to final 100 ml volume
Resolving gel buffer	Acetic acid–KOH pH 4.3: 17.2 ml glacial acetic acid and 48.0 ml 1 M KOH mixed. Adjusted to final 100 ml volume
Electrophoresis buffer	Acetic acid–β-alanine pH 4.5: 8.0 ml glacial acetic acid and 31.2 g β-alanine dissolved in water. Adjusted to final 1 liter volume
Neutral pH discontinuous (116)	Stacks at pH 7.0, separates at pH 8.0
Stacking gel buffer	Tris–phosphate pH 5.5: 4.95 g Tris dissolved in 40 ml water. Titrated to pH 5.5 with 1 M orthophosphoric acid. Adjusted to final 100 ml volume
Resolving gel buffer	Tris–HCl pH 7.5: 6.85 g Tris dissolved in 40 ml water. Titrated to pH 7.5 with 1 M HCl. Adjusted to final 100 ml volume
Electrophoresis buffer	Tris–diethylbarbiturate pH 7.0: 5.52 g diethylbarbituric acid and 10.0 g Tris dissolved in water. Adjusted to final 1 liter volume
High pH discontinuous (117)	Stacks at pH 8.3, separates at pH 9.5
Stacking gel buffer	Tris–HCl pH 6.8: 6.0 g Tris dissolved in 40 ml water. Titrated to pH 6.8 with 1 M HCl ($\sim$48 ml). Adjusted to final 100 ml volume
Resolving gel buffer	Tris–HCl pH 8.8: 36.3 g Tris and 48.0 ml 1 M HCl mixed. Titrated to pH 8.8 with HCl if needed. Adjusted to final 100 ml volume
Electrophoresis buffer	Tris–glycine pH 8.3; 3.0 g Tris and 14.4 g glycine dissolved in water. Adjusted to final 1 liter volume

Table 19. Recipe for gel preparation using nondenaturing discontinuous buffer systems

Stock solution	Stacking gel[a]	% Acrylamide in resolving gel[a]			
		20.0	**15.0**	**12.5**	**10.0**
Acrylamide–bisacrylamide (30:0.8)	2.5	20.0	15.0	12.5	10.0
Stacking gel buffer stock[b]	5.0	–	–	–	–
Resolving gel buffer stock[b]	–	3.75	3.75	3.75	3.75
Water	11.5	4.75	9.75	12.25	14.75
1.5% (w/v) APS	1.0	1.5	1.5	1.5	1.5
TEMED[c]	0.015	0.015	0.015	0.015	0.015

Electrophoresis buffer: as described in *Table 18*.
Sample loading buffer: use stacking gel buffer stock diluted 1/4–1/8, containing 10% (w/v) sucrose (or glycerol) and 0.002% tracking dye.
%Acrylamide–polyacrylamide gel concentration expressed in terms of *total* monomer (i.e. acrylamide and crosslinker).
APS, ammonium persulfate; bisacrylamide, *N,N'*-methylene bisacrylamide; TEMED, *N,N,N',N'*-tetramethylethylenediamine.
[a]The columns represent volumes (ml) of the various constituents required to prepare the gel mixtures.
[b]As described in *Table 18*.
[c]Increase volume of TEMED to 0.15 ml for the resolving gel when low pH discontinuous buffer system is used. Adjust volume of water accordingly.

mobility which is inversely proportional to their size. However, as some proteins bind less SDS and so migrate anomalously, the accuracy of size estimates by SDS–PAGE is generally taken to be about 10%. The Laemmli system (113), is a discontinuous SDS system and is probably the most widely used electrophoretic system today. *Table 20* gives recipes for discontinuous systems that are likely to be useful for most types of separation.

23.3. Determination of isoelectric points of enzymes using isoelectric focusing

Isoelectric focusing (IEF) is a method by which enzymes are separated in a pH gradient according to their isoelectric points. The most popular method for generating pH gradients for IEF is the incorporation of low molecular weight amphoteric compounds, synthetic carrier ampholytes, into a polyacrylamide gel matrix. When an electric field is applied the ampholyte molecules 'migrate' to one or other of the electrodes, depending on their net charge. Analysis of the position of the protein band of interest after running the gel will give an indication of its isoelectric point. *Table 21* lists solutions for IEF electrodes and *Table 22* gives recipes for IEF gels.

Table 20. Recipe for gel preparation using the SDS–PAGE discontinuous buffer system (113)

Stock solution	Stacking gel[a]	% Acrylamide in resolving gel[a]			
		20.0	15.0	12.5	10.0
Acrylamide–bisacrylamide (30:0.8)	2.5	20.0	15.0	12.5	10.0
0.5 M Tris–HCl pH 6.8	5.0	–	–	–	–
3.0 M Tris–HCl pH 8.8	–	3.75	3.75	3.75	3.75
10% (w/v) SDS	0.2	0.3	0.3	0.3	0.3
Water	11.3	4.45	9.45	11.95	14.45
1.5% (w/v) APS	1.0	1.5	1.5	1.5	1.5
TEMED	0.015	0.015	0.015	0.015	0.015

Electrophoresis buffer: 0.025 M Tris, 0.192 M glycine, pH 8.3, 0.1% (w/v) SDS.
Sample loading buffer: 2% (w/v) SDS, 5% (v/v) 2-mercaptoethanol, 10% (w/v) sucrose (or glycerol) and 0.002% bromophenol blue, in 0.01 M phosphate buffer pH 7.2 or 0.0625 M Tris–HCl pH 6.8.
Final concn of buffers
    stacking gel: 0.125 M Tris–HCl pH 6.8
    resolving gel: 0.375 M Tris–HCl pH 8.8.
%Acrylamide, APS, bisacrylamide, TEMED, see footnotes to *Table 19*.
[a]The columns represent volumes (ml) of stock solutions required to prepare the gel mixtures.

Table 21. Electrolytes used in IEF

Anolyte	Concn (M)	Used for	Catholyte	Concn (M)	Used for
CH₃COOH	0.5	Alkaline pH ranges (pH > 7)	Histidine	0.2	Acidic pH ranges (pH < 5)
H₂SO₄	0.1	Very acidic pH ranges (pH < 4)	NaOH	1.0	All pH ranges
H₃PO₄	1.0	All pH ranges	Tris	0.5	Acidic and neutral pH ranges (pH < 5)

Table 22. Recipes for preparation of IEF gels (4–7% T and 4–8 M urea)

Gel vol. (ml)	30% T monomer solution[a] (ml)				2% Carrier ampholytes[b] (ml)		Urea (g)		TEMED (μl)	40% APS[c] (μl)
	4% T	5% T	6% T	7% T	A	B	4 M	8 M		
30	4.00	5.00	6.00	7.00	1.50	1.88	7.20	14.40	9.0	30
25	3.34	4.17	5.00	5.83	1.25	1.56	6.00	12.00	7.5	25
20	2.66	3.34	4.00	4.67	1.00	1.25	4.80	9.60	6.0	20
15	2.00	2.50	3.00	3.50	0.75	0.94	3.60	7.20	4.5	15
10	1.33	1.66	2.00	2.33	0.50	0.63	2.40	4.80	3.0	10
5	0.66	0.83	1.00	1.16	0.25	0.31	1.20	2.40	1.5	5

$\%T$ = g monomers per 100 ml; $\%C$ = g crosslinker per 100 g monomers.
[a]Monomer solution
30 %T, 2.5 %C; 29.25 g acrylamide and 0.75 g bisacrylamide mixed. Water added to final 100 ml volume.
30 %T, 3.0 %C; 29.10 g acrylamide and 0.90 g bisacrylamide mixed. Water added to final 100 ml volume.
30 %T, 4.0 %C; 28.80 g acrylamide and 1.20 g bisacrylamide mixed. Water added to final 100 ml volume.
[b]A, for 40% solution (Ampholine, Resolyte®, Servalyte®); B, for Pharmalyte®.
[c]To be added after degassing the solution and just before pouring it into the gel mold.

IEF using immobilized pH gradients (IPG), an alternative to IEF using carrier ampholytes, is often preferred. This is achieved by the incorporation of Immobiline® reagents (Pharmacia Biosystems). The Immobiline® reagents are a series of seven acrylamide derivatives, forming a series of buffers with different pK values distributed throughout the pH 3–10 range. *Table 23* list volumes of Immobiline® required to prepare broad pH unit gradients, and *Table 24* gives recipes for forming two IPG gels. IPG electrode solutions are given in *Table 25*.

23.4. Two-dimensional gel electrophoresis

The most common two-dimensional electrophoresis method for analysing polypeptides is to separate them in the first dimension on the basis of charge by IEF and then to separate the polypeptides in the second dimension in the presence of SDS primarily on the basis of molecular mass of the polypeptides (*Table 26*).

An alternative to IEF–SDS is nonequilibrium pH gradient electrophoresis (NEPHGE)–SDS, used for analysing basic proteins (114). Most of the components, both chemical and apparatus, are the same as in IEF–SDS. The major difference between the two systems is in the first dimension. Instead of applying the sample to the basic end of the gel, it is applied at the acidic end. All the enzymes are thus positively charged and migrate towards the basic end of the gel. Care must be taken not to run the gels too long or the enzymes will migrate too far (*Table 27*).

In contrast, two-dimensional gel electrophoresis can also be achieved under native conditions in the absence of urea or detergents and hence this is suitable for very labile enzymes (*Table 28*).

Table 23. Volume of Immobiline® required to prepare broad pH gradients (118)

pH range	Control pH at 20°C	Acidic dense solution Volume (μl) 0.2 M Immobiline® pK^a						Control pH at 20°C	Basic light solution Volume (μl) 0.2 M Immobiline® pK^a					
		3.6	4.6	6.2	7.0	8.5	9.3		3.6	4.6	6.2	7.0	8.5	9.3
3.5–5.0	3.53±0.06	299	223	157	–	–	–	5.06±0.07	212	310	465	–	–	–
4.0–6.0	4.00±0.06	569	99	439	–	–	–	6.09±0.14	390	521	276	–	–	722
4.0–7.0	4.01±0.05	578	110	450	–	–	–	7.02±0.14	302	738	151	269	–	876
5.0–7.0	5.08±0.03	69	428	414	–	–	–	7.01±0.06	–	474	270	219	–	320
5.0–8.0	5.03±0.12	702	254	416	133	346	–	8.12±0.07	175	123	131	345	346	–
6.0–8.0	6.06±0.08	435	–	323	208	44	–	8.11±0.09	286	–	174	325	329	–
6.0–9.0	6.04±0.14	779	–	402	93	364	80	9.01±0.06	241	–	161	449	237	225
7.0–9.0	7.03±0.24	1349	–	–	272	372	845	8.94±0.07	484	–	–	232	189	546
7.0–10.0	6.98±0.07	542	–	–	378	351	–	9.88±0.06	90	–	–	324	237	225
8.0–10.0	8.10±0.07	399	–	–	364	355	94	9.89±0.05	91	–	–	329	366	289

aVolume of Immobiline® for 15 ml of each starting solution.

Table 24. Recipe for the preparation of immobilized pH gradient gels (119)

Stock solutions	Acidic dense solution	Basic light solution
0.2 M Immobiline®	_[a]	_[a]
Water	To 7.5 ml	To 7.5 ml
1 M Acetic acid[b]	–	Titrate to pH 6.8
1 M NaOH[b]	Titrate to pH 6.8	–
Acrylamide–bisacrylamide (28.8 : 1.2)[c]	2 ml	2 ml
100% Glycerol	3.8 g	–
Water	To 15 ml	To 15 ml
TEMED	10 μl	10 μl
40% (w/v) APS	16 μl	16 μl

[a]Volumes (μl) of 0.2 M Immobiline® as given in *Table 23*.
[b]Neutralization is not necessary when the mixture contains ampholytes in a pI range (e.g. 4–9.5) exhibiting a neutral pH prior to electrofocusing.
[c]These concentrations provide a 4%*T*, 4%*C* gel (%*T* = % acrylamide + % bisacrylamide; %*C* = % bisacrylamide). Polyacrylamide stock solution must be modified to obtain alternative gel concentrations.

Table 25. Electrolytes used in immobilized pH gradients

Solution	Concn	Use
Carrier ampholytes[a,b]	0.3–1.0%	Anolyte and catholyte
Distilled water[b]	–	Anolyte and catholyte
Glutamic acid	10 mM	Anolyte
Lysine	10 mM	Catholyte

[a]Of the same or of a narrower range than the IPG.
[b]For mixed-bed gels or for samples with high salt concentration.

23.5. Immunoelectrophoresis

Immunoelectrophoresis is a procedure in which proteins and other antigenic substances are characterized by both their electrophoretic migration in a gel and their immunological properties. There are many variations of the technique but they are all based on the electrophoretic migration of antigens in an antibody-containing gel and specific immunoprecipitation of the antigens by means of corresponding precipitating antibodies. *Table 29* gives the reagents necessary for immunoelectrophoresis.

24. SAMPLE PREPARATION

In order to calibrate gels, whether they are SDS–PAGE or IEF gels, standard proteins are used as markers. *Tables 30* and *31* list standard marker proteins used in nondenaturing or denaturing conditions of electrophoresis, respectively. It is usual to select three or four proteins that migrate similarly to the enzyme of interest.

25. ANALYSIS OF GELS

Enzymes separated on gels can be visualized using a number of different procedures

Table 26. Recipes for first-dimensional IEF gel and second-dimensional SDS–PAGE gel (120)

First dimension
  Gel mixture: 1.33 ml 28.38% acrylamide, 1.62% bisacrylamide; 5.5 g ultrapure urea; 2 ml 10% NP-40; 0.4 ml 40% Ampholines (pH 5–7); 0.1 ml 40% Ampholines (pH 3.5–10); 1.95 ml water; 10 μl 10% (w/v) APS; and 5 μl TEMED

  Electrolytes: anolyte, 10 mM H_3PO_4; catholyte, 20 mM NaOH
  Sample loading buffer: 9.5 M urea, 5% 2-mercaptoethanol, 2% NP-40, 1.6% Ampholines® (pH 5–7), and 0.4% Ampholines (pH 3.5–10)
  Equilibration buffer: 2.5% (w/v) SDS, 5 mM DTT, 125 mM Tris–HCl pH 6.8, 10% (w/v) glycerol, 0.05% bromophenol blue

Second dimension

Stock solution	Stacking gel[a]	Resolving gel	
		Light solution[a]	Dense solution[a]
29.2% acrylamide, 0.8% bisacrylamide	0.75	5.3	4.3
0.5 M Tris–HCl pH 6.8, 0.4% (w/v) SDS	1.25	–	–
1.5 M Tris–HCl pH 8.8, 0.4% (w/v) SDS	–	4.0	2.0
75% glycerol	–	–	1.7
Water	3.0	6.7	–
10% (w/v) APS	0.015	0.025	0.010
TEMED	0.005	0.008	0.004

Electrophoresis buffer: 25 mM Tris, 0.192 M glycine, 0.1% (w/v) SDS
Sealing gel: 0.1% (w/v) agarose in 0.125 M Tris–HCl pH 6.8. Melt the agarose in Tris buffer and allow to cool to 55°C for normal agarose, or 45°C low gelling temperature (LGT) agarose, before adding a tenth volume 20% (w/v) SDS

[a]Columns represent volumes (ml) of stock solutions required to prepare the gel mixtures.

Table 27. Recipe for the first-dimensional gel for nonequilibrium pH gradient electrophoresis (NEPHGE) (114)

Gel mixture: 1.33 ml 28.38% acrylamide, 1.62% bisacrylamide (30%T, 5.7%C); 5.5 g ultrapure urea; 2.0 ml 10% NP-40; 0.5 ml 40% carrier ampholytes[a]; 1.93 ml water; 20 μl 10% (w/v) APS and 14 μl TEMED

The electrophoresis buffers and the second dimension are the same as that described in *Table 26*.
%T = g monomers per 100 ml; %C = g crosslinker per 100 g monomers.
[a]Use Pharmacia Biosystems Ampholine pH 7–10 if basic proteins are to be analysed. Use pH 3.5–10 if a wider range of pIs is to be studied.

(*Tables 32–35*). Generally, staining using Coomassie blue is the most popular method. However, if a more sensitive stain is required, silver staining is recommended.

If a polypeptide separation is transferred from a gel to a membrane such as nitrocellulose, it becomes extremely sensitive to detection techniques. All of the sample is bound to the surface of the membrane and is therefore available for the binding of specifically labeled probes or for

▶ p. 53

Table 28. Two-dimensional electrophoresis of enzymes under native conditions in the absence of urea or detergents (121)

Stock solutions	First dimension	Second dimension	
	4.0%	4.0%	21.0%
Acrylamide–bisacrylamide (16:0.8)	4.0	10.0	–
Acrylamide–bisacrylamide (42:0.4) and 20% (w/v) sucrose	–	–	20.0
40% (w/v) Ampholine	0.8	–	–
0.3 M Tris–HCl pH 8.9, 0.23% TEMED	–	5.0	5.0
Water	1.2	15.0	10.0
0.1% (w/v) APS	8.0	10.0	5.0
0.23% (v/v) TEMED	2.0	–	–

Electrophoresis buffer
  first dimension: anode, 0.01 M H_3PO_4; cathode, 0.04 M NaOH;
  second dimension: 0.05 M Tris–0.38 M glycine pH 8.3.
Sample loading buffer: 40% (w/v) sucrose.

Table 29. Reagents for immunoelectrophoresis

Gel mixture	1% (w/v) agarose[a] in Tris–barbital buffer pH 8.6
Tris–barbital buffer pH 8.6	2.4 g 5,5-Diethylbarbituric acid[b]; 44.3 g Tris; distilled water to final 1 liter volume. Dilute fivefold before use

Electrophoresis buffer: Tris–barbital buffer pH 8.6.
Agarose is dissolved by boiling for 5 min. Kept molten in a water bath at 50–60°C. Once cool and set, gel can be stored at 4°C, and is ready for use once more after a short period of boiling.
[a]Use agarose with electroendosmosis M_r = 0.13 for normal procedures.
[b]Caution, diethylbarbituric acid is highly poisonous, **handle with care!**

Table 30. Standard marker proteins used in nondenaturing gels (122, 123)

Polypeptide	Species	Tissue	Isoelectric point (pI)	No. of subunits	Subunit $M_r(\times 10^3)$
Acetylcholinesterase	*Electrophorus*	–	4.5	4	70.0
Adenine phosphoribosyl transferase	Human	Erythrocyte	4.8	3	11.0
Adenylate kinase	Rat	Liver (cytosol)	7.5	3	23.0
Alcohol dehydrogenase	Yeast	–	5.4	4	35.0
Aldolase	Yeast	–	5.2	2	40.0
Alkaline phosphatase	Calf	Intestine	4.4	2	69.0
Arginase	Human	Liver	9.2	4	30.0
Catalase	Cow	Liver	5.4	4	57.5
Ceramide trihexosidase	Human	Plasma	3.0	4	22.0
Chymotrypsinogen A	Bovine	Pancreas	9.2	1	25.7
Deoxyribonuclease I	Cow	Pancreas	4.8	1	31.0
Deoxyribonuclease II	Pig	Spleen	10.2	1	38.0
Galactokinase	Human	Erythrocyte	5.7	2	27.0

Table 30. Continued

Polypeptide	Species	Tissue	Isoelectric point (pI)	No. of subunits	Subunit $M_r(\times 10^3)$
β-Glucuronidase	Rat	Liver	6.0	4	75.0
Glyceraldehyde 3-phosphate dehydrogenase	Rabbit	Muscle	8.5	2	72.0
Glycerol 3-phosphate dehydrogenase	Rabbit	Kidney	6.4	2	34.0
Glycogen synthetase	Pig	Kidney	4.8	4	92.0
Hemoglobin	Rabbit	Erythrocyte	7.0	4	16.0
Hexokinase	Yeast	–	5.3	2	51.0
β-Lactoglobulin	Bovine	Serum	5.2	2	17.5
Lipoxidase	Soybean	–	5.7	2	54.0
Lysine decarboxylase	E. coli	–	4.6	10	80.0
Malate dehydrogenase	Pig	Heart	5.1	2	35.0
Micrococcal nuclease	S. aureus	–	9.6	1	16.8
Pepsinogen	Pig	Stomach	3.7	1	41.0
Phosphoenolpyruvate carboxylase	E. coli	–	5.0	4	99.6
Phosphoenolpyruvate carboxylase	Spinach	Leaf	4.9	2	130.0
Ribonuclease	Bovine	Pancreas	7.8	1	13.7
Triosephosphate isomerase	Rabbit	Muscle	6.8	2	26.5
Trypsinogen	Cow	Pancreas	9.3	1	24.5
Urease	Jack bean	–	4.9	2	240.0
Uricase	Pig	Liver	6.3	4	32.0

Table 31. Standard marker proteins used in denaturing gels (122, 124–127)

Polypeptide	Species	Tissue	Isoelectric point (pI)	No. of subunits	Subunit $M_r(\times 10^3)$
Alcohol dehydrogenase	Horse	Liver	–	–	41.0
Aldolase	Rabbit	Muscle	–	–	40.0
Bovine serum albumin	Bovine	Serum	4.7	1	68.0
Carbonic anhydrase	Parsley	Leaf	–	6	28.7
Catalase	Bovine	Liver	–	–	57.5
Enolase	Rabbit	Muscle	8.8	2	42.0
Fumarase	Pig	Heart	–	4	48.5
β-Galactosidase	E. coli	–	–	4	130.0
Glutamate dehydrogenase	Bovine	Liver	–	–	53.0
Glyceraldehyde 3-phosphate dehydrogenase	Rabbit	Muscle	–	–	36.0
α-Lactalbumin	Bovine	Milk	–	–	14.4
Lactate dehydrogenase	Pig	Heart	–	4	36.0
Lysozyme	Chicken	Egg white	10.7	1	14.3
Myoglobin	Horse	Heart	6.8	1	16.9

Table 31. Continued

Polypeptide	Species	Tissue	Isoelectric point (pI)	No. of subunits	Subunit $M_r (\times 10^3)$
Myosin heavy chain	Rabbit	Muscle	–	2	212.0
Phosphorylase a	Rabbit	Muscle	5.8	4	92.5
Pyruvate kinase	Rabbit	Muscle	6.6	4	57.2
Trypsin inhibitor	Soybean	–	–	–	20.1

Several proteases, such as trypsin, chymotrypsin and papain, have been used as molecular mass standards but these may sometimes cause proteolysis of other polypeptide standards and thus are omitted here.

Table 32. Staining and detection methods used in IEF

Method	Application	References
Alcian blue	Glycoproteins	128
Coomassie blue G-250	General use	129
Coomassie blue G-250/urea/perchloric acid	In presence of detergents	130
Copper stain	General use	131
Fast green FCF	General use	132
Immunoprecipitation *in situ*	Antigens	133
Silver stain	General use	134
Sudan black	Lipoproteins	135
Zymograms[a]	Enzymes	136

[a]The concentration of buffer in the assay medium usually needs to be increased to counteract the buffering action by carrier ampholytes.

Table 33. Staining procedures for two-dimensional polyacrylamide gels

Staining technique	Destaining technique	Comments
1. 3–4 h in 0.1% Coomassie blue in methanol : water : acetic acid (5 : 5 : 1)	Overnight by diffusion against methanol : water : acetic acid (5:5:1)	–
2. 20 min in 0.1% Coomassie blue in 50% trichloroacetic acid	Several changes of 7% (v/v) acetic acid	Removal of commercial ampholytes
3. 3 h in 0.25% Coomassie blue in methanol : water : acetic acid (5 : 5 : 1)	Several changes of 5% (v/v) methanol, 10% (v/v) acetic acid	–
4. 1–4 h in 0.1% Coomassie blue R-250 in 7.5% (v/v) acetic acid, 50% (v/v) methanol in water	Overnight in 7.5% (v/v) acetic acid, 50% (v/v) methanol in water	–
5. Overnight in 25% (v/v) isopropyl alcohol, 10% (v/v) acetic acid, 0.025–0.05% Coomassie blue, followed by 6–9 h in 10% (v/v) isopropyl alcohol, 10% (v/v) acetic acid, 0.0025–0.005% Coomassie blue	Several changes of 10% (v/v) acetic acid	An additional optional staining overnight in 10% (v/v) acetic acid containing 0.0025%

Table 33. Continued

Staining technique	Destaining technique	Comments
		Coomassie blue helps to intensify the gel pattern
6. 15 min in 0.55% Amido black in 50% (v/v) acetic acid	40 h in 1% (v/v) acetic acid	–
7. 3 h at 80°C, or overnight at RT in 0.1% Amido black in 0.7% (v/v) acetic acid, 30% (v/v) ethanol in water	Several changes of 7% (v/v) acetic acid, 20% (v/v) ethanol in water	–
8. 1 h in 50% (v/v) methanol, stain with fresh alkaline 0.8% AgNO$_3$ solution. Wash with water for 5 min. To develop soak in fresh 0.02% HCHO, 0.005% citric acid for 10 min	Wash gel in water, transfer to 50% (v/v) methanol	Much more sensitive than Coomassie blue methods

Table 34. Staining procedures used after immunoelectrophoresis

Stain	Staining procedure	Destaining procedure
Coomassie blue R-250	Stain gel for 5 min in 0.5% Coomassie blue R-250 in 96% ethanol : glacial acetic acid : water (4.5 : 1 : 4.5)	Destain 2–3 times, each for 5 min in 96% ethanol : glacial acetic acid : water (4.5:1:4.5)
Nigrosin	Stain gel until precipitates are stained sufficiently in 0.14% Nigrosin in glacial acetic acid : 0.1 M sodium acetate : methanol : glycerol (2.15 : 5.75 : 1.4 : 0.7)	Destain in 20% (v/v) methanol in 5% (v/v) acetic acid followed by 5% (v/v) acetic acid and water

Table 35. Detection of specific enzymes on gels

Enzyme	References	Enzyme	References
Acid phosphatase	136–139	Alkaline phosphatase	136–139
Aconitase	136, 138, 139	Amine oxidase	137
Adenine phosphoribosyl transferase	136	Amino acid oxidase	137, 141
		AMP deaminase	140
Adenosine deaminase	136, 138	Amylase	136, 137
Adenosine kinase	140	Arginase	140
Adenylate kinase	136, 138, 139	Arginosuccinase	140
ADP–glycogen transferase	137	Aromatic amino acid transaminase	139
Alanine aminotransferase	136, 138		
Alcohol dehydrogenase	136–139	Arylsulfatases	140
Aldolase	136, 138, 139	Aspartate aminotransferase	136, 139

Table 35. Continued

Enzyme	References	Enzyme	References
Aspartate carbamoyl transferase	136	Glycogen phosphorylase	137
		Glycolate oxidase	140
Carbonic anhydrase	136, 139	Glycosidases	137
Cellobiose phosphorylase	137	Glyoxalase	136, 141
Cholinesterase	137	Guanine deaminase	136, 137
Citrate synthase	136	Guanylate kinase	137
Creatine kinase	136, 137, 139	Hexokinase	136, 137, 139
3', 5'-cyclic AMP phosphodiesterase	136, 137		
		Hyaluronidase	146
Cysteine S-conjugate	142	Hydroxyacyl-CoA dehydrogenase	141
Cytidine deaminase	136		
DNA polymerase	137	D(-)3-Hydroxybutyrate dehydrogenase	139
DNase	143		
Enolase	136, 138	Hydroxysteroid dehydrogenase	137
Esterases	136–139	Hypoxanthine–guanine phosphoribosyl transferase	136, 137, 139
Folate reductase	137		
β-D-Fructofuranosidase	138	Inorganic pyrophosphatase	136, 137, 139
Fructose 1,6-bisphosphate	139		
α-Fucosidase	136, 137	Inosine triphosphatase	137
Fumarase	136, 138, 139	Isocitrate dehydrogenase	136–139
Galactokinase	136, 137	α-Ketoglutarate semialdehyde dehydrogenase	137
Galactose 6-phosphate dehydrogenase	139		
		α-Ketoisocaproate dehydrogenase	137
Galactose 6-phosphate uridyltransferase	136, 137		
		α-Keto-β-methyl valerate dehydrogenase	137
α-Galactosidase	136, 138		
β-Galactosidase	136, 137	Lactate dehydrogenase	136–139
Glucanases	144, 145	Leucine aminopeptidase	139
Glucose oxidase	137	Lipase	137
Glucose 6-phosphate dehydrogenase	136, 138, 139	Lipoyl dehydrogenase	137
		Malate dehydrogenase	136–139
Glucose phosphate isomerase	138	Malic enzyme	136, 138
		Mannose-phosphate isomerase	136, 138
Glucose 1-phosphate uridylyltransferase	136, 137	α-Mannosidase	136, 138
		β-N-acetyl-hcxosaminidase	136, 138
α-Glucosidase	136, 137	NAD(P) nucleosidase	141
β-Glucosidase	136, 137	Nitrate reductase	137
β-Glucuronidase	138, 139, 141	Nucleoside phosphorylase	136, 138
Glutamate dehydrogenase	137, 139, 141	Nucleotidase	137
Glutamine transaminase K	142	Ornithine carbamoyltransferase	136
Glutathione peroxidase	136	Oxidases	137
Glutathione reductase	136, 139	Peptidases	136, 138, 139
Glyceraldehyde 3-phosphate dehydrogenase	136–139		
		Peroxidase	138, 139, 147
Glycerol kinase	137	Phosphofructokinase	136, 138
Glycerol 3-phosphate dehydrogenase	136–139	Phosphoglucoisomerase	136, 138, 139

Table 35. Continued

Enzyme	References	Enzyme	References
Phosphoglucomutase	136, 138, 139, 141	RNase	137, 139
		Sorbitol dehydrogenase	136, 139
Phosphogluconate dehydrogenase	136, 138, 139, 148	Sucrose phosphorylase	137
		Superoxide dismutase	136
Phosphoglycerate kinase	136, 138	Thymidine kinase	139
Phosphoglyceromutase	136, 138	Triose-phosphate isomerase	136, 138, 139
Phosphoglycollate phosphatase	140	UDPG dehydrogenase	137
Phosphatases	149	UDPG pyrophosphorylase	136, 137
Polynucleotide phosphorylase	137	UMP kinase	136
Pyridoxine kinase	141	Urease	137
Pyruvate kinase	136, 138	Xanthine dehydrogenase	139
Retinol dehydrogenase	139		

autoradiography without quenching. *Table 36* lists the many blotting matrices available. *Table 37* gives blocking solutions to prevent nonspecific binding of probe to matrix. Transfer buffers for the transfer of protein to the blotting matrix are given in *Table 38*.

Detection procedures of proteins on blots are many, of which some are exemplified in *Table 39*. If the polypeptides are radioactive, they can be located by using autoradiography (*Tables 40* and *41*).

Table 36. Common blotting membranes

Blotting matrix	Additional information
Glass fiber	Recommended for amino acid sequence analysis of separated proteins. Protein-binding capacity between $10–20$ μg cm^{-2}. The proteins can be blotted from SDS–PAGE gels and then acid-hydrolysed directly while still immobilized on the filter
Nitrocellulose membrane	Pure nitrocellulose membranes have good protein-binding capacity ($\sim 80–100$ μg cm^{-2}). Nitrocellulose membranes with 0.45 μm pore size are typically used but $0.22–0.1$ μm has been recommended for lower molecular mass proteins. Recommended for immunodetection analyses due to reduced nonspecific binding of antibodies, and for the analysis of basic proteins. Mixed ester membranes which contain cellulose acetate have reduced capacity
Nylon membrane	Far stronger and more robust than conventional pure nitrocellulose sheets. Substantially higher protein binding capacity, e.g. Zeta-Probe® (Bio-Rad) has approx. sixfold higher protein binding capacity (~ 480 μg cm^{-2}) than nitrocellulose. Due to its highly cationic nature, recommended for electroblotting of SDS–PAGE gels to maximize binding of highly anionic SDS–polypeptide complexes. Charged nylon filters such as Zeta-Probe® are recommended for electroelution of SDS–PAGE gels. Binding is also much stronger than in the case of nitrocellulose. Uncharged nylon membranes should give higher binding

Table 36. Continued

Blotting matrix	Additional information
	of basic proteins. A major disadvantage of all nylon membranes is the lack of simple general staining procedures
Polyvinyldifluoride (PVDF)	Hydrophobic in nature. Compatible with commonly used protein stains as well as standard immunodetection methods

Table 37. Blocking solutions to prevent nonspecific binding of the probe to the matrix when blotting (150)

Blocker	Concn (% w/v)	Blocker	Concn (% w/v)
Bovine serum albumin (BSA)	0.5–10	Milk	5
Casein	1–2	Newborn calf serum (NCS)	5
Ethanolamine	10	Ovalbumin	1–5
Fetal calf serum (FCS)	10	Polyvinylpyrrolidone	2
Gelatin	0.25–3	Tween 20	0.05–0.5

Table 38. Transfer buffers used in blotting

Transfer buffer	Additional information	Reference
25 mM Tris, 192 mM glycine, 20% (v/v) methanol, pH 8.3	0.05–0.1% (w/v) SDS can be included	151
48 mM Tris, 39 mM glycine, 20% (v/v) methanol, pH 9.2	0.0375% (w/v) SDS can be included	152
10 mM $NaHCO_3$, 3 mM $NaCO_3$, 20% (v/v) methanol, pH 9.9		153
0.7% (v/v) acetic acid		154

Table 39. General stains for blot transfers

Protein stain	Blotting matrix	Sensitivity
Amido black 10B	Nitrocellulose, PVDF	1.5 µg
Colloidal gold	Nitrocellulose, PVDF	4 ng
Colloidal iron	Nitrocellulose, nylon, PVDF	30 ng
Coomassie brilliant blue R-250[a]	Nitrocellulose, PVDF	1.5 µg
Fast green FC	Nitrocellulose, PVDF	–
India ink	Nitrocellulose, PVDF	100 ng
In situ biotinylation + HRP-avidin	Nitrocellulose, nylon, PVDF	30 ng
Ponceau S	Nitrocellulose, PVDF	–
Silver enhanced copper	Nitrocellulose, nylon	–

HRP, horseradish peroxidase.
[a]Results obtained in high background.

Table 40. *In situ* detection of radioactive proteins in gels

Detection method	References
Double-label detection using X-ray film	155, 156
Electronic data capture	157, 158
Fluorography using PPO in DMSO	159, 160
Fluorography using PPO in glacial acetic acid	161
Fluorography using sodium salicylate	160, 162, 163
Fluorography using commercial reagents	164, 165
Image intensification	166
Indirect autoradiography using an X-ray intensifying screen	167, 168
Quenching of radiolabeled proteins by gel conditions	169
Radiolabeling proteins *in vivo* prior to electrophoresis	170, 171
Radiolabeling proteins *in vitro* prior to electrophoresis	170
Radiolabeling proteins after gel electrophoresis	172, 173

DMSO, dimethylsulfoxide; PPO, 2,5-diphenyloxazole.

Table 41. Sensitivities of methods for radioisotope detection in polyacrylamide gels (174)

Isotope	Method	Detection limit d.p.m. cm^{-2} for 24 h	Relative performance compared to direct autoradiography
^{3}H	Direct autoradiography	$> 8 \times 10^6$	1
	Fluorography using PPO	8000	> 1000
^{14}C or ^{35}S	Direct autoradiography	6000	1
	Fluorography using PPO	400	15
^{32}P	Direct autoradiography	525	1
	Intensifying screen	50	10.5
^{125}I	Direct autoradiography	1600	1
	Intensifying screen	100	16

The data are for exposure at $-70°C$ using X-ray film pre-exposed to $A_{540} = 0.15$ above the background absorbance of unexposed film.

26. TROUBLESHOOTING

This is described in *Table 42*.

Table 42. Troubleshooting guide

Symptom	Cause	Remedy
1. Lack of polymerization. The gel consistency is not firm, gel does not hold its shape after removal from the mold. Too fast or too slow polymerization	Incorrect concentrations of prepared reagents; omission of reagent from gel mixture; impurities in reagents; old APS stock solution	Discard solutions and prepare fresh batch using pure reagents. Check and vary concentrations of polymerization catalysts

Table 42. Continued

Symptom	Cause	Remedy
2. Cracking of gel during polymerization	Excessive heat production by polymerization reaction	Use cooled solutions. For rod gels, siliconizing the tubes may also help
3. Detachment of slab gels from glass plates during gel electrophoresis	Inadequately cleaned plates; low concn gels sometimes detach from rod gel tubes even though these are clean	Thoroughly clean plates and tubes, a solution of a strong decontaminating detergent such as Contrad 70, DECON 90, or RBS 35 should be used. Rinse extensively with distilled water, and finally with ethanol prior to air-drying. For latter cause, attach a piece of nylon mesh to bottom of the tube
4. Failure of the sample to form a layer at the well bottom when applied to slab gels	Omission of sucrose or glycerol from sample buffer; use of sample comb where teeth do not form a snug fit with the glass plates, permitting gel to polymerize between the teeth and glass plates	Prepare fresh batch of sample buffer. Use better-fitting comb, but, in the short term, remove excess gel from wells using syringe needle
5. Insoluble material in the sample.	Provided that the ionic strength of nondissociating buffers is high enough to prevent aggregation of native proteins, denatured protein represents the insoluble material	Remove by centrifugation prior to electrophoresis
6. Insolubility in SDS-containing buffers	Too little SDS; too little reducing agent; too low a pH, especially after trichloroacetic acid precipitation of proteins	Adjust concentrations of the said reagents; add urea in addition to SDS to ensure solubilization
7. Protein streaking along individual rod gels or slab gel tracks, and protein at the gel origin	Protein precipitation followed by dissolution of the protein precipitates during electrophoresis; overloading of the gel	See symptoms no. 5 and no. 12; amount of sample loaded should be decreased
8. Protein bands observed in all tracks of a slab gel or all rod gels; continuous stained region from the gel origin to near the buffer front, even in tracks which have not been loaded with	Contamination of the sample buffer; contaminated electrophoresis buffer; sample from one well has contaminated adjacent wells, usually by overflowing	Prepare fresh buffers; reduce the volume of sample loaded

Table 42. Continued

Symptom	Cause	Remedy
sample; same protein bands observed in several neighboring lanes of a slab gel		
9. High background of protein staining along individual rod gels or slab gel tracks with indistinct bands. Irreproducibility of protein band pattern; reduction in the staining intensity or complete loss of individual components or appearance of previously unobserved fast migrating bands	Caused by problems of sample preparation; extensive sample proteolysis; use of impure grades of SDS	Work at low temperature and use protease inhibitors during sample preparation. If problem occurs with SDS–PAGE, check that sample is heated to at least 90°C for 2 min during dissociation; use purified SDS
10. Distorted bands	Insoluble material; bubbles in the gel; inconsistent pore size throughout the gel; uneven heating of the gel	Filter gel reagents before use and ensure gel mixture is well mixed and degassed before pouring the gel; use a cooled apparatus or reduce the current during electrophoresis
11. Variations in staining density along the width of a stained band	Uneven gel surface, resulting in sample accumulating at the low points prior to electrophoresis	Overlay gels carefully and prepare gels in vibration-free places
12. Heavily stained band at the gel origin	Inability of a substantial portion of the protein to enter the resolving gel; may be aggregated protein in sample prior to electrophoresis or, in the case of nondenaturing discontinuous buffer systems, precipitation of the proteins due to the formation of highly concentrated zones during electrophoresis in the stacking gel	Aggregated protein (see symptom no. 5). In the case of the nondissociating discontinuous buffer systems, use less concentrated samples with a continuous buffer system
13. Particulate material in the sample and streaking in both dimensions	May be using too low a ratio of solubilizing agent to sample	Try using larger ratios of extraction buffer to sample

Table 42. Continued

Symptom	Cause	Remedy
14. Reduction in sharpness, streaking of spots, and smearing at the alkaline end of the gel during sample loading	Interaction of nucleic acids with proteins and carrier ampholytes; can form a precipitate	In some cases nucleic acids can be removed by selective extraction or precipitation procedures (175–178)
15. Poor staining after SDS–PAGE; uneven staining with any buffer system; stained bands lost on destaining		Increase volume of staining solution to dilute out the SDS present; allow more time for staining step to permit the dye to penetrate fully; reduce destaining time or use a better fixative
16. Metallic sheen on gels after staining with Coomassie blue R-250	Solvent has been allowed to evaporate, causing dye to dry on the gel at that point	Slight films of Coomassie blue sometimes observed on the gel surface after destaining can be removed by a quick rinse in 50% methanol or by gently swabbing the gel surface with methanol-soaked tissue paper
17. High background after silver-staining for proteins	Acrylic acid contamination in the acrylamide and/or bisacrylamide. Nucleic acids may also stain with silver stains	Highest quality reagents, including the purest (deionized) water, should always be used
18. Proteins left in gel after blotting	Protein probably precipitated in the gel	Add <0.01% SDS to transfer buffer
19. After blotting, proteins not in gel, and not present on blot	Protein lost	Ensure polarity of blotting used is correct (proteins should move towards positive pole)
20. After blotting no protein left in gel, but protein bands stain only faintly on blot. The front and back sides of the blot are equally stained	Protein moved through the filter, or protein was never present in gel; low binding capacity	Check sample protein concentration in gel; reduce detergent load in the blotting buffer; check pH of the blotting buffer and whether methanol was added to the transfer buffer
21. Staining intensity of protein in bands on nitrocellulose is weaker than expected	Not enough protein loaded; fraction of protein moved through the membrane	Determine protein concentration; also see symptom no. 20

Table 42. Continued

Symptom	Cause	Remedy
22. Protein bands appear fuzzy on the filter	Inefficient electrophoresis system; air bubbles between gel and nitrocellulose during transfer	Check electrophoresis system; remove air bubbles by passing glass pipette over membrane before blotting sandwich is assembled
23. Artifactual blackening of X-ray film during fluorography	Inadequate removal of DMSO	Ensure sufficient soaking in water before drying the gel

E. ACTIVE SITE TITRATION

K. Brocklehurst

27. INTRODUCTION

Kinetic characterization of enzyme reactions (involving, for example, determination of k_{cat}, via V_{max}, and k_{cat}/K_m) requires knowledge of the molarity of functional active sites ('operational' molarity). The use of a rate assay for this purpose (see, for example, refs 179, 180) is unsatisfactory, mainly because it requires a sample of 100% pure active enzyme as a primary standard (which is difficult, if not impossible, to obtain) but also because of additional uncertainties related to variables such as temperature, ionic strength, pH and concentration of substrate and possibly of cofactor that need to be controlled accurately in a rate assay. Concentrations of active enzyme are determined instead by making use of stoichiometric reactions or interactions of the intact enzyme active site with an inhibitor or substrate (i.e. an 'all or none' assay) to produce a product that is readily observable by a physico-chemical technique. This is known as an active site titration. ('Active site' is in common usage in this context and is retained in this section. More generally it is more informative to distinguish catalytic site from binding sites and to refer to the combination of the two as the active center.)

Techniques for active site titration were developed initially for a variety of proteolytic enzymes, and titration of this class of enzyme, particularly by using analog substrates, still dominates the literature. This section describes the principles of active site titration, illustrates developments aimed at improvements in selectivity, sensitivity, generality and reliability in data analysis, and provides a selection of examples.

The characteristics of a good titrant are that it should:

(i)   recognize structural features of the active site to provide specificity for active enzyme.
(ii) Permit titration over a short interval of time (a few minutes at most) to obviate problems arising from denaturation and for convenience.
(iii) Be usable in the pH range that maintains conformational integrity of the enzyme.
(iv) Be a stable and water-soluble compound whose stoichiometric reaction with the active site is easily detectable by a convenient and sensitive physico-chemical technique, notably electronic absorption or fluorescence spectroscopy.
(v)   Provide a reliable titration over a wide range of enzyme concentration.

These characteristics are similar, but not identical, to those proposed by Bender *et al.* (181) in their seminal paper on the use of chromogenic substrates for the active site titration of various hydrolytic enzymes.

28. PRINCIPLE OF TITRATION USING (CHROMOGENIC ANALOG) SUBSTRATES

Titration using substrates is described in terms of the well-known minimal kinetic model that applies to some proteinases, notably serine proteinases such as chymotrypsin and cysteine proteinases such as papain (i.e. the three-step acylenzyme model):

$$E + S \underset{}{\overset{K_s}{\rightleftharpoons}} ES \xrightarrow{k_{+2}} P_1 + ES' \xrightarrow{k_{+3}} E + P_2$$

where ES is the adsorptive complex; ES', the acylenzyme intermediate; P_1, the chromophoric leaving group (e.g. 4-nitrophenolate if substrate S is a 4-nitrophenyl ester); P_2, the carboxylate product; and K_s, the equilibrium dissociation constant of ES.

Different approaches to titration may be used (181) which vary according to relative values of $[S]_o$, $[E]_T$ (concentration of enzyme active sites), k_{+2}, k_{+3}, K_s, and K_m (the value of $[S]_o$ when $v_i = V_{max}/2 = k_{+3} (k_{-1} + k_{+2})/k_{+1} (k_{+2} + k_{+3})$ which becomes $k_{+3} K_s/(k_{+2} + k_{+3})$ when quasi-equilibrium around ES, i.e. $k_{+2} << k_{-1}$ or $k_{cat}/K_m << k_{+1}$ (see ref. 182) may be assumed; for the three-step model $k_{cat} = k_{+2} k_{+3}/(k_{+2} + k_{+3})$).

The most commonly used approach specifies $[S]_o >> [E]_T$ and records $[P_1]$ as a function of time as a biphasic progress curve: a pre-steady-state phase (acylation) which is usually essentially instantaneous on the time-scale of a conventional spectrophotometer and a slow steady-state phase (deacylation) in the early stage of which increase in $[P_1]$ with increase in t is essentially linear. $[P_1]$ produced in time t is given by *Equation 1* (ref. 183); quasi-equilibrium around ES is assumed and thus $(k_{-1} + k_{+2})/k_{+1}$ is written as K_s in the expression for a.

$$[P_1] = \frac{k_{cat}[E]_T[S]_o t}{K_m + [S]_o} + \left(\frac{k_{cat}/k_{+3}}{1 + K_m/[S]_o} \right) \left(1 - e^{-at} \right) [E]_T \qquad \text{Equation 1}$$

where $a = \dfrac{(k_{+2} + k_3)[S]_o + k_{+3} K_s}{K_s + [S]_o}$.

After completion of the pre-steady state, when the value of t is large, the exponential term approaches zero and *Equation 1* becomes *Equation 2* in which π, the concentration of P_1 formed at completion of the pre-steady state, is given by *Equation 3*.

$$[P_1] = \pi + \frac{k_{cat}[E]_T[S]_o t}{K_m + [S]_o} \qquad \text{Equation 2}$$

$$\pi = \left(\frac{k_{cat}/k_{+3}}{1 + K_m/[S]_o} \right)^2 \cdot [E]_T. \qquad \text{Equation 3}$$

π is determined experimentally as the intercept on the $[P_1]$ axis when the linear steady-state phase of the plot of $[P_1]$ versus t is extrapolated to $t = 0$.

The significance of the value of π depends on the relative values of $[S]_o$ and K_m, with $k_{+2} >> k_{+3}$ assumed in order to produce the 'initial burst' of P_1 in the biphasic $[P_1]$ versus t plot.

Equation 3 may be transformed into *Equation 4*. With the condition $k_{+2} >> k_{+3}$ this becomes *Equation 5*, which it is useful to write also as *Equation 6*.

$$\frac{1}{\sqrt{\pi}} = \left(\frac{k_{+2} + k_{+3}}{k_{+2}}\right) \cdot \frac{1}{\sqrt{[E]_T}} + \left(\frac{k_{+2} + k_{+3}}{k_{+2}}\right) \cdot \frac{K_m}{[S]_o} \cdot \frac{1}{\sqrt{[E]_T}}$$

Equation 4

$$\frac{1}{\sqrt{\pi}} = \left(1 + \frac{K_m}{[S]_o}\right) \cdot \frac{1}{\sqrt{[E]_T}}$$

Equation 5

$$\frac{1}{\sqrt{\pi}} = \frac{1}{\sqrt{[E]_T}} + \frac{K_m}{\sqrt{[E]_T}} \cdot \frac{1}{[S]_o}$$

Equation 6

When the titration is carried out with $[S]_o >> K_m$ *Equation 5* shows that $\pi = [E]_T$ and the active site concentration may be read directly as the intercept on the $[P_1]$ axis.

If it is not practicable to use $[S]_o >> K_m$, π is determined at several values of $[S]_o$ and a plot of $1/\sqrt{\pi}$ versus $1/[S]_o$ provides the value of $1/\sqrt{[E]_T}$ as the ordinate intercept when $1/[S]_o = 0$ (*Equation 6*).

29. DEVELOPMENT OF THE TITRATION TECHNIQUE USING CHROMOGENIC AND FLUOROGENIC SUBSTRATES

Techniques for active site titration have been developed and improved over the years as indicated below.

(i) Observation of a biphasic progress curve (initial burst followed by zero-order steady state) in the α-chymotrypsin-catalysed hydrolysis of 4-nitrophenyl acetate providing evidence of an acylenzyme intermediate and the possibility of an active site titration (184).

(ii) Derivation of the complex rate equations rigorously describing the hydrolysis of 4-nitrophenyl acetate catalysed by α-chymotrypsin (183, 185–187).

(iii) Development of an active site titration of α-chymotrypsin using the chromophoric acylating agent, *N-trans*-cinnamoylimidazole (188).

(iv) Development of titration theory and of practical methodology for the active site titration of a variety of hydrolytic enzymes: α-chymotrypsin, trypsin, elastase, subtilisin, papain and acetylcholinesterase (181, 189).

(v) Variation in the structure of the titrant to provide a slower steady-state reaction while maintaining a rapid 'burst' phase. A problem with using specific substrates is that the steady-state reaction that occurs subsequent to the initial 'burst' of P_1 may be too rapid to allow a reliable extrapolation to provide an accurate value of π. This is the case, for example, with the use of *N*-benzyloxycarbonyl L-tyrosine 4-nitrophenylester as a titrant for papain (181) where deacylation is rapid, necessitating titration at pH 3. By changing the acyl moiety of the substrate to that of *N*-benzyloxycarbonyl D-norleucine while retaining 4-nitrophenolate as the electronically activated leaving group, deacylation is retarded (due to the D-configuration of the norleucine moiety), which permits a convenient titration in approximately neutral media (190).

(vi) Variation in the structure of the leaving group to provide increased sensitivity. The titration of trypsin is used as an example. Chase and Shaw (191) introduced 4-nitrophenyl 4'-guanidinobenzoate (NPGB) as a valuable titrant for trypsin and trypsin-like enzymes which exert a high specificity for cationic substituents that can bind in the S_1-subsite. Acylation of the hydroxy group of the catalytic site Ser residue produces a 'burst' of 4-nitrophenolate, which is recorded as an increase in A_{410} and quantified by using a pH-dependent value of ε_{410} (e.g. $\varepsilon_{410} = 1.68 \times 10^4$ $M^{-1}cm^{-1}$ at pH 8.3 and $\varepsilon_{410} = 1.26 \times 10^4$ M^{-1} cm^{-1} at pH 7.5). The sensitivity of the titration is greatly increased by replacing the 4-nitrophenolate leaving group of NPGB by a fluorogenic leaving group. The sensitivity is increased by c. two orders of magnitude by using the 4-methylumbelliferone substrate (192) or by 5–6 orders of magnitude by using the fluorescein substrate (193). Unless the fluorescein titrant is considerably in excess of the enzyme, however, complications in both kinetic and end-point analysis arise because monoacylfluorescein is fluorescent. This difficulty, which is a general criticism of fluorescein-based substrates, is overcome by using as the fluorophore a fluorescein derivative containing only a single hydroxy substituent on the xanthene structure, 3'-hydroxyspiro[isobenzofuran-1(3H), 9'-[9H] xanthen]-3-one (3-hydroxyfluoran, 3HF) which can be measured at 10^{-12} M. The titrant containing the guanidinobenzoyl recognition feature and the 3HF leaving group can titrate 10^{-11}–10^{-12} M trypsin (194).

(vii) Variation in the structure of the titrant to provide increased specificity. In principle it should be possible to provide specificity for a particular member of an enzyme family by making use of known molecular recognition features of the active sites. In some cases, however, specificities are not realized as effectively as might have been predicted. For example, active site titrants designed for trypsin based on recognition of the 4-guanidinobenzoate moiety react also with α-chymotrypsin (195). Similarly, some active site titrants designed for α-chymotrypsin (e.g. N-benzyloxycarbonyl-L-tyrosine-4-nitrophenylester) react also with trypsin (196). Specificity is increased in some cases by use of azapeptide 4-nitrophenyl esters (197). For example Ac-Ala-Aphe-ONp reacts with α-chymotrypsin but not with either trypsin or elastase.

30. TITRATION USING NONCHROMOGENIC SUBSTRATES

Although the use of chromophores or fluorophores provides the most convenient version of active site titration using substrates, it is sometimes necessary to resort to more laborious methods using other types of detection (e.g. the use of a radiolabel). Examples are:

(i) Titration of penicillin-binding protein 4 of *Escherichia coli* by using [^{14}C]benzylpenicillin (198).

(ii) Titration of the *Drosophila* kinesin motor domain containing both the microtubule and ATP binding sites using [α-^{32}P]ATP (199).

(iii) Titration of aminoacyl-tRNA synthetases by using either saturating amounts of [^{14}C]-labeled amino acid and ATP (200, 201) or [γ-^{32}P]ATP (202). On mixing tRNA synthetase, the appropriate amino acid, [γ-^{32}P]ATP and inorganic pyrophosphatase under suitable conditions there is an initial rapid 'burst' of depletion of ATP (rate constant, k_1) stoichiometric with the formation of enzyme-bound amino acid adenylate. Following the 'burst' an initially linear further decrease in [ATP] occurs as the complex undergoes hydrolysis (rate constant, k_2) to regenerate active enzyme. The initial burst provides the stoichiometry of aminoacyl adenylate formation, provided that $k_2 < < k_1$ Complexes too unstable to be isolated by gel or nitrocellulose disk filtration may be assayed in this way.

31. TITRATION USING CHROMOGENIC TIME-DEPENDENT INHIBITORS: GENERAL PRINCIPLES

An alternative to titrations using chromogenic substrates, where rapid reaction subsequent to the initial burst of product can be problematic, is the use of chromogenic time-dependent ('irreversible') inhibitors.

The requirement is for a reagent that reacts specifically and stoichiometrically with the active site of the enzyme to produce a stable inactive derivative with consequent spectral change.

The use of an inhibitor rather than a substrate obviates the difficulties in the titration associated with rapid turnover.

The chief potential disadvantage of using an inhibitor is that the reaction may depend on a chemical reaction of, most commonly, an active site nucleophile with an electrophilic reagent that may not easily distinguish reaction of an intact functional active site from reactions of protein fragments, denatured enzyme or contaminants containing an analogous nucleophilic center.

This potential difficulty may be overcome in two ways: (i) by making use of special features of particular active sites that provide unusual reactivities towards group-specific reagents that are not formally substrate analogs and which result in site-selectivity; and (ii) by using site-directed or mechanism-based inhibitors that make use of known specific recognition features of the active site. Examples of titrants that make use of each of these approaches are given below.

32. EXAMPLES OF THE TWO APPROACHES TO PROVISION OF SITE SPECIFICITY IN REACTIONS OF TIME-DEPENDENT INHIBITOR TITRANTS

32.1. Exploitation of special features of active sites

This is exemplified by the use of 2-pyridyl disulfides (R-S-S-2Py) as two-protonic-state electrophiles and their application as active site titrants (and other types of protein chemical tool) for cysteine proteinases.

The major common feature of cysteine proteinase active sites is a $(Cys)\text{-}S^-/(His)\text{-}Im^+H$ ion-pair state with nucleophilic character formed by protonic dissociation, with pK_a $c.$ 3.0–3.5 in many cases (203, 204).

2-Pyridyl disulfides (such as $CH_3\text{-}S\text{-}S\text{-}2\text{-}Py$ and $2\text{-}Py\text{-}S\text{-}S\text{-}2\text{-}Py$; the latter available from Aldrich as Aldrithiol 2) are valuable titrants for cysteine proteinase active sites. They are absolutely thiol-specific in reactions with proteins and, in acidic media, highly selective for catalytic site thiol groups because of nucleophilic character in the $(Cys)\text{-}S^-/(His)\text{-}Im^+H$ ion-pair. This is particularly useful for enzymes such as chymopapains A and B (205) that contain additional thiol groups.

In reactions with thiol groups they release stoichiometric amounts of pyridine-2-thione (λ_{max} 343 nm) over a wide range of pH (206). This stoichiometric release of a chromophoric leaving group does not occur in reactions of all aromatic disulfides. Reactions of the well-known general nonselective thiol titrant, DTNB [5,5'-dithiobis(2-nitrobenzoic acid)], need to be carried out in alkaline solution where all thiol groups exist substantially as nucleophilic uncomplicated thiolate anions. In acidic media reactions of DTNB reach adverse pH-dependent equilibrium positions with consequent loss of stoichiometry (207).

The special property that makes 2-pyridyl disulfides particularly valuable as titrants for cysteine proteinases is their ability to increase their reactivity by $c.$ 1000 times consequent upon protonation of the pyridyl N atom. This property accounts for the ability of $2\text{-}Py\text{-}S\text{-}S\text{-}2\text{-}Py$, for example, to discriminate between active papain and both low M_r mercaptans and denatured enzyme (208, 209). At pH values around 4, rapid reaction with cysteine proteinase active site thiol groups results from reaction of $(Cys)\text{-}S^-/(His)\text{-}Im^+H$ with $R\text{-}S\text{-}S\text{-}2\text{-}Py^+H$. At this pH isolated thiol groups exist essentially as undissociated nonnucleophilic SH.

The theoretical basis for the use of two-protonic-state electrophiles, such as R-S-S-2-Py $\rightleftharpoons$ R-S-S-2-Py$^+$H, is described in refs 210 and 211, and their applications, not only as active site titrants but also as reactivity probes, delivery vehicles for spectroscopic reporter groups, heterobifunctional cross-linking agents and in covalent chromatography, have been reviewed (203, 210–214). Recent studies involving 2-pyridyl disulfides are described in refs 204, 205, 215–217.

32.2. Site-directed or mechanism-based inhibitors

Site selectivity can be achieved also by providing known recognition features in the chromogenic titrant to produce an 'active site-directed' inhibitor (218) or a 'mechanism-based' enzyme inactivator (219, 220). The latter, sometimes referred to as a suicide substrate, is a relatively unreactive compound with structural similarity to the substrate for a particular enzyme that, via its normal catalytic mechanism, converts the potential inactivator molecule into a species that reacts chemically with the enzyme without prior release from the active site. Two examples of the use of specific recognition features in titrants are given below.

(i)   Incorporation into R-S-S-2-Py of specific recognition features for the active site of papain (a P_1-P_2 amide bond and a hydrophobic residue at P_2, as in Ac-[L-Phe]-NH-(CH$_2$)$_2$-S-S-2-Py) produces a titrant that discriminates between papain and a low M_r mercaptan even in approximately neutral media ($k_{papain}/k_{lowM_r\ RSH}$ c. 10 000). This effect arises from the possibility of activating the 2-mercaptopyridine leaving group not only by formal protonation at low pH but also by association with a hydrogen-bond donor when this is provided for by specific interactions involving the extended enzyme binding site and the nonpyridyl part of the titrant (221). The existence of this additional mechanism of activation provides opportunities to define the nature of binding site–catalytic site signaling mechanisms, which might be an important facet of molecular recognition processes, by using this type of titrant also as a reactivity probe (for example, refs 216, 222).

(ii) One approach to producing highly specific titrants for individual proteinases is to use protein proteinase inhibitors as a basis (223, 224). High specificity for individual serine proteinases has been achieved by use of derivatives of bovine pancreatic trypsin inhibitor (BPTI, aprotinin) (225). The starting material is modified BPTI in which the reactive-site peptide bond, Lys15–Ala16, has been cleaved and the two peptide chains are held together by two cystinyl disulfide bonds. Lys15 is then substituted, for example by Arg-, Phe- or Val-4-nitroanilides or by Val-7-amido 4-methyl coumarin, to provide specificities for different serine proteinases. Reaction with the catalytic site Ser hydroxy group results in formation of the chromophoric or fluorescent product and the resynthesis of the Lys15–Ala16 bond. The enzyme remains strongly complexed with the modified BPTI which prevents further reaction with more of the chromogenic BPTI derivative and thus the titration is stoichiometric.

33. TITRATION USING SPECIFIC INHIBITORS WITHOUT CHROMOGENIC LEAVING GROUPS

Four approaches using tight binding or covalently bonded inhibitors are exemplified below: (i) use of spectroscopic labels, (ii) use of radiolabels, (iii) use of antibodies, and (iv) use of activity loss measurements.

33.1. Use of spectroscopic labels and radiolabels

Active sites can be complexed with, or derivatized by, specific ligands that provide quantifiable physico-chemical signals.

(i)   Phenylhydrazine inhibits bovine serum amino acid oxidase by imine formation with the carbonyl group of the quinonoid form of the covalently bonded 2,4,5-trihydroxyphenylalanine cofactor (225). Active sites may be titrated by using (a) an increase in A_{450} consequent upon imine formation, (b)[^{14}C]phenylhydrazine (and (c) activity loss data) (226).

(ii) A recently reviewed strategy combines site-specific inhibitor affinity labeling with the versatility of group selective chemical modication to permit active site labeling of enzymes with a wide variety of spectroscopic probes by use of a single affinity-labeling reagent (227). The active site is alkylated by a chloromethyl ketone containing a free thiol group as a substituent which is subsequently alkylated by a reagent bearing a spectroscopic reporter group.

33.2. Use of antibodies

The use of antibodies in active site assays may be illustrated by a recent description of the approach as applied to titration of enzymes of coagulation and fibrinolytic pathways (228).

(i)   The enzyme of interest is labeled covalently at the active site by a reagent that incorporates an antibody recognition site, exemplified below in terms of biotinyl-ε-aminocaproyl-D-Phe-Pro-Arg-chloromethylketone.
(ii) Following active site labeling, one of two approaches may be selected.
(iii) The biotinylated enzyme may be captured by solid-phase avidin and detected by a peroxidase-conjugated specific antibody in a standard ELISA. This has the advantage of specificity for active enzyme but the presence of excess biotinylated chloromethylketone might compete for avidin binding sites.
(iv) Alternatively, capture is by the antibody and detection by peroxidase-conjugated avidin. This method protects against excess biotinylated reagent but immunoreactive, inactive enzyme species, if present in large quantities, might compete with the capturing reagent.
(v)   The use of antibodies promises the possibility of carrying out active site titrations *in vivo*, for example the determination of circulating tissue plasminogen activator (t-PA) in connection with the use of recombinant t-PA in thrombolytic therapy.

33.3. Activity loss measurements

Measurement of loss of catalytic activity during titration of separate samples of enzyme with different sub-stoichiometric amounts of inhibitor is a general method of active site titration which requires rapid inhibition but no special spectral characteristics or labeling of the titrant.

In the simplest cases the plot of catalytic activity (as ordinate) versus molar concentration of titrant is linear and the intercept on the abscissa provides the active site molarity.

An example of a well-known titrant of this type is the naturally occurring compound *trans*-expoxysuccinyl-L-leucylamido (4-guanidino)-butane (E-64) isolated from *Aspergillus japonicus* and reviewed in ref. 229. This compound has generally been considered to be a specific inhibitor of cysteine proteinases although it has recently been shown to inhibit trypsin (230). Examples of linear plots of activity versus [E-64] are provided by the active site titrations of human cathepsin B (231) and cruzipain (232). It should be noted that the crystal structure at 2.4 Å resolution of the product of covalent modification of Cys25 of papain by E-64 (233) shows that a binding mode involving the S'-subsites proposed previously (231) is incorrect (see ref. 230 for further comment).

In more complex situations where the titrant is a group-specific (but not site-specific) chemical modification reagent and the enzyme contains a number of potential modification sites, only one or a smaller number of which are sites essential for catalytic activity, the method of analysis suggested by Tsou Chen-Lu (234) and applied, for example, to the study of pepsin (235), urease

(236) and chymopapains A and B (237) may be used. The enzyme is allowed to react with a chemical reagent that causes activity loss and is specific for a particular type of functional group. The number of essential groups per molecule is determined by measuring both the residual catalytic activity and the number of groups modified, in samples of partially modified enzyme. A plot of (activity)$^{1/i}$, where i ($= 1, 2$, etc.) is the number of essential groups per molecule, versus the number of groups modified per molecule becomes linear only when the correct value of i is used.

ACKNOWLEDGMENTS

I thank Hasu Patel for searching the literature, Rita Dobrin and Joy Smith for rapid production of the typescript and SERC, MRC and AFRC for project grants and earmarked studentships.

F. ACTIVE ENZYME CENTRIFUGATION
S.E. Harding and A.J. Rowe

Sedimentation coefficients (s values; units, seconds or 'Svedbergs' $= 10^{-13}$ sec) can yield important information concerning mass and shape properties of both native and engineered biopolymers. In most circumstances their estimation calls for the isolation of the substance under investigation in either purified or, at least, paucidisperse form. For enzymes an elegant alternative possibility exists: their velocity of migration in an applied field can be monitored by means of spectral change linked to their distinctive catalytic property. In essence, their movement leaves a spectral 'footprint', and the migration of this 'footprint' can be followed. First devised by Cohen (238) and analysed in detail by Cohen and Mire (239), the method has since that time seen a modest degree of application. With the advent of the new generation of analytical ultracentrifuges (AUCs) with full on-line computer-based data analysis, this method, which is uniquely valuable in enabling quality physical information to be obtained about systems which have been perhaps only partially purified, and which can target the *active* component of a polydisperse system, can expect to find extended use.

34. BASIC PRINCIPLE OF ACTIVE ENZYME CENTRIFUGATION

The method is quite simple to apply, but a limited number of criteria must be satisfied if valid results are to be obtained. A solution column is set up in one sector of an AUC double-sector cell which contains all the substrates and cofactors necessary for spectral change to result from the introduction of the enzyme under study. For example, if the enzyme-catalysed reaction results in the uptake of protons, then a pH indicator such as phenol red will be present, and the solution will be (weakly) buffered to a pH (around 6.5) below the pK of the phenol red (7.3). The layering of a very small volume of enzyme on to the top of this solution column while the cell is being accelerated to speed causes a very thin zone of the enzyme to be formed, which at a sufficiently high centrifuge speed will sediment progressively down the solution column. As it does, the spectral color of the solution will be changed by the interaction of the reaction product with the indicator – in the example given above, the region above the migrating zone will become red, and the migration of the interface between the red upper region of the cell and the paler, more yellow, lower region can be followed and analysed to give a sedimentation rate (s value). In simple theory, this s value can be regarded as the sedimentation rate of the (fastest) active species of the enzyme. It is not necessary for the enzyme-catalysed reaction to yield a measurable spectral change directly: biochemical ingenuity can be used to couple almost any reaction to a spectrally significant change (e.g. the $NAD^{+}/NADH$ couple; see Chapter 3).

Figure 6 shows the relationship between the change in extinction and the location of the zone of enzyme for a simple case. The change in the extinction caused by the enzyme can, of course, be either positive or negative.

35. PRACTICAL USE OF AN AUC TO GIVE ACTIVE ENZYME CENTRIFUGATION

It is assumed that the reader has the use of an AUC equipped with scanning absorption optics (e.g. a Beckman XL-A) and is competent in its basic use. The type of cell often called a 'Vinograd' cell (240) must be employed – or rather a centerpiece of this type. In this double-sector centerpiece are located one or more small drilled-out holes (*Figure 7*). The two sectors contain the full assay solution in one case, and an appropriate reference solution in the other. Both solutions will normally contain enough additional salt (e.g. 50–100 mM excess) or glycerol to ensure freedom from density inversion when the enzyme solution is overlayered. It is generally convenient for automated analysis to ensure that the change in extinction caused by the enzyme is *positive* (i.e. not as shown in *Figure 6*). This can be arranged by appropriate choice of reference solution or choice of channel.

Figure 6. Diagram showing the association between a migrating zone of enzyme and the presence of a boundary in the radial scan of the AUC cell at an appropriate wavelength.

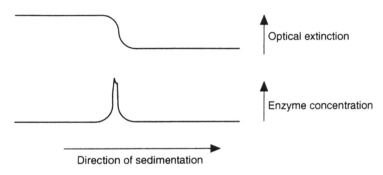

Figure 7. Diagram of a 'Vinograd' type AUC cell, with two drilled holes adjacent to the sectors.

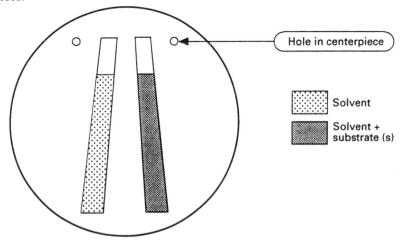

The hole immediately adjacent to the sector containing the assay solution is filled *prior to assembly of the cell* with a small quantity of enzyme solution (usually 10 µl), using an accurate ('Hamilton') syringe to place the enzyme solution at the bottom of the hole. This is essential to avoid a trapped air bubble which would probably lead to loss of this solution during cell assembly and tightening. After the cell has been carefully assembled and torqued up, the assay and reference solutions are loaded, the rotor loaded and accelerated to a speed (usually 10 000 r.p.m. (8000 *g*) +) at which transfer of enzyme solution from the hole to the top of the column occurs. A quick radial scan is then performed at the selected wavelength to check for the presence of a level plateau region, indicating no leakage or convection. The rotor is then accelerated to operational speed, and automatic scans taken at intervals defined by the presumed range of *s* value under investigation.

After completion of the experiment, the *s* value associated with the migration of the active species can then be estimated by the normal methods – either by following the second moment of the migrating boundary, or (more simply) its first moment (point of inflection). For most cases the latter suffices.

36. CONDITIONS THAT MUST BE SATISFIED

The reader is referred to full treatments given in refs 239 and 241. The most important conditions to be met are as follows:

(i)   all components of the assay system, including the agent of the spectral change and any coupling enzymes, must sediment much more slowly than the enzyme under study. Normally only coupling enzymes must be considered seriously here.

(ii) The enzyme must be loaded at a concentration such that a substantial *but not total, and generally not more than 50% maximal* spectral change occurs during passage of the zone. Although ultimately this is a matter of trial and error, a good start will be made if in dummy experiments in a spectrophotometer the enzyme in the cuvette *at its initial loading concentration* causes a spectral change of this magnitude within 5–10 min.

(iii) If the enzyme solution contains any substances (e.g. reaction products) which might themselves induce spectral change, then it is vital that the velocity of movement of the boundary exceeds by a reasonable margin the diffusion rate of such substances.

(iv) Care must be taken to ensure that the conditions employed do not inhibit any coupling enzymes used; for example, one could not use EDTA in the PK/LDH system described below, since Mg^{2+} is a vital cofactor.

37. EXAMPLES OF THE USE OF ACTIVE ENZYME CENTRIFUGATION

37.1. Native myosin filaments

These have been followed by their ATPase activity being linked (via pyruvate kinase and lactic dehydrogenase) to the NAD/NADH couple, at 340 nm (242; *Figure 8*). This was an early use of a multienzyme coupling system. The *s* value of these filaments (132S) had to be measured at above 12 000 r.p.m. (11 000 *g*) to avoid problems from diffusion of the ADP present in the preparation.

37.2. Solubilized (α+β) chains of Na⁺/K⁺ ATPase

Here active enzyme centrifugation has been employed to show that the monomeric species was active, no dimer needed to be formed (244). The activity was coupled to phenol red in this case, and scanned at 550 nm. The *s* value found (6.5±0.2S) was confirmed as being identical to that measured using the purified enzyme in conventional velocity analysis. A similar study has been performed on the monomer of the sarcoplasmic reticulum (SR) calcium pump in $C_{12}E_8$ solution (245).

Figure 8. Active enzyme sedimentation diagram of native thick filaments from vertebrate skeletal muscle (243).

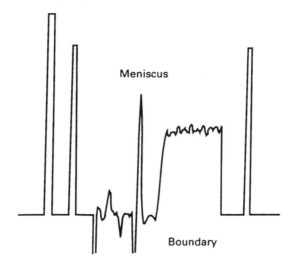

37.3. Glutamate dehydrogenase

The activity of the wild-type enzyme was compared with that of two point mutants (R61E and F187D) to show that only the hexameric form of the enzyme is active (246). The activity was coupled to NADH *production* (i.e. the opposite to that used above (242)) and scanned at 340 nm.

Figure 9. Active enzyme sedimentation diagram of glutamate dehydrogenase in the presence of glutamate and NAD^+. Rotor speed, 25 000 r.p.m. (50 000 g); scan interval, 5 min. The line fitted follows the point of inflection of the sedimenting boundary in the same way as a conventional sedimenting boundary would be followed.

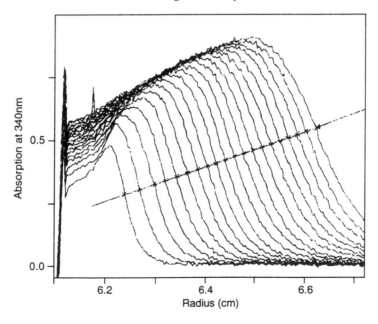

Figure 9 shows the active enzyme velocity profile for the R61E mutant in phosphate–chloride buffer (pH 7.0), yielding an $s_{20,w} = 14.3 \pm 0.2S$. In a buffer of higher pH (8.8) where the enzyme is entirely in its trimeric form, no active enzyme sedimenting boundary was observed.

38. REFERENCES

1. Pai, S.B. (1983) *Biochem. Biophys. Res. Commun.*, **110**, 412.

2. Weitzman, P.D.J. and Ridley, J. (1983) *Biochem. Biophys. Res. Commun.*, **112**, 1021.

3. Yamamoto, H., Tanaka, M., Okochi, T. and Kishimoto, S. (1983) *Biochem. Biophys. Res. Commun.*, **111**, 36.

4. Farmer, E.E. and Easterby, J.S. (1982) *Anal. Biochem.*, **123**, 373.

5. Farmer, E.E. and Easterby, J.S. (1984) *Anal. Biochem.*, **141**, 79.

6. Léonil, J., Langrené, S., Sicsic, S. and Le Goffie, F. (1985) *J. Chromatogr.*, **347**, 316.

7. Inagaki, K., Miwa, I. and Okuda, J. (1982) *Arch. Biochem. Biophys.*, **216**, 337.

8. Fulton, S. (1980) *Dye–Ligand Chromatography*. Amicon Corporation, MA.

9. Hey, J. and Dean, P.D.G. (1981) *Chem. Ind. (Lond.)*, **20**, 726.

10. Smolin, E.M. and Rapoport, L. (1959) *The Chemistry of Heterocyclic Compounds, 13:* s-*Triazine and Derivatives*. Interscience Publishers, London.

11. Clonis, Y.D. and Lowe, C.R. (1981) *Biochim. Biophys. Acta*, **659**, 86.

12. Hughes, P., Lowe, C.R. and Sherwood, R.F. (1982) *Biochim. Biophys. Acta*, **700**, 90.

13. Hughes, P., Lowe, C.R. and Sherwood, R.F. (1982) *Biochem. J.*, **205**, 453.

14. Hughes, P. (1989) in *Protein–Dye Interactions, Developments and Applications* (M.A. Vijayalakshmi and O. Bertrand, eds). Elsevier-Applied Science, Cambridge, p. 337.

15. Lowe, C.R., Burton, S.J., Burton, N.P., Alderton, W.K., Pitts, J.M. and Thomas, J.A. (1992) *Trends Biotechnol.* **10**, 442.

16. Lindner, N.M., Jeffcoat, R. and Lowe, C.R. (1989) *J. Chromatogr.*, **473**, 227.

17. Burton, S.J., McLoughlin, S.B., Stead, C.V. and Lowe, C.R. (1988) *J. Chromatogr.*, **435**, 127.

18. Burton, S.J., Stead, C.V. and Lowe, C.R. (1988) *J. Chromatogr.*, **455**, 210.

19. Burton, N.P. and Lowe, C.R. (1992) *J. Mol. Recognit.*, **5**, 55.

20. Burton, N.P. and Lowe, C.R. (1993) *J. Mol. Recognit.*, **6**, 31.

21. Burton, S.J., Stead, C.V. and Lowe, C.R. (1990) *J. Chromatogr.*, **508**, 109.

22. Thomas, J.A. and Lowe, C.R. (1993) in *Biotechnology of Blood Proteins* (C. Rivat and J.-F. Stoltz, eds). INSERM/JohnLibbey Eurotext Ltd., *22*, p. 25.

23. Hermanson, G.T., Mallia, A.K. and Smith, P.K. (1992) *Immobilised Affinity Ligand Techniques*. Academic Press, New York.

24. Pearson, J.C. and Lowe, C.R. (1989) in *Protein–Dye Interactions. Developments and Applications* (M.A. Vijayalakshmi and O. Bertrand, eds). Elsevier Applied Science, Cambridge, p. 337.

25. Stewart, D.J., Purvis, D.R., Pitts, J.M. and Lowe, C.R. (1992) *J. Chromatogr.*, **623**, 1.

26. Boyer, P.M. and Hsu, J.T. (1992) *Chem. Eng. Sci.*, **47**, 241.

27. Lowe, C.R. (1984) in *Topics in Enzyme and Fermentation Biotechnology* (A. Wiseman, ed.). Ellis Horwood, Chichester, p. 78.

28. Kopperschläger, G., Böhme, H.-J. and Hofmann, E. (1982) *Adv. Biochem. Engin.*, **25**, 101.

29. Lowe, C.R., Small, D.A.P. and Atkinson, A. (1981) *Int. J. Biochem.*, **13**, 33.

30. Scawen, M.D. and Atkinson, A. (1987) in *Reactive Dyes in Protein and Enzyme Technology* (Y.D. Clonis, T. Atkinson, C.J. Bruton and C.R. Lowe, eds). Macmillan, Basingstoke, P. 51.

31. Qadri, F. (1985) *Trends Biotechnol.*, **3**, 7.

32. McLoughlin, S.B. and Lowe, C.R. (1988) *Rev. Prog. Coloration*, **18**, 16.

33. Brehm, R.D., Tranter, H.S., Hambleton, P. and Melling, J. (1990) *Appl. Environ. Microbiol.*, **56**, 1067.

34. Vanderlinden, M.P.G., Mottle, H. and Keck, W. (1992) *Eur. J. Biochem.*, **204**, 197.

35. Li, X.X. and Chen, K.C. (1993) *Dyes Pigments*, **22**, 27.

36. Travis, J. and Pannell, R. (1973) *Clin. Chim. Acta*, **49**, 49.

37. Monaghan, C., Holland, S. and Dale, J.W. (1982) *Biochem. J.*, **205**, 413.

38. Lowe, C.R., Burton, S.J., Burton, N.P., Stewart, D.J., Purvis, D.R., Pitfield, I. and Eapen, S. (1990) *J. Mol. Recognit.*, **3**, 117.

39. Watson, D.H., Harvey, M.J. and Dean, P.D.G. (1978) *Biochem. J.*, **173**, 591.

40. Turner, A.J. and Hryszko, J. (1980) *Biochim. Biophys. Acta*, **613**, 256.

41. Lowe, C.R., Hans, M., Spibey, N. and Drabble, W.T. (1980) *Anal. Biochem.*, **104**, 23.

42. Clonis, Y.D., Goldfinch, M.J. and Lowe, C.R. (1981) *Biochem. J.*, **197**, 203.

43. McFarthing, K.G., Angal, D. and Dean, P.D.G. (1982) *Anal. Biochem.*, **122**, 186.

44. Hammond, P.M., Atkinson, T. and Scawen, M.D. (1986) *J. Chromatogr.*, **366**, 79.

45. Scopes, R.K. (1986) *J. Chromatogr.*, **376**, 131.

46. Warburg, O. and Christian, W. (1941) *Biochem. Z.*, **310**, 384.

47. Hamilton, R.J. and Hamilton, S.E. (eds) (1992) *Lipid Analysis: a Practical Approach*. IRL Press, Oxford.

48. Chaplin, M.F. and Kennedy, J.F. (eds) (1986) *Carbohydrate Analysis: a Practical Approach*. IRL Press, Oxford.

49. Hooper, N.M. and Turner, A.J. (eds) (1992) *Lipid Modification of Proteins: a Practical Approach*. IRL Press, Oxford.

50. Scopes, R.K. (1982) *Protein Purification: Principles and Practice*. Springer, New York.

51. Laue, T.M. and Rhodes, D.G. (1990) *Methods Enzymol.*, **182**, 566.

52. Garfin, D.E. (1990) *Methods Enzymol.*, **182**, 425.

53. Garfin, D.E. (1990) *Methods Enzymol.*, **182**, 459.

54. Lunney, A., Chrambach, A. and Rodbard, D. (1971) *Anal. Biochem.*, **40**, 158.

55. Righetti, P.G. and Drysdale, J.W. (1976) in *Laboratory Techniques in Biochemistry and Molecular Biology* (T.S. Work and E. Work, eds). North-Holland, Amsterdam, Vol. 5, p. 337.

56. Vesterberg, O. (1971) *Methods Enzymol.*, **22**, 389.

57. Tiselius, A. (1937) *Trans. Faraday Soc.*, **33**, 524.

58. Ackers, G.K. (1975) in *The Proteins* (H. Neurath and R.L. Hill, eds). Academic Press, New York., Vol. 1, p. 2.

59. Fischer, L. (1980) in *Laboratory Techniques in Biochemistry and Molecular Biology* (T.S. Work and R.H. Burdon, eds). North-Holland, Amsterdam, Vol. 1 part II, p. 157.

60. Himmelhoch, S.R. (1971) *Methods Enzymol.*, **22**, 273.

61. An excellent series of books on the principles and methods of ion-exchange chromatography, affinity chromatography, hydrophobic interaction chromatography and chromatofocusing are available from Pharmacia LKB Biotechnology.

62. Cuatrecasas, P. and Anfinsen, C.B. (1971) *Methods Enzymol.*, **22**, 345.

63. Lowe, C.R. and Dean, P.D.G. (1974) *Affinity Chromatography*. John Wiley & Sons, London.

64. Lowe, C.R. (1979) in *Laboratory Techniques in Biochemistry and Molecular Biology* (T.S. Work and E. Work, eds). North-Holland, Amsterdam, Vol. 7, p. 267.

65. Dean, P.D.G., Johnson, W.S. and Middle, F.A. (1985) *Affinity Chromatography: a Practical Approach*. IRL Press, Oxford.

66. Sluyterman, L.A.A. and Wijdenes, J. (1981) *J. Chromatogr.*, **206**, 441.

67. Schachman, H.K. (1959) *Ultracentrifugation in Biochemistry*. Academic Press, New York.

68. Williams, J.W. (1972) *Ultracentrifugation of Macromolecules*. Academic Press, New York.

69. Birnie, J.D. and Rickwood, D. (1978) *Centrifugal Separations in Molecular and Cell Biology*. Butterworths, London.

70. Chervenka, C.H. (1970) *A Manual of Methods for the Analytical Ultracentrifuge*. Spinco Div., Beckman Instruments, Palo Alto, CA.

71. Johnson, M.L., Correia, J.J., Yphantis, D.A. and Halvorson, H.R. (1981) *Biophys. J.*, **36**, 575.

72. Meng, C.K., Mann, M. and Fenn, J.B. (1988) *Z. Phys. D.*, **10**, 361.

73. Karas, M. and Hillenkamp, F. (1988) *Anal. Chem.*, **60**, 2299.

74. Loo, J.A., Udseth, H.R. and Smith, R.D. (1989) *Anal. Biochem.*, **179**, 404.

75. Karas, M., Bahr, U., Ingendoh, A. and Hillenkamp, F. (1989) *Angew. Chem. Int. Ed. Engl.*, **28**, 760.

76. Smith, R.D., Loo, J.A., Barinaga, C.J., Edmonds, C.G. and Udseth, H.R. (1990) *J. Am. Soc. Mass Spectrom.*, **1**, 53.

77. Hirayama, K., Akashi, S., Furuya, M. and Fukuhara, K. (1990) *Biochem. Biophys. Res. Commun.*, **173**, 639.

78. Poulter, L., Green, B.N., Kaur, S. and Burlingame, A.L. (1990) in *Biological Mass Spectrometry* (A.L. Burlingame and J.A. McCluskey, eds). Elsevier, Amsterdam, p. 477.

79. Page, M.J., Aitken, A., Cooper, D.J., Magee, A.I. and Lowe, P.N. (1990) in *Methods: a Companion to Methods in Enzymology*, Vol. 1, p. 221.

80. Harbour, G.C., Garlick, R.L., Lyle, S.B., Crow, F.W., Robins, R.H. and Hoogerheide, J.G. (1992) in *Techniques in Protein Chemistry III* (R.H. Angeletti, ed.). The Protein Society, Academic Press, New York, p. 487.

81. Reinhold, B.B., Reinherz, E.L. and Reinhold, V.N. (1992) in *Techniques in Protein Chemistry III* (R.H. Angeletti, ed.). The Protein Society, Academic Press, New York, p. 287.

82. Wang, Y.K., Liao, P.C., Allison, J., Gage, D.A., Andrews, P.C., Lubman, D.M., Hanash, S.M. and Strahler, J.R. (1993) *J. Biol. Chem., 268*, 14269.

83. Papac, D.I., Oatis, J.E., Crouch, R.K. and Knapp, D.R. (1993) *Biochemistry, 32*, 5930.

84. Sanger, F. (1945) *Biochem. J., 39*, 507.

85. Gray, W.R. and Hartley, B.S. (1963) *Biochem. J., 89*, 59.

86. Edman, P. (1950) *Acta Chem. Scand., 4*, 283.

87. Levy, W.P. (1981) *Methods Enzymol., 79*, 27.

88. Chang, J.Y. and Creaser, E.H. (1976) *Biochem. J., 157*, 77.

89. Chang, J.Y., Brauer, D. and Wittmann-Liebold, K.W. (1978) *FEBS Lett., 93*, 205.

90. Kézdy, F.J. and Kaiser, E.T. (1970) *Methods Enzymol., 19*, 3.

91. Fersht, A.R. (1985) *Enzyme Structure and Mechanism (2nd edn)*. W.H. Freeman, New York.

92. Peterson, G.L. (1983) *Methods Enzymol., 91*, 95.

93. Darbre, A. (1986) in *Practical Protein Chemistry: a Handbook* (A. Darbre, ed.). John Wiley & Sons, Chichester.

94. Stoscheck, C.M. (1990) *Methods Enzymol., 182*, 50.

95. Blakeley, R.L. and Zerner, B. (1975) *Methods Enzymol., 35*, 221.

96. Stevens, L. (1992) in *Enzyme Assays: a Practical Approach* (R. Eisenthal and M.J. Danson, eds). IRL Press, Oxford.

97. Kirschenbaum, D.M. (1975) *Anal. Biochem., 68*, 465.

98. Scopes, R.K. (1987) *Protein Purification: Principles and Practice (2nd edn)*. Springer, New York.

99. Gill, S.C. and von Hippel, P.H. (1989) *Anal. Biochem., 182*, 319.

100. Layne, E. (1957) *Methods Enzymol., 3*, 447.

101. Kalb, V.F., Jr and Bernlohr, R.W. (1977) *Anal. Biochem., 82*, 362.

102. Whitaker, J.R. and Granum, P.E. (1980) *Anal. Biochem., 109*, 156.

103. Scopes, R.K. (1974) *Anal. Biochem., 59*, 277.

104. Waddell, W.J. (1956) *J. Lab. Clin. Med., 48*, 311.

105. Chang, J.-Y., Knecht, R. and Braun, D.G. (1983) *Methods Enzymol., 91*, 41.

106. Ozols, J. (1990) *Methods Enzymol., 182*, 587.

107. Smith, P.K., Krohn, R.I., Hermanson, G.T., Mallia, A.K., Gartner, F.H., Provenzano, M.D., Fujimoto, E.K., Goeke, N.M., Olson, B.J. and Klenk, D.C. (1985) *Anal. Biochem., 150*, 76.

108. Harris, D.A. (1987) in *Spectrophotometry and Spectrofluorimetry: a Practical Approach* (D.A. Harris and C.L. Bashford, eds). IRL Press, Oxford.

109. Bradford, M.M. (1976) *Anal. Biochem., 72*, 248.

110. Sedmak, J.J. and Grossberg, S.E. (1977) *Anal. Biochem., 79*, 544.

111. Congdon, R.W., Muth, G.W. and Splittgerber, A.G. (1993) *Anal. Biochem., 213*, 407.

112. Butcher, E.C. and Lowry, O.H. (1976) *Anal. Biochem., 76*, 502.

113. Laemmli, U.K. (1970) *Nature, 227*, 680.

114. O'Farrell, P.Z., Goodman, H.M. and O'Farrell, P.H. (1977) *Cell, 12*, 1133.

115. Reisfeld, R.A., Lewis, V.J. and Williams, D.E. (1962) *Nature, 195*, 281.

116. Williams, D.E. and Reisfeld, R.A. (1964) *Ann. NY Acad. Sci., 121*, 373.

117. Davis, B.J. (1964) *Ann. NY Acad. Sci., 121*, 404.

118. Pharmacia Biosystems Application Note 324 (1984).

119. Görg, A., Fawcett, J.S. and Chrambach, A. (1988) in *Advances in Electrophoresis* (A. Chrambach, M.J. Dunn and B.J. Radola, eds). VCH, Weinheim, Vol. 2, p. 1.

120. O'Farrell, P.H. (1975) *J. Biol. Chem., 250*, 4007.

121. Manabe, T., Tachi, K., Kojima, K. and Okuyama, T. (1979) *J. Biochem., 85*, 649.

122. Fasman, G.D. (ed.) (1976) *Handbook of Biochemistry and Molecular Biology, Proteins (3rd edn)*. CRC Press, Cleveland, OH, Vol. II.

123. Malamud, D. and Drysdale, J.W. (1978) *Anal. Biochem., 86*, 620.

124. Hames, B.D. and Rickwood, D. (eds) (1990) *Gel Electrophoresis of Proteins: a Practical Approach (2nd edn)*. IRL Press, Oxford.

125. Dautrevaux, M., Boulanger, Y., Han, K. and Biserte, G. (1969) *Eur. J. Biochem., 11*, 267.

126. Lambin, P.C. (1978) *Anal. Biochem., 85*, 114.

127. Weber, K. and Osborn, M. (1969) *J. Biol. Chem., 244*, 4406.

128. Cowman, M.K., Slahetka, M.F., Hittner, D.M., Kim, J., Forino, M. and Gadelrad, G. (1984) *Biochem. J., 221*, 707.

129. Blakesly, R.W. and Boezi, J.A. (1977) *Anal. Biochem., 82*, 580.

130. Vesterberg, O. (1972) *Biochim. Biophys. Acta,* **257,** 11.

131. Lee, C., Levin, A. and Branton, D. (1987) *Anal. Biochem.,* **166,** 308.

132. Allen, R.E., Masak, K.C. and McAllister, P.K. (1980) *Anal. Biochem.,* **104,** 494.

133. Richtie, R.F. and Smith, R. (1976) *Clin. Chem.,* **22,** 497.

134. Merril, C.R., Goldman, D., Sedman, S.A. and Ebert, M.H. (1981) *Science,* **211,** 1438.

135. Godolphin, W.J. and Stinson, R.A. (1974) *Clin. Chim. Acta,* **56,** 97.

136. Harris, H. and Hopkinson, D.A. (1976) *Handbook of Enzyme Electrophoresis in Human Genetics.* North-Holland, Amsterdam.

137. Gabriel, O. (1971) *Methods Enzymol.,* **22,** 578.

138. Siciliano, M.J. and Shaw, C.R. (1976) in *Chromatographic and Electrophoretic Techniques* (I. Smith, ed.). William Heinemann, London, Vol. 2, p. 185.

139. Shaw, C.R. and Prasad, R. (1980) *Biochem. Genet.,* **4,** 297.

140. Harris, H. and Hopkinson, D.A. (1978) *Handbook of Enzyme Electrophoresis in Human Genetics.* North-Holland, Amsterdam.

141. Harris, H. and Hopkinson, D.A. (1977) *Handbook of Enzyme Electrophoresis in Human Genetics.* North-Holland, Amsterdam.

142. Abraham, D.G. and Cooper, A.J.L. (1991) *Anal. Biochem.,* **197,** 421.

143. Blank, A., Suigiyama, R.H. and Dekker, C.A. (1982) *Anal. Biochem.,* **130,** 267.

144. Mathew, R. and Rao, K.K. (1992) *Anal. Biochem.,* **206,** 50.

145. Grenier, J. and Asselin, A. (1993) *Anal. Biochem.,* **212,** 301.

146. Steiner, B. and Cruce, D. (1992) *Anal. Biochem.,* **200,** 405.

147. Mittler, R. and Zilinskas, B.A. (1993) *Anal. Biochem.,* **212,** 540.

148. Akyama, S. (1990) *Electrophoresis,* **11,** 509.

149. Queiroz-Clapet, C. and Meunier, J.-C. (1993) *Anal. Biochem.,* **209,** 228.

150. Gershoni, J.M. (1987) in *Advances in Electrophoresis* (A. Chrambach, M.J. Dunn and B.J. Radola, eds). VCH, Weinheim, Vol. 1, p. 141.

151. Towbin, H., Staehelin, T. and Gordon, J. (1979) *Proc. Natl Acad. Sci. USA,* **76,** 4350.

152. Bjerrum, O.J. and Schafer-Nielsen, C. (1986) in *Electrophoresis '86* (M. Dunn, ed.). VCH, Weinheim, p. 315.

153. Dunn, S.D. (1986) *Anal. Biochem.,* **157,** 144.

154. In *Protein Blotting, a Guide to Transfer and Detection* (1993) Bio-Rad Laboratories, Bulletin 1721. Bio-Rad Laboratories Ltd, Hemel Hempstead, UK.

155. Walton, K.E., Styer, D. and Gruenstein, E. (1979) *J. Biol. Chem.,* **254,** 795.

156. Cooper, P.C. and Burgess, A.W. (1982) *Anal. Biochem.,* **126,** 301.

157. Davidson, J.B. and Case, A. (1982) *Science,* **215,** 1398.

158. Burbeck, S. (1983) *Electrophoresis,* **4,** 127.

159. Laskey, R.A. and Mills, A.D. (1975) *Eur. J. Biochem.,* **56,** 335.

160. Bonner, W.M. (1984) *Methods Enzymol.,* **104,** 461.

161. Skinner, K. and Griswold, M.D. (1983) *Biochem. J.,* **209,** 281.

162. Chamberlain, J.P. (1979) *Anal. Biochem.,* **98,** 132.

163. Heegard, N.H.H., Hebsgaard, K.P. and Bjerrum, O.J. (1984) *Electrophoresis,* **5,** 230.

164. Roberts, P.L. (1985) *Anal. Biochem.,* **147,** 521.

165. McConkey, E.H. and Anderson, C. (1984) *Electrophoresis,* **5,** 230.

166. Laskey, R.A. (1981) *Amersham Research News No. 23.*

167. Laskey, R.A. and Mills, A.D. (1977) *FEBS Lett.,* **82,** 314.

168. Bonner, W.M. (1983) *Methods Enzymol.,* **96,** 215.

169. Harding, C.R. and Scott, I.R. (1983) *Anal. Biochem.,* **129,** 371.

170. Dunbar, B.S. (1987) *Two-dimensional Gel Electrophoresis and Immunological Techniques.* Plenum Press, New York.

171. Latter, G.I., Burbeck, S., Fleming, S. and Leavitt, J. (1984) *Clin. Chem.,* **30,** 1925.

172. Christopher, A.R., Nagpal, M.L., Carrol, A.R. and Brown, J.C. (1978) *Anal. Biochem.,* **85,** 404.

173. Zapolski, E.J., Gersten, D.M. and Ledley, R.S. (1982) *Anal. Biochem.,* **123,** 325.

174. Rickwood, D., Patel, D. and Billington, D. (1993) in *Biochemistry Labfax* (J.A.A. Chambers and D. Rickwood, eds). BIOS Scientific Publishers, Oxford, p. 37.

175. Johns, E.W. (1976) in *Subcellular Components: Preparation and Fractionation* (G. D. Birnie, ed.). Butterworths, London, p. 202.

176. Kruh, J., Schapira, G., Lareau, J. and Dreyfus, J.C. (1964) *Biochem. Biophys. Acta,* **87,** 669.

177. Adamietz, P. and Hiltz, H. (1976) *Hoppe-Seylers Z. Physiol. Chem.*, **357**, 527.

178. Sinclair, J.H. and Rickwood, D. (1985) *Biochem. J.*, **229**, 771.

179. Laskowski, M. (1955) *Methods Enzymol.* **2**, 8.

180. Bergmeyer, H.-U. (ed.) (1963) *Methods of Enzymatic Analysis*. Academic Press, New York.

181. Bender, M.L., Begué-Cantón, M.L., Blakely, R.L., Brubacher, L.J., Feder, J., Gunter, C.R., Kézdy, F.J., Killheffer, J.V., Marshall, T.H., Miller, C.G., Roeske, R.W. and Stoops, J.K. (1966) *J. Am. Chem. Soc.*, **88**, 5890.

182. Brocklehurst, K. (1979) *Biochem. J.*, **181**, 775.

183. Gutfreund, H. and Sturtevant, J.M. (1956) *Biochem., J.*, **63**, 656.

184. Hartley, B.S. and Kilby, B.A. (1954) *Biochem. J.*, **56**, 288.

185. Gutfreund, H. and Sturtevant, J.M. (1956) *Proc. Natl Acad. Sci. USA*, **42**, 719.

186. Ouellet, L. and Stewart, J.A. (1959) *Can. J. Chem.*, **37**, 737.

187. Sturtevant, J.M. (1960) *Brookhaven Symp. Biol.*, **13**, 164.

188. Schonbaum, G.R., Zerner, B. and Bender, M.L. (1961) *Biol. Chem.*, **236**, 2930.

189. Bender, M.L., Kézdy, F.J. and Wedler, F.C. (1967) *J. Chem. Educ.*, **44**, 84.

190. Lucas, E.C. and Williams, A. (1969) *Biochemistry*, **8**, 5125.

191. Chase, T. and Shaw, E. (1967) *Biochem. Biophys. Res. Commun.*, **29**, 508.

192. Jameson, D.M., Williams, J.F. and Wehrly, J.A. (1977) *Anal. Biochem.*, **79**, 623.

193. Livingston, D.C., Brocklehurst, J.R., Cannon, J.F., Leytus, S.P., Wehrly, J.A, Peltz, S.W., Peltz, G.A. and Mangel, W.F. (1981) *Biochemistry*, **20**, 4298.

194. Green, D.P.L. (1985) *Anal. Biochem.*, **147**, 487.

195. Chase, T. and Shaw, E. (1970) *Methods Enzymol.*, **XIX**, 3.

196. Kézdy, F.J. and Kaiser, E.T. (1970) *Methods Enzymol.*, **XIX**, 3.

197. Powers, J.C., Boone, R., Carroll, D.L., Gupton, B.F., Kam, C.M., Nishino, N., Sakamoto, M. and Tuhy, P.M. (1984) *J. Biol. Chem.*, **259**, 4288.

198. Gilbert, S.P. and Johnson, K.A. (1993) *Biochemistry*, **32**, 4677.

199. Mottl, H. and Keck, W. (1991) *Eur, J. Biochem.*, **200**, 767.

200. Söll, D. and Schimmel, P.R. (1974) in *The Enzymes (3rd edn)* (P.D. Boyer, ed.). Academic Press, New York, Vol. 10, p. 489.

201. Johnson, D.L. and Yang, C.H. (1981) *Proc. Natl Acad. Sci. USA*, **78**, 4059.

202. Fersht, A.R., Ashford, J.S., Bruton, C.J., Jakes, R., Koch, G.L.E. and Hartley, B.S. (1975) *Biochemistry*, **14**, 1.

203. Brocklehurst, K., Willenbrock, F. and Salih, E. (1987) in *Hydrolytic Enzymes* (A. Neuberger and K. Brocklehurst, eds). Elsevier, Amsterdam, p. 39.

204. Mellor, G.W., Patel, M., Thomas, E.W. and Brocklehurst, K. (1993) *Biochem. J.*, **294**, 201.

205. Thomas, M.P., Topham, C.M., Kowlessur, D., Mellor, G.W., Thomas, E.W., Whitford, D. and Brocklehurst, K. (1994) *Biochem. J.*, **300**, 805.

206. Grassetti, D.R. and Murray, J.F. (1967) *Arch. Biochem. Biophys.*, **119**, 41.

207. Brocklehurst, K., Kierstan, M. and Little, G. (1972) *Biochem. J.*, **128**, 811.

208. Brocklehurst, K. and Little, G. (1970) *FEBS Lett.*, **5**, 63.

209. Brocklehurst, K. and Little, G. (1973) *Biochem. J.*, **133**, 67.

210. Brocklehurst, K. (1974) *Tetrahedron*, **30**, 2397.

211. Brocklehurst, K. (1982) *Methods Enzymol.*, **87C**, 427.

212. Brocklehurst, K. (1979) *Int. J. Biochem.*, **10**, 259.

213. Brocklehurst, K., Carlsson, J., Kierstan, M.P.J. and Crook, E.M. (1974) *Methods Enzymol.*, **34B**, 531.

214. Shipton, M., Stuchbury, T. and Brocklehurst, K. (1976) *Biochem. J.*, **159**, 235.

215. Topham, C.M., Salih, E., Frazao, C., Kowlessur, D., Overington, J.P., Thomas, M., Brocklehurst, S.M., Patel, M., Thomas, E.W. and Brocklehurst, K. (1991) *Biochem. J.*, **280**, 79.

216. Patel, M., Kayani, S., Templeton, W., Mellor, W., Thomas, E. and Brocklehurst, K. (1992) *Biochem. J.*, **287**, 881.

217. Mellor, G.W., Sreedharan, K., Kowlessur, D., Thomas, E.W. and Brocklehurst, K. (1993) *Biochem. J.*, **290**, 75.

218. Baker, B.R. (1967) *Design of Active-Site Directed Irreversible Inhibitors*. John Wiley & Sons, New York.

219. Silverman, R.B. (1988) *Mechanism-Based Enzyme Inactivation Chemistry and Enzymology*. CRC Press, Boca Raton, FL, Vol. I.

220. Silverman, R.B. (1988) *Mechanism-Based Enzyme Inactivation Chemistry and Enzymology*. CRC Press, Boca Raton, FL, Vol. II.

221. Brocklehurst, K., Kowlessur, D., O'Driscoll, M., Patel, G., Quenby, S., Salih, E., Templeton, W., Thomas, E.W. and Willenbrock, F. (1987) *Biochem. J.*, **244**, 173.

222. Kowlessur, K., Topham, C.M., Thomas, E.W., O'Driscoll, M., Templeton, W. and Brocklehurst, K. (1989) *Biochem. J.*, **258**, 755.

223. Laskowski, M. and Sealock, R.W. (1971) in *The Enzymes (3rd edn)* (P.D. Boyer, ed.). Academic Press, New York, Vol. III, p. 375.

224. Birk, Y. (1987) in *Hydrolytic Enzymes* (A. Neuberger and K. Brocklehurst, eds). Elsevier, Amsterdam, p. 257.

225. Janes, S.M., Mu, D., Wemmer, D., Smith, A.J., Kaur, S., Maltby, D., Burlingame, A.L. and Klinman, J.P. (1990) *Science*, **248**, 981.

226. Janes, S.M. and Klinman, J.P. (1991) *Biochemistry*, **30**, 4559.

227. Bock, P.E. (1993) *Methods Enzymol.*, **222**, 478.

228. Tracy, R.P., Jerry, R., Williams, E.B. and Mann, K.G. (1993) *Methods Enzymol.*, **222**, 574.

229. Shaw, E. (1990) *Adv. Enzymol.*, **63**, 271.

230. Sreedharan, S.K., Shah, H.N. and Brocklehurst, K. (1994) *Biochem. Soc. Trans.*, **22**, 212S.

231. Barrett, A.J., Kembhavi, A.A., Brown, M.A., Kirscke, H., Knight, C.G., Tamai, M. and Hanada, K. (1982) *Biochem. J.*, **201**, 189.

232. Cazzulo, J.J., Franke, M.C.C., Martinez, J. and de Cazzulo, B.M.F. (1990) *Biochim. Biophys. Acta*, **1037**, 186.

233. Varughese, K.I., Ahmed, F.R., Carey, P.R., Hasnain, S., Huber, C.P. and Storer, A.C. (1989) *Biochemistry*, **28**, 1330.

234. Tsou Chen-Lu (1962) *Sci. Sin.*, **11**, 1535.

235. Paterson, A.K. and Knowles, J.R. (1972) *Eur. J. Biochem.*, **31**, 510.

236. Norris, R. and Brocklehurst, K. (1976) *Biochem. J.*, **159**, 245.

237. Baines, B.S. and Brocklehurst, K. (1982) *J. Prot. Chem.*, **1**, 119.

238. Cohen, R. (1963) *C. R. Acad. Sci.*, **256**, 3513.

239. Cohen, R. and Mire, M. (1971) *Eur. J. Biochem.*, **23**, 276.

240. Vinograd, J., Bruner, R., Kent, R., and Weigle, J. (1963) *Proc. Natl Acad. Sci. USA*, **49**, 902.

241. Kemper, D.I. and Everse, J. (1973) *Methods Enzymol.*, **XVII**, 67.

242. Emes, C.H. and Rowe, A.J. (1978) *Biochim. Biophys. Acta*, **537**, 125.

243. Emes, C.H. (1977) Ph.D. Thesis, University of Leicester.

244. Madden, C.S., Ward, D.G., Walton, T.J.H., Washbrook, R.F., Rowe, A.J. and Cavieres, J.D. (1994) in *The Sodium Pump: Structure, Mechanism, Hormonal Control and its Role in Disease* (E. Bamberg and W. Schoner, eds). Springer, Darmstadt, p. 445.

245. Martin, D.W. (1983) *Biochemistry*, **22**, 2276.

246. Pasquo, A. Britton, K.L., Stillman, T.J., Rice, D.W., Cölfen, H., Harding, S.E., Scandurra, R. and Engel, P.C. (1995) *Biochim. Biophys. Acta*, in press.

CHAPTER 3
ENZYME KINETICS

A. TYPES OF ASSAY PROCEDURE

P.C. Engel

1. INTRODUCTION

Fundamental to virtually all enzyme work is the measurement of catalytic activity – either because factors affecting activity may be the central issue under study or because, in other contexts (e.g. monitoring purification) measurement of activity is the most sensitive way to get a quantitative estimate of how much enzyme is present.

One approach to determining the absolute number of active enzyme molecules present is active-site titration (see Chapter 2E), but for most purposes it is sufficient and much more sensitive to measure the *units of activity* per unit volume (see Chapter 1) in a steady-state turnover assay. By also measuring total protein concentration (see Chapter 2) this may be converted to a *specific activity*, expressed as units per mg protein. This provides an indication of purity or integrity if the corresponding figure for the pure, native enzyme is known. If the specific activity of the pure enzyme and its molecular weight are both known, the activity measurement also provides an estimate of the *molar concentration* of active enzyme molecules.

2. EXAMPLE

Activity	0.05 U ml^{-1}
Protein concentration	5 mg ml^{-1}
$\therefore$ Specific activity	0.01 U mg^{-1}

If specific activity of pure enzyme $= 10$ U mg^{-1} then enzyme at this stage is only 0.1% pure (the 99.9% could include inactivated enzyme as well as other proteins).

If molecular weight $= 50\,000$ per active site, then 5 mg ml^{-1} of pure enzyme $\equiv 10^{-4}$ M active sites. But this sample is only 0.1% pure. Thus the concentration of active sites is 10^{-7} M.

It is always necessary to fix a set of conditions for a *standard assay*. These are usually chosen on grounds of sensitivity, convenience, economy and reliability. If the enzyme has been investigated by earlier workers, there will be a published assay procedure. For ease of comparison it is often best to adopt the existing procedure unless there are compelling reasons for rejecting it.

3. PRODUCT OR SUBSTRATE?

Consider a reaction

$$A + B \rightarrow C + D.$$

Since the reaction is stoichiometric, in principle the rate may be measured equally well by monitoring the appearance of C or D or the disappearance of either A or B. Depending on the method of measurement, there may be some advantage in opting for measurement of product formation: in principle it is easier to measure precisely a finite increase from zero than a small decrease from a large starting concentration.

4. PHYSICAL OR CHEMICAL PROPERTY TO MEASURE

Following a reaction depends on using a property of one of the reactants which is changed measurably by the reaction. Potentially, a wide range of properties may be used (*Table 1*) and the choice will be governed by availability of equipment, cost of materials and the quality of the assay procedure.

Table 1. A summary of the main enzyme assay methods (adapted from Chambers, J.A.A. and Rickwood, D. (1993) *Biochemistry Labfax*. BIOS Scientific Publishers, Oxford, Table 8, p. 140)

Assays	Principles and comments	Examples
Spectrophotometry	Suitable when there is a difference between substrate and product in the absorbance of light of a particular wavelength according to the Beer–Lambert law. Often used with an absorbance less than 0.8. Can use either natural or synthetic chromogenic substrates	Lipoxidase, (deoxy)ribonucleases, pepsin, trypsin, α-chymotrypsin(ogen), lysozyme and most dehydrogenases, kinases and phosphatases
Spectrofluorimetry	Used when there is a difference in fluorescence between substrate and product. More sensitive than spectrophotometry. Main problems: (i) fluorescence varies with temperature; (ii) quenching	Firefly luciferase, $NAD(P)^+$-dependent dehydrogenases
Manometry	For reactions where there is gas uptake or output, and also those involving the production or consumption of acid or alkaline substances, if they are carried out in the presence of bicarbonate in equilibrium with a gas mixture containing a defined percentage of CO_2	Oxidases, decarboxylases
Electrode methods	For enzymes catalysing redox reactions. Potential generated can either be measured by potentiometric techniques or be related to the concentration of a specific substance, using ion-selective electrodes. Advantages: contaminating substances can be present in suspension without affecting the analysis	Lipase, chymopapain, papain, carbonic anhydrase, butyrylcholinesterase

Table 1. Continued

Assays	Principles and comments	Examples
Microcalorimetry	For reactions in which heat (enthalpy) is gained or lost. Advantages: sensitivity, freedom from interference and wide range of applications	Hexokinase
Radiochemical labeling	By using radioactively labeled substrate, the product concentration can be determined indirectly by measuring the radioactivity of the product fraction. Extremely sensitive. Disadvantages: (i) possible health hazard; (ii) multi-steps; (iii) difficult to monitor reactions continuously; (iv) quenching	Many types of enzyme, restricted only by the availability of radio-labeled compounds
Polarimetry	For enzymes involved in conversion of optically inactive substrates to optically active products and for racemases. Not very convenient	Lactate dehydrogenase
Chromatographic methods	Chromatography can be used to detect the formation of a product when other methods fail. Disadvantage: time consuming	Chloramphenicol acetyl transferase (CAT)

4.1. Alternatives

Sometimes none of the reactants offers a convenient direct assay method. *Table 2* lists possible alternatives in this situation.

5. DEPENDENCE ON TIME AND ENZYME CONCENTRATION

A basic assumption in an assay procedure is that doubling the enzyme addition will double the rate. However, in practice this can only be true over a limited range of enzyme concentrations. In the extreme case, addition of a large amount of enzyme could bring the assay reaction to equilibrium before observation even commenced. Paradoxically, therefore, a zero rate would be observed and doubling the enzyme addition would make no difference. Good linearity with enzyme concentration can only be achieved if the reaction conditions are such that product formation is *linear with time*, and this needs to be directly observed or established in setting up new assay conditions.

Frequently a curved time-course is obtained, so that attempting to measure the initial rate necessitates drawing a tangent. Under these circumstances adding much less enzyme should result in a much decreased extent of progress towards equilibrium and therefore a more linear time-course. Ideally one should also aim to increase the sensitivity of observation; for example,

Table 2. Alternative assay strategies when reactants do not offer useful properties for direct measurement

Strategy	Advantage	Disadvantage	Examples
Secondary chemical reactions to yield a measurable product	Reagents usually cheap	Usually multistep procedure requiring stopped assay	2,4-Dinitrophenyl hydrazine reaction with oxoacids, forming hydrazone, Lipmann and Tuttle procedure for thioesters and acyl phosphates: $NH_2OH \rightarrow$ hydroxamate acid $FeCl_3 \rightarrow$ colored complex
Artificial chromogenic substrates	High sensitivity is built in. Usually suitable for continuous assay	For a new assay this approach requires synthesis of the substrate	BAPNA assay for trypsin, releasing p-nitroaniline. X-gal assay for β-galactosidase
Coupled assay involving a second enzyme step with a convenient indicator substrate	Suitable for continuous assay	Large number of reagents. Expense of large amount of coupling enzyme for each assay – coupling reaction must not limit the rate	LDH assay for pyruvate produced in pyruvate kinase reaction – detection of NADH

BAPNA, benzoyl arginine p-nitroanilide; LDH, lactate dehydrogenase; X-gal, 5-bromo-4-chloro-3-indoyl-β-D-galactopyranoside.

if a spectrophotometric assay trace is curved on the 0–0.5 absorbance scale, one could add 1/10 as much enzyme and observe on the 0–0.05 absorbance scale (see *Figure 1*).

Even when the reaction is linear with time, proportionality to enzyme concentration needs to be checked. In some cases the concentration of *active* enzyme may not equal the concentration of enzyme protein; for example, an oligomeric enzyme may dissociate upon dilution (e.g. yeast glucose 6-phosphate dehydrogenase).

6. CONTINUOUS OR STOPPED ASSAY?

A major issue is whether it is possible to establish a *continuous assay* – one that provides a continuous read-out of the course of reaction as it proceeds. This is closely linked to the choice of the physical or chemical parameter to be measured (*Table 1*). Thus a spectrophotometer or fluorimeter may be linked to a chart recorder or provide a digitized output to a computer and a display screen for the purposes of continuous assay. However, an assay based on high performance liquid chromatography (HPLC) or measurement of radioactivity must inevitably be a *stopped assay*. In the latter case a time-course may be constructed by taking multiple samples at timed intervals. In each case the enzyme reaction must be promptly terminated by change of pH, heating, addition of inhibitor, etc. Alternatively, *if linearity with time and enzyme addition can be confidently established*, it may be possible to use a *single time* of incubation, say 30 min, on the assumption that the time-course is linear up to and beyond that time.

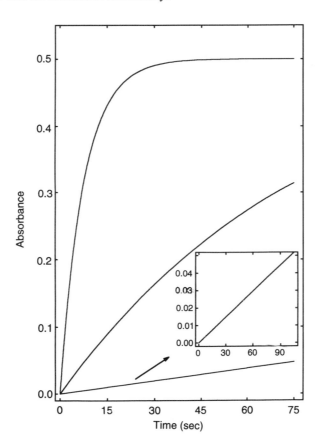

Figure 1. Choosing the amount of enzyme to add. The upper curve shows a reaction trace where so much enzyme has been added that the reaction proceeds all the way to equilibrium during the observation period. Indeed, if observation does not commence promptly after initiating the reaction, the rise phase of the absorbance trace could be entirely missed. The middle curve shows the same reaction with 10 times less enzyme added. This is still so curved that it is impossible to get a true initial rate by conventional manual mixing methods. Typically one might miss the first 10–15 sec of the trace and then measure the tangent as a false 'initial' rate. The lowest line shows the result of adding 10 times less enzyme than in the middle trace (i.e. 100 times less than in the upper trace). This slows down the reaction so that even in 60 sec the absorbance change is only 0.03 (approx.). This can, however, be made to extend over the full vertical scale simply by altering the range setting on the spectrophotometer to 0–0.05, as in the inset, instead of 0–0.5. The resulting trace is so close to linearity that the initial rate can be estimated accurately.

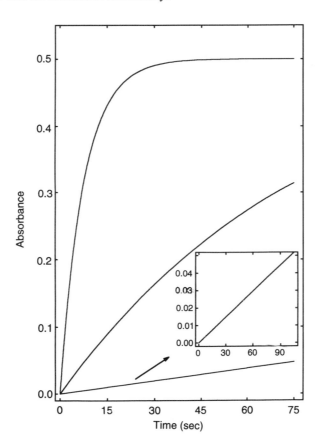

7. ENZYME INSTABILITY

An early consideration in any enzyme project is the choice of buffer conditions that favor long-term stability of the enzyme. This should be checked by sampling the enzyme stock solution periodically, and using the standard assay procedure. In some cases temperature (high or low), pH, or choice of buffer salt may be critical. Another possibility that arises is that an enzyme that is acceptably stable over hours or days in the stock solution may nevertheless be very unstable at high dilution. Typically an enzyme stock solution will be diluted 50- to 500-fold into an assay solution. If, therefore, a continuous assay shows marked curvature (decreasing rate with time) it is important to establish whether this reflects rapid progress to equilibrium/depletion of substrates/inhibition by products, or whether possibly the enzyme is losing activity to a significant extent over the observed time-course of the assay.

This may be tested by one or other of the following procedures (*Figure 2*):

(i)   Second enzyme addition. Suppose that the measured rate drops to 40% of the initial value after 2 min. A second addition of enzyme, identical to the first, is made. If the decline is due directly to the progress of reaction then the rate should exactly double, to 80% of the initial value. If, on the other hand, the decline is entirely due to enzyme inactivation, the addition of a fresh portion of enzyme should raise the rate to 140% of the initial value (*Figure 2a*).

(ii) Double enzyme addition. This approach is ideally suited to continuous recording of a spectrophotometric assay. If, say, 5 µl of enzyme solution gives a curved trace on the 0–0.1 scale of a spectrophotometer, the assay may be repeated, adding 10 µl but recording on the 0–0.2 scale. If the curvature is entirely due to first-order inactivation of the enzyme, then the second trace should be superimposable on the first (i.e. show identical curvature). Conceivably if the inactivation process is dependent on enzyme concentration the curvature might even be less pronounced in the second case. On the other hand, if the curvature is mainly due to the progress of the reaction towards equilibrium, then clearly the second assay will proceed further towards equilibrium than the first in the same period of time. Accordingly (*Figure 2b*), the second trace should then be much more curved.

(iii) Initiation with substrate. If one of the substrates can be prepared in concentrated solution, a set of reaction mixtures may be prepared with this substrate missing. Enzyme is then added and reaction is initiated after varying lengths of time by adding the missing substrate. If the enzyme is unstable under the conditions used, initial rates will decline with increasing time of pre-incubation (*Figure 2c*).

8. CHOICE OF REACTION MIXTURE

If an assay procedure is being designed for the first time, it is desirable to choose conditions close to the pH optimum and to use substrate concentrations reasonably close to saturation. Clearly, these conditions give higher rates and accordingly better sensitivity. More important, however, they make for a more robust assay in the sense that small errors (e.g. in the weighing of reagents or adjustment of buffer pH) will cause minimal changes in rate. One wants differences in rates, not only from sample to sample on one day, but from day to day, and from lab. to lab., to be determined, as far as possible, only by the different amounts of enzyme added.

9. VOLUMETRIC ACCURACY

Once a reaction mixture has been optimized for assay purposes, and assuming adequate instrumentation, the main limitation on meaningful enzyme assays is volumetric accuracy in delivering the enzyme. It is important to choose a reliable physical method of delivery, and to practice and directly demonstrate repeatability. Among the more reliable procedures are constriction micropipettes and Hamilton syringes. If automatic pipettes are used, it is vital to check that they are delivering reliably. The volume they deliver is not visibly correct and reproducible, and O-rings have a finite life. It is therefore essential to validate their reliability at regular intervals. Clearly it is also essential to use the appropriate pipette. Using a 100 µl pipette to deliver 10 µl cannot be expected to give acceptable accuracy. With care and practice a routine repeatability of ±1–2% should be expected.

Figure 2. Testing for enzyme instability. (a) Second enzyme addition. The reaction trace is very curved and the rate, measured by the tangential slope, decreases from $0.20\ \text{min}^{-1}$ at zero time to $0.08\ \text{min}^{-1}$ after 75 sec. Depending on the reason for the curvature, a second, equal enzyme addition at this point should increase the rate *either* to $0.16\ \text{min}^{-1}$ (2×0.08) (curve 1) *or* to $0.28\ \text{min}^{-1}$ ($0.08 + 0.20$) (curve 2). (b) Double enzyme addition. The reaction trace shown by the solid line in the left panel is very curved (curve 1). The right panel shows three possible consequences of doubling the enzyme addition and halving the sensitivity of measurement (0–0.2 scale instead of 0–0.1). Curve 2 is the expected result if curvature is entirely due to time-dependent inactivation of the enzyme during the assay. This curve is superimposable on curve 1 – see dashed curve in left panel. (c) Initiation with substrate. When enzyme and the missing substrate are added together to complete the reaction mixture, the initial rate is measured as $0.4\ \text{min}^{-1}$. When the substrate is added 1 min after the enzyme the rate is only $0.16\ \text{min}^{-1}$, and when the reaction is initiated at 2 min the initial rate is still lower, at $0.064\ \text{min}^{-1}$. The design of the assay means that in this case the rapid decline cannot be attributed to progress of the reaction towards equilibrium. It must reflect instability of one of the components in the initial mixture, most likely the enzyme.

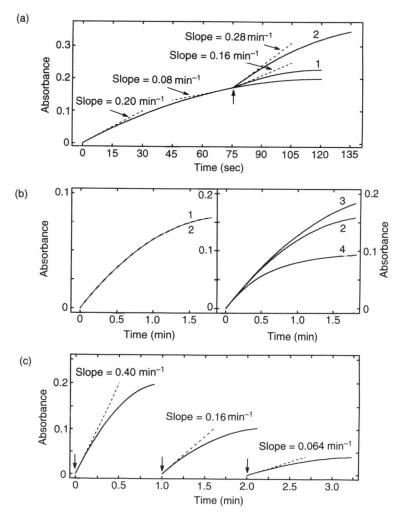

B. ENZYME KINETICS

F.M. Dickinson

Initial-rate measurements of enzyme-catalysed reactions are of interest for two principal reasons:

(i) To characterize an individual enzyme and provide data for deciding how the enzyme functions under physiological conditions.
(ii) To provide information about the mechanism of catalysis.

10. ANALYSIS OF INITIAL-RATE DATA

10.1. Two-substrate systems

Enzyme-catalysed reactions commonly involve two substrates (A, B) and two products (P, Q) as indicated in *Equation 1*:

$$A + B \rightleftharpoons P + Q. \qquad \qquad \text{Equation 1}$$

As the reaction is, in general, reversible it will be clear that P and Q become substrates if the reaction is studied from the right-hand side. Experience has shown that, for many reactions of this type, the initial-rate equation for the forward reaction at fixed pH and temperature, in the absence of products is of the form:

$$\frac{v}{[E_t]} = \frac{[A][B]}{\phi_0[A][B] + \phi_A[B] + \phi_B[A] + \phi_{AB}} \qquad \text{Equation 2}$$

$v/[E_t]$ is the specific initial rate (the rate per unit active site concentration), [A] and [B] are the substrate concentrations and ϕ_0, etc., the initial-rate parameters which characterize the enzyme under the chosen conditions. Conventions and symbols in *Equation 2* are based on those of Dalziel (1).

The reciprocal form of the rate equation is shown in *Equation 3*:

$$\frac{[E_t]}{v} = \phi_0 + \frac{\phi_A}{[A]} + \frac{\phi_B}{[B]} + \frac{\phi_{AB}}{[A][B]}. \qquad \text{Equation 3}$$

The initial-rate parameters are measured by making initial-rate measurements at varying [A] at a series of fixed [B] and plotting the data in double reciprocal (Lineweaver–Burk) plots. Re-arrangement of *Equation 3* with 1/[A] as the variable shows that the intercepts of the primary plots (*Figure 3*) are given by $[\phi_0 + \phi_B/[B]]$ and the slopes by $[\psi_A + \phi_{AB}/[B]]$. The secondary plots (*Figure 4*) of the slopes and intercepts of the primary plots with 1/[B] as the variable provide estimates for the individual initial-rate parameters. Advantages of expressing the rate equation in the form of *Equation 3* rather than as *Equation 6* (see below) are:

(i) The initial-rate parameters arise naturally in the derivation of initial-rate equations.
(ii) The initial rate parameters are usually simpler functions of the rate constants than are Michaelis constants. In some cases they are functions of single velocity constants (see below).

10.2. Single-substrate/pseudo single-substrate systems

If one substrate is always present in high and invariable concentration (for example for hydrolytic reactions B = H_2O) the variation of v with change of [B] cannot readily be studied and *Equation 2* reduces to *Equations 4* and *5*, or, more correctly to *Equation 7*.

Figure 3. Analysis of a two-substrate system which conforms to *Equation 2*. Data are obtained by varying [A] (four or five different concentrations) at a series of fixed concentrations of B. Accurate estimation of intercepts and slopes requires reproducible data with points close to the ordinate (i.e. within 10% of the extrapolated value) and a change in rate over the substrate range of at least twofold.

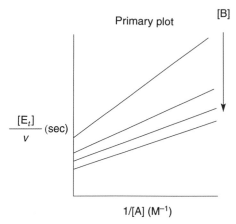

Figure 4. Plot of the intercepts and slopes of *Figure 3* vs. 1/[B]. The initial rate parameters are obtained from the intercepts and slopes of the secondary plot, as indicated.

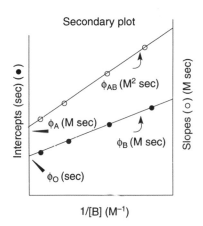

$$\frac{\nu}{[E_t]} = \frac{[A]}{\phi_0[A] + \phi_A}.$$

Equation 4

The reaction is effectively treated as a one-substrate system. *Equation 4* is one form of the Michaelis–Menten equation (2), which is often written as:

$$\frac{\nu}{[E_t]} = \frac{V_{max}[A]}{K_m + [A]} \text{ or } \frac{k_{cat}[A]}{K_m + [A]}.$$

Equation 5

Comparison of *Equations 4* and *5* yields the following relationships:

$$\phi_0 = \frac{1}{V_{max}} = \frac{1}{k_{cat}} \quad K_m = \frac{\phi_A}{\phi_0}.$$

ENZYME KINETICS

85

V_{max} (k_{cat}) and K_m can be obtained from the intercept and slope of a Lineweaver–Burk plot (3). V_{max} is the initial rate achieved as $[A] \to \infty$ and K_m is the concentration of A giving $V_{max}/2$. The various possible mechanisms for one-substrate reactions cannot be distinguished on the basis of initial-rate measurements. All mechanisms, whether they contain one or more intermediate complexes and whether they are treated by equilibrium or steady-state methods, predict the same kinetic features and share the same Haldane (see below) relationship. The details of such mechanisms can only be elucidated by recourse to other methods and are not further considered here.

The traditional Michaelis–Menten symbols can also be applied to two-substrate systems (4) but the definitions are modified. V_{max} is now the initial rate achieved as *both* [A] and $[B] \to \infty$. K_{mA} is the value of [A] giving $V_{max}/2$ when $[B] \to \infty$, and, similarly, for K_{mB} *Equation 2* now becomes:

$$\frac{v}{[E_t]} = \frac{V_{max}[A][B]}{[A][B] + K_{mA}[B] + K_{mB}[A] + K_{AB}}.$$

Equation 6a

The term K_{AB} has no counterpart in the Michaelis–Menten equation. Comparison of *Equations 2* and *6a* shows that:

$$V_{max} = \frac{1}{\phi_0} \qquad K_{mA} = \frac{\phi_A}{\phi_0} \qquad K_{mB} = \frac{\phi_B}{\phi_0} \qquad K_{AB} = \frac{\phi_{AB}}{\phi_0}.$$

Cleland (5) writes *Equation 6a* in a slightly different form:

$$\frac{v}{[E_t]} = \frac{V_1[A][B]}{[A][B] + K_a[B] + K_b[A] + K_{ia}K_b}.$$

Equation 6b

This equation highlights the fact that the last two terms in the denominator differ only by the factor K_{ia} which is the dissociation constant for the binary EA complex (*Table 4*). *Equation 6a* can also be written as:

$$\frac{v}{[E_t]} = \frac{\left(\frac{V_{max}[B]}{[B]+K_{mB}}\right)[A]}{[A] + \left(\frac{K_{mA}[B]+K_{AB}}{[B]+K_{mB}}\right)} = \frac{V_{max.app}[A]}{[A] + K_{mA.app}}$$

Equation 7

if a fixed, finite value of [B] is used in the experiments (compare with *Equation 5*; the subscript 'app' denotes 'apparent'). Obviously

$V_{max.app} \to V_{max}$ and $K_{m.app} \to K_{mA}$ as $[B] \to \infty$, ($[B] \gg K_{mB}$ and $K_{mA}[B] \gg K_{AB}$).

10.3. Three-substrate systems

For reactions involving three substrates:

$A + B + C \rightleftharpoons P + Q + (R)$,

the general reciprocal initial-rate equation for the forward reaction is of the form of *Equation 8* (6).

$$\frac{[E_t]}{v} = \phi_0 + \frac{\phi_A}{[A]} + \frac{\phi_B}{[B]} + \frac{\phi_C}{[C]} + \frac{\phi_{AB}}{[A][B]} + \frac{\phi_{BC}}{[B][C]} + \frac{\phi_{AC}}{[A][C]} + \frac{\phi_{ABC}}{[A][B][C]}.$$

Equation 8

The initial-rate parameters are obtained from the intercepts and slopes of tertiary plots (6). *Figures 3* and *4* show how the primary and secondary plots would appear at each fixed value of [C]. Tertiary plots are obtained by plotting the slopes and intercepts of plots such as *Figure 4* versus $1/[C]$. By analogy with two-substrate reactions, $V_{max} = 1/\phi_0$ is defined as the initial rate achieved as [A], [B] and [C]$\rightarrow\infty$. $K_{mA} = \phi_A/\phi_0$ is the value of [A] giving $V_{max}/2$ as [B] and [C]$\rightarrow\infty$, and the Michaelis constants for B and C are defined in the corresponding way. Again, by analogy with *Equation 7*, apparent values of V_{max}, K_{mA}, etc., will be obtained if fixed concentrations of two substrates are used and the third substrate concentration is varied.

11. ASSIGNMENT OF MECHANISM

11.1. Two-substrate reactions

This section is largely based on the theoretical work of Alberty (4), Dalziel (1) and Cleland (5). Two-substrate reactions can occur by one of two fundamentally different sorts of mechanism. Mechanism (i) below occurs by *partial reactions* so that the first product is released before combination of the second substrate and a stable modified form of the enzyme is formed. Mechanism (ii) below involves reaction of the enzyme with both substrates before the release of any products.

$$E+A \underset{k_2}{\overset{k_1}{\rightleftharpoons}} \begin{pmatrix} EA \\ E'Q \end{pmatrix} \overset{k_3}{\underset{k_4}{\rightleftharpoons}} E' + Q$$

$$E' + B \underset{k_6}{\overset{k_5}{\rightleftharpoons}} \begin{pmatrix} E'B \\ EP \end{pmatrix} \overset{k_7}{\underset{k_8}{\rightleftharpoons}} E + P$$

$$E \underset{}{\overset{k_1 \overset{A}{\downarrow} k_2 \quad k_3 \overset{Q}{\uparrow} k_4 \quad k_5 \overset{B}{\downarrow} k_6 \quad k_7 \overset{P}{\uparrow} k_8}{\underline{}} E$$

$$\begin{pmatrix} EA \\ E'Q \end{pmatrix} \quad E' \quad \begin{pmatrix} E'B \\ EP \end{pmatrix}$$

Mechanism (i)

$$E + A \underset{k_2}{\overset{k_1}{\rightleftharpoons}} EA$$

$$EA + B \underset{k_4}{\overset{k_3}{\rightleftharpoons}} \begin{pmatrix} EAB \\ EPQ \end{pmatrix} \overset{k_5}{\underset{k_6}{\rightleftharpoons}} EP + Q$$

$$EP \underset{k_8}{\overset{k_7}{\rightleftharpoons}} E + P$$

$$E \underset{}{\overset{k_1 \overset{A}{\downarrow} k_2 \quad k_3 \overset{B}{\downarrow} k_4 \qquad k_5 \overset{Q}{\uparrow} k_6 \quad k_7 \overset{P}{\uparrow} k_8}{\underline{}} E$$

$$EA \quad \begin{pmatrix} EAB \\ EPQ \end{pmatrix} \quad EP$$

Mechanism (ii)

The representations on the left are conventional. In the convenient short-hand notation on the right (5), the vertical arrows represent substrate combination and product dissociation (reaction running from left to right). The enzyme intermediates having kinetically significant concentrations are shown below the horizontal line at the appropriate position in the reaction sequence. Where two species are indicated together in parentheses (i.e. (EAB/EPQ)) it means that while two species would be expected to participate, for reasons of symmetry, the rate equation has been derived assuming the formation of only one central complex. This simplifies the rate equations, but does not alter their form or any of the relationships between the initial rate parameters.

Initial-rate equations, derived by using steady-state methods, predict quite different kinetic characteristics for the two mechanisms. The *ternary complex mechanism* (mechanism (ii)) predicts equations of the type shown in *Equations 2* and *3*, but for the *enzyme-substitution mechanism* (mechanism (i)) only the ϕ_0, ϕ_A and ϕ_B terms are present (*Table 3*). For this case $\phi_{AB} = 0$ so that the primary plots are strictly parallel. The same situation applies for the reverse reaction. Additional criteria can be used to decide between two-substrate mechanisms which conform to *Equation 2*. Some of these (maximum rate relationships, use of alternative substrates, product inhibition patterns), like the case described above, rely solely on the availability of detailed initial-rate data. Others (e.g. Haldane relationships) require additional data from other types of experiment. The application of the various criteria is illustrated by *Tables 3* and *4* and allows firm conclusions to be drawn regarding the mechanism of any particular two-substrate enzyme system.

Table 3. Mechanisms and initial-rate parameters for two-substrate reactions conforming to *Equation 2*

Mechanism	Initial rate parameters			
	ϕ_0	ϕ_A	ϕ_B	ϕ_{AB}

(i) Enzyme substitution/group transfer/ping-pong Bi-Bi

$$
\begin{array}{cccc}
 & A & B & P & Q \\
E & {\scriptstyle k_1\downarrow k_2}\ & {\scriptstyle k_3\uparrow k_4}\ {\scriptstyle k_5\downarrow k_6}\ & {\scriptstyle k_7\uparrow k_8}\ & E \\
 & \begin{pmatrix}EA\\EP\end{pmatrix} & E' & \begin{pmatrix}E'B\\EQ\end{pmatrix} &
\end{array}
$$

ϕ_0	ϕ_A	ϕ_B	ϕ_{AB}
$\dfrac{1}{k_3}+\dfrac{1}{k_7}$	$\dfrac{k_2+k_3}{k_1k_3}$	$\dfrac{k_6+k_7}{k_5k_7}$	0

(ii) Compulsory order ternary complex/ordered Bi-Bi

$$
\begin{array}{cccc}
 & A & B & Q & P \\
E & {\scriptstyle k_1\downarrow k_2}\ {\scriptstyle k_3\downarrow k_4}\ & {\scriptstyle k_5\uparrow k_6}\ {\scriptstyle k_7\uparrow k_8}\ & E \\
 & EA & \begin{pmatrix}EAB\\EPQ\end{pmatrix} & EP &
\end{array}
$$

ϕ_0	ϕ_A	ϕ_B	ϕ_{AB}
$\dfrac{1}{k_5}+\dfrac{1}{k_7}$	$\dfrac{1}{k_1}$	$\dfrac{k_4+k_5}{k_3k_5}$	$\dfrac{k_2(k_4+k_5)}{k_1k_3k_5}$

Reverse reaction

$\phi_0' = \dfrac{1}{k_2}+\dfrac{1}{k_4}$	$\phi_P = \dfrac{1}{k_8}$	$\phi_Q = \dfrac{k_3+k_6}{k_4k_6}$	$\phi_{PQ} = \dfrac{k_7(k_3+k_6)}{k_8k_4k_6}$

(iiia) Theorell–Chance/ordered Bi-Bi, no ternary complex

$$
\begin{array}{cccc}
 & A & B & Q & P \\
E & {\scriptstyle k_1\downarrow k_2}\ {\scriptstyle k_3}\ & {\scriptstyle k_4}\ {\scriptstyle k_5\uparrow k_6}\ & E \\
 & EA & & EP &
\end{array}
$$

ϕ_0	ϕ_A	ϕ_B	ϕ_{AB}
$\dfrac{1}{k_5}$	$\dfrac{1}{k_1}$	$\dfrac{1}{k_3}$	$\dfrac{k_2}{k_1k_3}$

(iiib) Theorell–Chance with isomeric binary complexes

$$
\begin{array}{ccccc}
 & A & B & Q & P \\
E & {\scriptstyle k_1\downarrow k_2}\ {\scriptstyle k_3}\ & {\scriptstyle k_5}\ {\scriptstyle k_6}\ & {\scriptstyle k_7}\ {\scriptstyle k_9\uparrow k_{10}}\ & E \\
 & EA & {\scriptstyle k_4}\,EA' & EP' & {\scriptstyle k_8}\,EP
\end{array}
$$

ϕ_0	ϕ_A	ϕ_B	ϕ_{AB}
$\dfrac{1}{k_7}+\dfrac{1}{k_3}+\dfrac{1}{k_9}+\dfrac{k_8}{k_7k_9}$	$\dfrac{k_2+k_3}{k_1k_3}$	$\dfrac{k_3+k_4}{k_3k_5}$	$\dfrac{k_2k_4}{k_1k_3k_5}$

(iva) Rapid equilibrium random order/random Bi-Bi

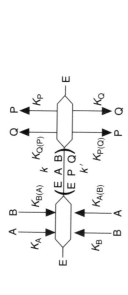

$$\frac{1}{k} \qquad \frac{K_{A(B)}}{k} \qquad \frac{K_{B(A)}}{k} \qquad \frac{K_A K_{B(A)}}{k}$$

(ivb) Rapid equilibrium random order with independent binding of substrates, i.e. $K_A = K_{A(B)}$ $K_B = K_{B(A)}$

$$\frac{1}{k} \qquad \frac{K_A}{k} \qquad \frac{K_B}{k} \qquad \frac{K_A K_{B(A)}}{k}$$

For random mechanisms, rapid equilibrium is assumed between the complexes formed prior to the catalytic step. This leads to equations of the form of *Equation 2*. K_A, etc., are the dissociation constants of the complexes to which they are assigned for these mechanisms, $K_A K_{B(A)} = K_B K_{A(B)}$. If steady-state methods are used, nonlinear reciprocal equations are obtained (7). These equations approach linearity as the variable substrate concentration is progressively reduced. In these circumstances the mechanism tends more and more to a compulsory mechanism as the bulk of the reaction passes through the preferred pathway. Detailed studies have been made of the conditions under which a general nonequilibrium random mechanism gives rise to an initial-rate equation of the form of *Equation 2* (7,8). These results have not, however, found wide application.

Mechanism (ii) is a special case of a random mechanism and mechanisms (iiia and iiib) are variants of mechanism (ii) with the central complexes being no longer kinetically significant.

The mechanisms shown are all symmetrical for convenience but this may not apply in a particular laboratory situation. It is possible to envisage, for example, a ternary complex mechanism in which the binding of A and B is random but where the dissociation of products (P and Q) is effectively compulsory. Mechanism (iiib) should not be confused with the iso-Theorell–Chance mechanism discussed by Cleland (5). The latter mechanism involves isomerization of the free enzyme and not the EA and EP complexes as here. Mechanism (iiib) was originally proposed to explain difficulties in getting data for liver alcohol dehydrogenase to fit the requirements of the Theorell–Chance mechanism (9). The problems were later shown to be due to impurities in coenzyme preparations (10). However, with sheep liver aldehyde dehydrogenase for example, a compulsory mechanism is found, ternary complex interconversions are not rate-limiting (11) and there is strong direct evidence for isomerization of the terminal E–NADH complex (12, 13). The mechanism, or simple variants of it, may thus be applicable in certain cases.

Table 4. Tests to allow discrimination between the mechanisms listed in *Table 3*

Mechanism	Relationships specified by the initial-rate parameters	Haldane relationship $K_{eq}=$	Tests to be applied if additional data are available from other types of experiment		
			Substrate binding studies	Stopped-flow relaxation methods	Other
(i)	$\phi_{AB}=0$	$\dfrac{\phi_P\phi_Q}{\phi_A\phi_B}$	—	—	Isotope exchange. Two stable forms of enzyme
(ii)	$\dfrac{\phi_A\phi_B}{\phi_{AB}} < \phi_0'$	$\dfrac{\phi_{PQ}}{\phi_{AB}}$	$\dfrac{\phi_{AB}}{\phi_B}=K_A=K_{ia}$	$\phi_A=\dfrac{1}{k_1}$	
(iiia)	$\dfrac{\phi_A\phi_B}{\phi_{AB}} = \phi_0'$	$\dfrac{\phi_{PQ}}{\phi_{AB}}=\dfrac{\phi_0'\phi_P\phi_Q}{\phi_0\phi_A\phi_B}$	$\dfrac{\phi_{AB}}{\phi_B}=K_A=K_{ia}$	$\phi_A=\dfrac{1}{k_1}$	
(iiib)	If $\dfrac{\phi_A\phi_B}{\phi_{AB}} < \phi_0'$, $\dfrac{\phi_P\phi_Q}{\phi_{PQ}} > \phi_0$	$\dfrac{\phi_{PQ}}{\phi_{AB}}$	$\dfrac{\phi_{AB}}{\phi_B}=K_{A,app}=K_{ia}$	Biphasic reaction of E with A (or P)	
(iva)	None	$\dfrac{\phi_{PQ}}{\phi_{AB}}$	$\dfrac{\phi_{AB}}{\phi_B}=K_A=K_{ia}$ $\dfrac{\phi_{AB}}{\phi_A}=K_B=K_{ib}$		
(ivb)	$\dfrac{\phi_A\phi_B}{\phi_{AB}} = \phi_0$	$\dfrac{\phi_{PQ}}{\phi_{AB}}=\dfrac{\phi_0\phi_P\phi_Q}{\phi_0'\phi_A\phi_B}$	$\dfrac{\phi_{AB}}{\phi_B}=K_A=K_{ia}$ $\dfrac{\phi_A}{\phi_0}=K_A=K_{ia}$ $\dfrac{\phi_B}{\phi_0}=K_B=K_{ib}$		

Values of K_A and K_B are obtained from binding studies. Values of K_{ia} and K_{ib} are obtained from product inhibition experiments with A or B as the competitive product inhibitor of the reaction for which P and Q are substrates (see *Table 5*).

Isotope exchange at equilibrium (14) is a particularly powerful method to discriminate between compulsory (ii) and random (iv) mechanisms. Complete suppression of isotope exchange at equilibrium between A and P as for lactate dehydrogenase (15) when B and Q are saturating indicates a strict compulsory mechanism. Incomplete suppression shows that a small fraction of the total reaction can proceed via an alternative pathway. Thus, a random mechanism exists, but only to an extent indicated by the residual exchange between A and P. If only a small proportion (< 10%) of the reaction proceeds via the alternative pathway, as for alcohol dehydrogenase (16), initial-rate criteria do not detect it (10).

$K_{A,app}$ (mechanism (iiib)) = [E][A]/([EA] + [EA']) and is the value of the dissociation constant obtained from direct binding studies of A by E if the method of analysis (i.e. equilibrium dialysis) distinguishes only between enzyme-bound and free A.

Relationships between the coefficients
For some mechanisms relationships exist which can easily be tested with detailed initial-rate data for forward and reverse reactions. *Table 3*, for example, shows that for mechanism (ii):

$$\frac{\phi_A \phi_B}{\phi_{AB}} = \frac{1}{k_2} < \frac{1}{k_2} + \frac{1}{k_4} = \phi_0'.$$

Since the mechanism is symmetrical, the reciprocal relationship also applies ($\phi_P \phi_Q / \phi_{PQ} < \phi_0$). As ϕ_0' and ϕ_0 are the reciprocal maximum velocities, the relationships are often called maximum rate relationships or, alternatively, Dalziel relationships. NB: These distinctive relationships can only be tested properly with accurate estimates of the initial-rate parameters.

Use of alternative substrates
For some mechanisms (e.g. mechanism (ii) where: $\phi_A = \frac{1}{k_1} \frac{\phi_{AB}}{\phi_B} = K_A$) use of an alternative substrate to B (say B'), if the specificity of the enzyme will allow, should leave the values of ϕ_A and ϕ_{AB}/ϕ_B unchanged, even though there may be dramatic changes to the values of ϕ_0, ϕ_B and ϕ_{AB}.

Product inhibition
Inclusion of one of the products in initial rate experiments causes inhibition, and the kind of inhibition (competitive, uncompetitive, noncompetitive) depends on which product is included and what the mechanism of reaction is. Cleland (5) has shown that many of the mechanisms in *Table 3* can be distinguished on the basis of their product inhibition patterns. For examples see *Table 5*.

One very valuable feature of product inhibition is that it allows identification of the substrate and product which react with the same enzyme species. Thus, for example, with mechanism (ii) discussed above, P should be strictly competitive towards A (and vice versa) under all conditions because these compounds compete for the free enzyme. A and P are the first substrates to combine for forward and reverse reactions. This information does not come out of conventional initial-rate studies because there is no way of deciding which of the two substrates is A. (There may of course be intuitive reasons for assigning identities to A and B, etc.) The inhibitor constants K_{ip} and K_{ia} obtained should of course $= K_P$ and K_A, respectively. For mechanism (i) the order of substrate addition is not critical, but product inhibition will show that Q is strictly competitive towards B, and P strictly competitive towards A. Thus Q and B combine with one form of the enzyme (E') and A and P combine with the other. This kind of information can also be obtained from direct measurements of substrate binding or by studying partial reactions (see below), but these would probably require substantial quantities (> 20 mg) of enzyme and specialized equipment. If enzyme supplies are very limited, the choice is limited to product inhibition experiments.

Ideally one would study the effect of products by independently varying [A] or [B] in the presence of varying [P] or [Q]. This allows a complete description of the effect of the inhibitors on the initial rate equation. Such a course would, however, involve a great deal of labor and the approach usually has been to study the effects of the products [P] or [Q] at various [B] with either 'unsaturating' or 'saturating' [A], and then to repeat the procedure with the alternative combinations. Unless the values of the initial rate parameters are known, the terms 'unsaturating' and 'saturating' are arbitrary and one cannot be sure that concentrations selected are satisfactory. Some of the possible distinctions between mechanisms might be blurred by an unfortunate choice of concentrations.

Table 5. Product inhibition patterns for some of the mechanisms shown in *Table 3*

Mechanism	Product inhibitor	Variable substrate			
		A		B	
		[B] non-saturating	[B] saturating	[A] non-saturating	[A] saturating
(i)	Q	Non-competitive	No inhibition	Competitive	Competitive
	P	Competitive	Competitive	Non-competitive	No inhibition
(ii)	Q	Non-competitive	Uncompetitive	Non-competitive	Non-competitive
	P	Competitive	Competitive	Non-competitive	No inhibition
(iv)	Q	Competitive	No inhibition	Competitive	No inhibition
	P	Competitive	No inhibition	Competitive	No inhibition

Many of the criteria used in *Table 4* require initial-rate data from forward and reverse reactions. If the reaction is essentially irreversible or the equilibrium constant is extremely unfavourable, a complete set of data will not be available. In these circumstances product inhibition by one or other product will be particularly valuable.

Addition of high concentrations of product can lead to the formation of additional complexes which were not envisaged in the basic mechanism. Thus if A and P are structurally similar (NAD$^+$ and NADH perhaps) the formation of complexes like EAQ and EPB is possible for mechanism (ii), for example. These complexes are 'abortive' and alter the characteristics of the mechanism. This obviously makes the task of elucidating the mechanism more difficult. Formation of abortive complexes will probably cause deviations from *Equation 2* because they may only be kinetically significant over a limited range of substrate concentration. An example of this is seen with liver alcohol dehydrogenase (17).

Since inclusion of products in reaction mixtures brings the initial mixture closer to equilibrium, the measurements of initial rates can be more difficult because of curvature of progress curves; a difficult problem if the equilibrium position is unfavourable.

Haldane relationships
If the value for the equilibrium constant is available under appropriate conditions, this can be combined with the initial rate parameters to produce a Haldane relationship. Thus for mechanism (ii):

$$K_{eq} = \frac{k_1 k_3 k_5 k_7 k_9}{k_2 k_4 k_6 k_8 k_{10}} = \frac{\phi_{PQ}}{\phi_{AB}}.$$

Similar relationships apply to other mechanisms (*Table 4*) and some mechanisms enjoy distinctive relationships. For the Theorell–Chance mechanism we have:

$$K_{eq} = \frac{\phi_0' \phi_P \phi_Q}{\phi_0 \phi_A \phi_B}.$$

This relationship arises because of the maximum rate relationship indicated in *Table 4*. It is not an independent test.

Dissociation constants of enzyme–substrate complexes
For some of the mechanisms in *Table 4*, $\frac{\phi_{AB}}{\phi_B} = K_A \left(\frac{\phi_{PQ}}{\phi_Q} = K_P\right)$, where K_A is the dissociation constant of the EA complex. Values of K_A can be determined in independent binding studies if sufficient enzyme is available. If only one substrate can be demonstrated to bind to free enzyme, it suggests that the mechanism is compulsory-order and that the leading substrate is A. If both substrates can be shown to bind independently, then a random mechanism would seem likely. If one substrate binds strongly and the other only weakly and below the detection limit of the experimental method used, binding studies might erroneously suggest a compulsory mechanism.

Rate constants for steps in the mechanism
For some mechanisms, such as mechanism (ii), $\phi_A = 1/k_1$. Again, if sufficient enzyme and a stopped-flow (or other rapid-reaction) apparatus are available, the value of k_1 might be obtained by direct study of the second-order (or pseudo first-order reaction if [A] >> [E]) reaction of E and A. In this way the predicted relationship could be checked. For the enzyme-substitution (group transfer) mechanism it should be possible to study one half reaction in the total absence of the alternative substrate/product; for example, with a rapid-reaction apparatus detecting the formation of E' in the absence of B. (For flavoprotein oxidases and dehydrogenases or aminotransferases, E and E' are spectrally distinct.) Alternatively, isotopically labeled compounds could be used to detect exchange between A and Q in the absence of B and P, or between B and P in the absence of A and Q. For mechanisms which are not of the group-transfer type, no exchange can be detected between A and Q unless B is present. Mechanisms involving isomerization of binary complexes (i.e. the Theorell–Chance mechanism with isomeric binary complexes) can possibly be best studied by a combination of initial-rate measurements and rapid-reaction studies. The rapid-reaction studies should detect the isomerization by seeing two 'relaxations' in the reaction of E and A.

11.2. Three-substrate reactions
This section is based on ref. 6, which gives a more complete discussion of the arguments. The approach is essentially as for the two-substrate reactions. *Table 6* shows five mechanisms which can be distinguished on the basis of a lack of terms in *Equation 8*. If accurate initial-rate data are available for forward and reverse reactions the mechanism should be assignable with confidence. Additional points are:

(i)   Several alternative mechanisms come under the general description of mechanism (ii) in *Table 6*. All can be distinguished from the other mechanisms in *Table 6* by initial-rate measurements, but they cannot be distinguished from each other. These finer distinctions require additional data (i.e. product inhibition, partial reactions, etc.).

(ii) For mechanism (iiia), a strictly compulsory mechanism, the lack of the ϕ_{AC} term identifies substrate B as the middle substrate to combine. Which substrate is identified as A and which as C can only be determined by product inhibition or binding studies.

(iii) For mechanisms (iiia) and (iiib) maximum-rate relationships are predicted (cf. two-substrate compulsory-order mechanism). If, for mechanism (iiia) the rate-limiting step under maximum-rate conditions is the dissociation of either the second or third product, more stringent maximum-rate relationships are obtained in the same way as for the Theorell–Chance mechanism (*Table 4*). These lead to additional Haldane relationships.

(iv) For some mechanisms, individual initial-rate parameters are functions of single rate constants and for others (e.g. rapid-random equilibrium mechanisms) they are measures of enzyme–substrate complex dissociation constants. In other cases the ratios of initial-rate parameters predict values for certain dissociation constants. Rapid-reaction or substrate binding studies can, in principle, be used to test such predicted relationships.

(v) Some authors have considered three-substrate mechanisms of the type shown below:

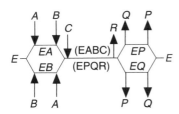

and, in deriving the initial-rate equation, have assumed that equilibrium exists between the free enzyme and the binary and ternary complexes. Such derivations lead to apparently distinctive initial-rate equations but the equations are not valid. In the examples shown, the equilibrium proposed can only apply when the second-order reaction of C with EAB is slow compared to the rates of dissociation of this complex. The increase in the concentration of C is bound to violate this condition at some point because the rate of reaction with C is directly proportional to the concentration of C. The precise conditions where the conditions would break down cannot be predicted, but that they would break down as $[C] \to \infty$ is inevitable. Accordingly, the equations derived in this way are generally invalid.

Table 6. Mechanisms and some diagnostic features of the initial-rate parameters for three-substrate reactions conforming to *Equation 8*

Mechanism	Distinguishing features
(i) Enzyme substitution/triple transfer Hexa Uni Ping-Pong 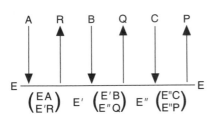	$\phi_{AB} = \phi_{AC} = \phi_{BC} = \phi_{ABC} = 0$
(ii) Concerted enzyme substitution/ Bi Uni Uni Bi Ping-Pong 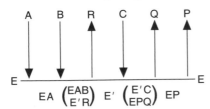	$\phi_{AC} = \phi_{BC} = \phi_{ABC} = 0$

Table 6. Continued

Mechanism	Distinguishing features

(iiia) Compulsory order, quaternary complex/
    Ordered Ter-Ter

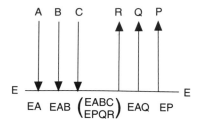

$\phi_{AC} = 0$

(iiib) Partially compulsory order, quaternary complex/
    Uni Random Bi-Bi

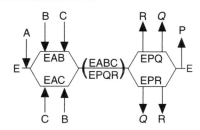

$\phi_{AB} = \phi_{AC} = 0$

(iv) Rapid equilibrium random order,   All parameters have finite values
    quaternary complex/Random Ter-Ter

If mechanism (iiib) is treated by steady-state methods a complex nonlinear reciprocal rate equation is obtained. However, if the rate of interconversion of the quaternary complex is much slower than the rates of dissociation, EABC is effectively in equilibrium with the ternary complexes and the rate equation reduces to the form of *Equation 8* (6).

Mechanism (iv) involves six different pathways for formation of EABC and is difficult to represent in the shorthand notation of Cleland (5). The rate equation is obtained by assuming equilibrium between all the enzyme–substrate complexes.

C. PLOTTING METHODS

A. Cornish-Bowden

12. GENERAL ASPECTS

It is sometimes claimed that the spread of computing has made plotting methods obsolete, but they remain in widespread use. As discussed in more detail elsewhere (18), this is not just conservatism: there are good reasons why plotting methods are essential, not only for displaying the results of experiments but also for their analysis. The human eye is much less easily deceived than any computer program currently available, and is capable of recognizing unexpected behavior even if nothing similar is described in existing textbooks (for an example, see ref. 19). Moreover, even when there is no unexpected behavior to be seen, statistical analysis as commonly applied incorporates several strong assumptions that are often not explicitly stated, and there is rarely any experimental information about whether they are true or not. Programs are now available that substantially decrease the dependence on such assumptions (18), but even then graphs remain essential for displaying the results of computations in such a way that they can be properly appreciated.

13. THE MICHAELIS–MENTEN EQUATION

A large number of equations encountered in enzyme kinetics fall into a small number of mathematical forms. Many have the same form as the Michaelis–Menten equation, $v = Va/(K_m + a)$, and so the methods of analysis worked out for it can often be adapted to other cases; Airas (20) gives many examples of the sorts of transformation needed.

Table 7 lists several plots that apply to the Michaelis–Menten equation. None of them is equally appropriate for all purposes, and in practice the enzymologist needs to use several of them. The direct plot of v against a (*Figure 5*) has an essential educational role for showing the mathematical nature of the dependence of rate on substrate concentration, but it is a poor plot for parameter estimation: it normally leads to substantial underestimation of both V and K_m, because the asymptote defining $v = V$ is nearly always drawn too close to the points. Of the three straight-line plots, the double-reciprocal plot (*Figure 6*) is by far the most widely used, but it has a severe distorting effect on the experimental error, especially if the original v values are homogeneous in variance. The plot of a/v against a (*Figure 7*) is much better in this respect. The plot of v against v/a (*Figure 8*) is complicated to analyse statistically, because errors in v affect both coordinates (so that deviations are towards or away from the origin, not parallel with the ordinate axis). However, in practice it gives quite satisfactory estimates and, more important, it has the excellent property of making it impossible to hide poor experimental design, because it makes it obvious how much of the observable range of v values from 0 to V has in fact been observed. The plot of v against log a (*Figure 9*) is not primarily useful for parameter estimation, though it was used for this purpose by Michaelis and Menten (2), but for comparing widely different kinetic behavior in a single plot.

All of these plots are drawn in observation space (i.e. each observation is shown as a point), but it is also possible to plot the Michaelis–Menten equation in parameter space, where each observation is drawn as a straight line and the parameters are represented by a point. The direct linear plot (*Figure 10*) is the simplest such case. It has the advantage for laboratory use of requiring no calculation, so that it can provide immediate feedback on the results of an experiment while it is proceeding. It has the opposite property from the double-reciprocal plot of making the effects of experimental error more obvious (*Figure 10b*), so that deviations that are barely noticeable in a plot in observation space lead to wide dispersion of the intersections corresponding to the parameter estimates. This is a desirable property for analysing data in the privacy of the laboratory, though one that makes it less appealing for subsequent publication.

The modified direct linear plot shown in *Figure 11* is somewhat more resistant to the effects of large observation errors than the original form, and the lines intersect at less acute angles, so that the points of intersection are more clearly defined. A third form of direct linear plot, with axes $1/K_m$ and V/K_m and observations represented by lines intersecting these at values of $-1/a$ and v/a respectively, appears to have no special advantages and is not used.

14. THE INTEGRATED MICHAELIS–MENTEN EQUATION

The prospect of estimating Michaelis–Menten parameters directly from the time course is one that has attracted numerous biochemists since the first years of enzyme kinetics, but it has

Table 7. Plots of the Michaelis–Menten equation[a]

Abscissa (x)	Ordinate (y)	Shape	Refs	Notes	Figure	Names[b]
a	v	Rectangular hyperbola through the origin		c	5	Michaelis–Menten plot
$1/a$	$1/v$	Straight line	3, 21	d–f	6	Lineweaver–Burk plot, double-reciprocal plot
a	a/v	Straight line	21, 22	d, g, h	7	Woolf plot, Hanes plot
v	v/a	Straight line	21, 23–25	d, i	8	Eadie plot, Hofstee plot, Eadie–Hofstee plot, Augustinsson plot
$\log a$	v	Sigmoid curve	2	j	9	Michaelis–Menten plot
K_m	V	Family of straight lines[k]	26, 27	l	10	Direct linear plot
K_m/V	$1/V$	Family of straight lines[m]	28	l	11	Direct linear plot

[a]The form of equation assumed throughout this table is $v = Va/(K_m + a)$.

[b]Many of the names in common use for enzyme kinetic plots are historically misleading.

[c]*Not* used by Michaelis and Menten (2), and so the name 'Michaelis–Menten plot' used by some authors is especially misleading. The difficulty of locating the asymptotes accurately makes it unsuitable for parameter estimation (both V and K_m are usually seriously underestimated when it is used).

[d]First described by Haldane and Stern (21) and attributed by them to B. Woolf.

[e]Gives a very distorted idea of the error in v used for calculating $1/v$, e.g. errors of $+0.1$ in true v values of 1 and 5 lead to errors of -0.091 and -0.0039, respectively, i.e. the errors in $1/v$ vary by a factor of 23 even though those in v are equal.

[f]Lineweaver and Burk (3) described both this plot and that of a/v against a.

[g]The error-distorting property of this plot is much less than that of the double-reciprocal plot.

[h]Hanes (22) gave the equation for this plot but did not use it to plot his data.

[i]As the entire observable range of v values maps into a finite range of plot, this plot encourages good experimental design by making it impossible to disguise bad design.

[j]Plot used by Michaelis and Menten (2). Rarely used for parameter estimation, but excellent for comparing properties of different isoenzymes or different substrates when the K_m values are widely different.

[k]The line for each observation intersects the K_m axis at $K_m = -a$ and the V axis at $V = v$.

[l]Each observation is plotted as a straight line, not as a point. The coordinates of the point of intersection of the family of lines defines the best-fit parameters. Because of experimental error the lines normally do not intersect at a unique point: each parameter is then estimated as the median of the values given by all possible intersection points. This approach depends on weaker statistical assumptions than methods requiring estimation of the slope and intercept of a straight line (27).

[m]The line for each observation intersects the K_m/V axis at $K_m/V = -a/v$ and the $1/V$ axis at $1/V = 1/v$.

Figure 5. Plot of rate (v) against substrate concentration (a) for an enzyme obeying the Michaelis–Menten equation. The inset shows that the part of the curve accessible to experimental observation is only a very small part of an infinite rectangular hyperbola with two limbs, one entirely and the other partly outside the potentially visible region.

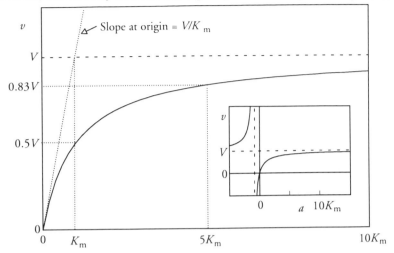

Figure 6. Double-reciprocal plot of $1/v$ against $1/a$. The straight line is the theoretical line, and the flanking curves show the uncertainty produced by errors in v of $\pm 0.05V$. In the inset (largely unlabeled to avoid overcrowding, but representing the same ranges of $1/a$ and $1/v$ as the main plot), the flanking curves are recalculated for errors of $\pm 0.1v$.

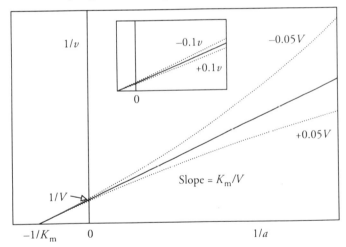

rarely, if ever, led to satisfactory results. Although in principle the Michaelis–Menten equation can be integrated to yield $V^{app}t = a_0 - a + K_m^{app}\ln(a_0/a)$, where a_0 and a are the substrate concentrations at times 0 and t respectively, the apparent parameters V^{app} and K_m^{app} that appear in this equation correspond to V and K_m, respectively, only under very special conditions that are rarely fulfilled: there must be no product inhibition at all, and no loss of enzyme activity during the reaction.

Despite this, the integrated Michaelis–Menten equation is useful as a means of obtaining more accurate estimates of the initial rate than are obtainable with a ruler and pencil when the rate

Figure 7. Plot of a/v against a. The flanking curves and the inset are as for *Figure 6*.

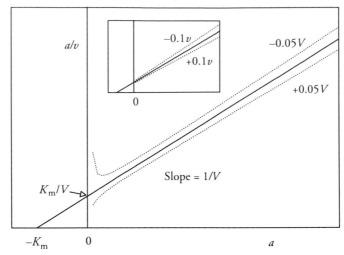

Figure 8. Plot of v against v/a. The flanking curves and the inset are as for *Figure 6*. Note that uncertainties in v produce deviations towards or away from the origin, as indicated by the broken lines.

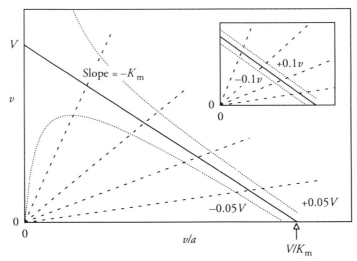

decreases rapidly during the period of measurement. This is because even though V^{app} and K_m^{app} may differ greatly from the true parameters, they normally provide an accurate estimate of the rate at time zero. The two plots listed in *Table 8* and illustrated in *Figures 12* and *13* exploit this property; they are representative of two families of plots in observation and parameter space that are related by the same sorts of algebraic transformations that relate the three straight-line plots of the Michaelis–Menten equation.

15. PLOTS FOR LINEAR INHIBITION DATA

Inhibition is said to be linear when the reciprocal rate is a linear function of the inhibitor concentration (i.e. when the rate equation can be written in the form $v = Va/[K_m(1 + i/K_{ic}) +$

Figure 9. Plot of v against $\log a$. Two curves are shown, for two activities with K_m values differing by an order of magnitude. The slope of the tangent at the point of inflection is equal to $0.576V$ (i.e. $0.25V \times \ln 10$), a relationship used by Michaelis and Menten (2) in their original study of invertase.

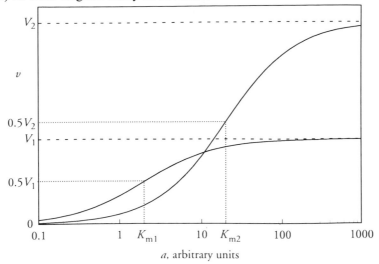

a, arbitrary units

$a(1 + i/K_{iu})]$). There are two basic ways to plot the data: one is to treat the rate as obeying the Michaelis–Menten equation at each inhibitor concentration and then estimate the inhibition constants from the dependence of the apparent parameters on the inhibitor concentration; the other is to analyse the dependence on inhibitor concentration from the outset, keeping the substrate concentration constant for each line. The first approach is illustrated in *Figure 14*. Note that what is actually plotted as ordinate in each secondary plot is an apparent Michaelis–Menten parameter. Although it may also be the slope or intercept of a primary plot, it is best not to refer to it as such because which it is will depend on which form of primary plot is used: the slopes in *Figure 14a* are the same as the ordinate intercepts in *Figure 14b*, and vice versa; moreover, it would be perfectly proper to obtain the apparent Michaelis–Menten parameters from primary plots of v against v/a at each i value, but then they would be neither slopes nor intercepts of these plots. For the same reason, it is best to refer to the two inhibition constants as competitive (K_{ic}) and uncompetitive (K_{iu}) inhibition constants, rather than using names and symbols that imply the use of particular primary plots.

The second approach is illustrated in *Figure 15*. If plots of both $1/v$ and a/v against i are made, the pattern of lines indicates the type of inhibition and both inhibition constants directly from the primary plots. The competitive inhibition constant K_{ic} is most conveniently obtained from the coordinates $(-K_{ic}, (1-K_{ic}/K_{iu})/V)$ of the point of intersection of the family of lines in the plot of $1/v$ against i, and the uncompetitive inhibition constant from the corresponding point $(-K_{iu}, (1-K_{iu}/K_{ic})/V)$ in the plot of a/v against i. However, if the limiting rate V is known, the second inhibition constant can be calculated in either case from the ordinate coordinate. It must be remembered, of course, that only in the case of mixed inhibition do both of these inhibition constants exist: K_{iu} is infinite in competitive inhibition, and K_{ic} is infinite in uncompetitive inhibition.

The approach illustrated in *Figure 14* assumes a reasonably small number of different inhibitor concentrations (say no more than five or six), with enough substrate concentrations at each to yield meaningful estimates of the apparent parameters. Conversely, that illustrated in *Figure 15* assumes a reasonably small number of different substrate concentrations, with enough inhibitor concentrations

Figure 10. Direct linear plot. Each line represents one observation and is drawn with intercepts $-a$ on the K_m axis and v on the V axis. (a) If the lines are drawn exactly (assuming no experimental error) they intersect at a unique point with coordinates that define the values of K_m and V. (b) Experimental errors that are barely perceptible in the original curve (which appears as a mirror image in the second quadrant) cause this unique point to degenerate into a family of points. In this case the best value of K_m can be defined by drawing a vertical line that passes to the right of exactly half the intersection points, and the best value of V likewise.

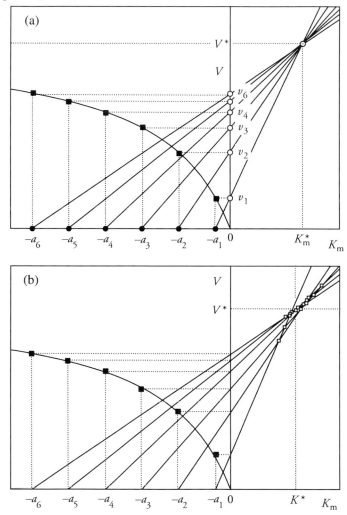

to define each line adequately. Although there are many experiments that satisfy both conditions, so the choice between the two approaches is purely a matter of personal preference, one may occasionally have to analyse data that satisfy neither condition even though the total number of observations is quite large. This may happen, for example, if the substrate and inhibitor concentrations are distributed haphazardly with no more than one or two inhibitor concentrations at each substrate concentration or vice versa. It is still possible to map all the observations on to a single curve provided that the kinetic behavior in the absence of inhibitor is well enough characterized for the uninhibited rate v_0 corresponding to each inhibited rate v to be either known or readily calculable.

Figure 11. Alternative form of direct linear plot.

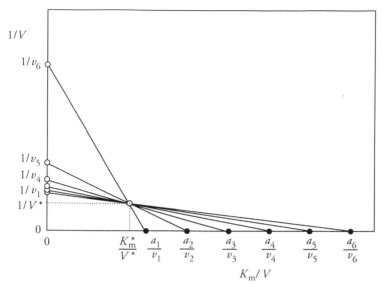

The method needed is illustrated in *Figure 16*. Although, in principle, it allows determination of both inhibition constants, in practice the curve does not usually approach the asymptote closely enough for its location to be estimated with any accuracy. In the event that the uncompetitive component is predominant ($K_{iu} < K_{ic}$), it is better to plot $(v_0 - v)/iv$ against a instead of $iv/(v_0 - v)$.

The plots discussed are listed in *Table 9*, together with another that is now little used, if at all, and which appears to have no special advantages that would justify reviving it.

16. PLOTS FOR TWO-SUBSTRATE DATA

In the simplest cases the rate of a two-substrate reaction follows an equation of the form:

$v = Vab/(K_{ia}K_{mB} + K_{mB}a + K_{mB}b + ab)$,

where a and b are the concentrations of the two substrates and the other symbols represent constants (given earlier, with different symbols, as *Equation 2*. Ed.). In a substituted-enzyme (ping-pong) mechanism the constant term in the denominator is missing, but in ternary-complex mechanisms all terms are needed (see Section 11.1). As this equation is algebraically similar to the equation for mixed inhibition used as the basis for discussion in the previous section, one might expect to find a variety of plots in use similar to those used for inhibition data. In practice, however, the only plotting method in widespread use involves obtaining apparent Michaelis–Menten parameters with respect to one substrate concentration from primary plots made at several constant values of the other concentration, followed by making secondary plots of these as functions of the second concentration (*Figure 17*). As illustrated, the appearance of the primary plots allows, in principle, discrimination between the two main classes of mechanism, though if K_{iA} is less than about $0.2K_{mA}$ the lines for the ternary-complex mechanisms may be difficult to distinguish in practice from those expected for a substituted-enzyme mechanism.

17. NON-MICHAELIS–MENTEN DATA

A full account of cooperativity and other aspects of non-Michaelis–Menten kinetics is beyond the scope of this chapter (but see, for example, ref. 36). However, there are two cases of general enough importance to be illustrated. The first occurs when the observed rate is actually the sum of two (or

▶ p. 106

Table 8. Plots of the integrated Michaelis–Menten equation[a]

Abscissa (x)	Ordinate (y)	Shape	Refs	Notes	Figure	Names
$t/\ln(a_0/a)$	$(a_0-a)/\ln(a_0/a)$	Straight line	29	b	12	Jennings–Niemann plot
$-a_0+p/2$	p/t	Family of straight lines	30	c	13	Integrated direct linear plot

[a]The form of equation assumed in this table is $V^{app}t = a_0 - a + K_m^{app}\ln(a_0/a)$.
[b]This plot has identical slope to the plot of a/v against a if the apparent Michaelis–Menten parameters are identified with the true Michaelis–Menten parameters, but this can be subject to huge (even infinite) errors if the product of the reaction binds significantly. It is better to regard it as an accurate (but more tedious) way to construct a plot of a/v against a. Similar analogs to the other common plots also exist (30).
[c]To produce a well-defined intersection point, the axes should be drawn at an angle of about 20–30° (the exact angle is not important), and the scales chosen so that $-a_0+p/2$ and p/t cover a similar range of values when measured in centimeters. Strictly the abscissa variable is $-p/\ln[a_0/(a_0-p)]$, but $-a_0+p/2$ is an excellent approximation that is much easier to calculate.

Table 9. Plots of inhibition data[a]

Abscissa (x)	Ordinate (y)	Shape	Refs	Notes	Figure	Names
i	K_m^{app}/V^{app}	Straight line		b	14c	
i	$1/V^{app}$	Straight line		b	14d	
i	$1/v$	Family of lines	31	c, d	15	Dixon plot
i	a/v	Family of lines	33	c	15	Cornish-Bowden plot
$iv/(v_0-v)$	a	See note[e]	34	e	16	Hunter–Downs plot
$v_0/(v_0-v)$	$1/i$	Straight line	35	f	–	Webb plot

[a]The form of equation assumed in this table is $v = Va/[K_m(1 + i/K_{ic}) + a(1 + i/K_{iu})]$, the rate in the absence of inhibitor, when needed, being $v_0 = Va/(K_m + a)$.
[b]Suitable for the case where apparent Michaelis–Menten parameters have been determined at each inhibitor concentration.
[c]Does not require separate determination of apparent Michaelis–Menten parameters.
[d]Not the same as the Dixon plots for pH dependences (32).
[e]Curved for mixed inhibition: limit as a approaches zero (intercept on ordinate axis) is K_{ic}; limit as a approaches infinity is K_{iu}. Although in principle this plot allows determination of both inhibition constants from a single plot, in practice they are unlikely to be both well defined unless they are approximately equal. When $K_{iu} < K_{ic}$ (predominantly uncompetitive inhibition) it is better to plot the reciprocals of both coordinates, when the characteristics are inverted. The Hunter–Downs plot maps all combinations of substrate and inhibitor concentrations on to a single line, not a family of lines, so it is suitable for plotting data that could not be plotted in other ways, e.g. if the substrate and inhibitor concentrations are distributed haphazardly.
[f]Gives a straight line with an ordinate intercept of 1 for all linear types, and slope $(K_m + a)/(K_m/K_{ic} + a/K_{iu})$.

Figure 12. Jennings–Niemann plot of time-course data. Each series of points represents a time course at a different initial substrate concentration, a_0. The individual lines do *not* offer a reliable way of estimating the Michaelis–Menten parameters, because even small amounts of product inhibition produce huge errors, but if they are extrapolated back to abscissa values of a_0 the resulting points (★) can be treated as accurate points on a plot of a_0/v_0 against a_0 (cf. *Figure 7*).

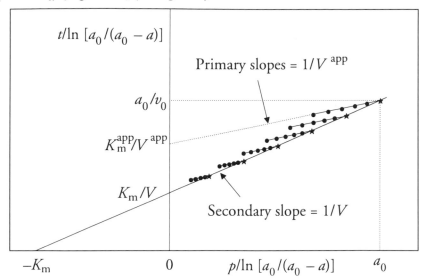

Figure 13. Direct linear plot for time-course data. The axes can, in principle, be drawn at any angle to one another, but values in the range 20–30° are best for giving a well-defined intersection point. Each experimental line represents one time point, and makes intersections $-a_0 + p/2$ (■) on one axis and p/t (●) on the other. Drawing an additional line through the common intersection point and making an intersection a_0 (□) with the first axis produces an intersection of v_0 (○) on the other. Scales of measurement should be chosen such that both series of intersections have similar values when expressed in centimetres.

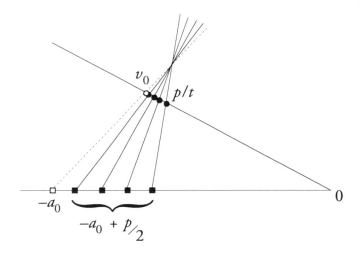

Figure 14. Determination of inhibition constants. Any kind of primary plot, such as (a) a/v against a (cf. *Figure 7*) or (b) $1/v$ against $1/a$ (cf. *Figure 6*) can be used to obtain apparent Michaelis–Menten parameters, from which the dependences on inhibitor concentration of the apparent values of (c) K_m/V and (d) $1/V$ yield the competitive (K_{ic}) and uncompetitive (K_{iu}) inhibition constants, respectively.

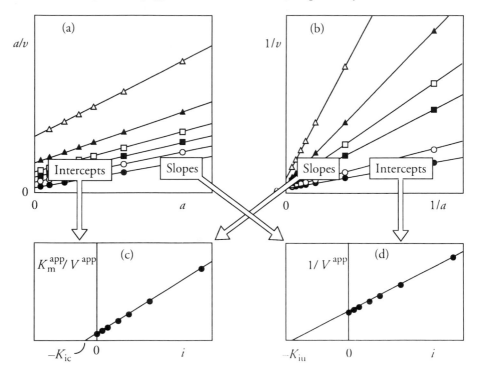

more) independent component activities that separately obey Michaelis–Menten kinetics (*Figure 18a*). In principle the ordinary straight-line plots of *Figures 6–8* become curved in this case, but in reality, as illustrated in *Figures 18b* and *18c*, the curvature is usually too slight to be easily recognized in the plots of $1/v$ against $1/a$ and a/v against a. It follows that neither of these plots provides a satisfactory way of recognizing deviations from Michaelis–Menten kinetics when they occur, and if either of them is used to obtain parameters when there is unnoticed curvature the resulting values will not, in general, correspond to the true parameters of either of the components. The simple plot of v against a is likewise unsatisfactory (*Figure 18a*) because the curve is not easily distinguishable from a rectangular hyperbola. Of the common plots, that of v against v/a (*Figure 18d*) is by far the best for recognizing deviations from Michaelis–Menten kinetics, as any curvature is usually obvious.

The relationship between the lines for the individual components and the resultant curve in such a plot of v against v/a is simple, but as it is very frequently misrepresented in published work it is important to understand it clearly: the components are additive along lines through the origin. This means that the resultant line is normally quite far from any of the lines for the individual components, so that if one tries to estimate the parameters for the components by drawing tangents to the experimental curve (as is done regrettably often) the resulting parameter estimates will be grossly in error. Essentially the same comments apply to the Scatchard plot (37), which is often used in the analysis of protein–ligand binding data and is algebraically equivalent to the plot of v against v/a with the axes reversed.

Figure 15. Determination of inhibition constants from primary plots. (a, c, e) Plots of $1/v$ against i at each substrate concentration yield a set of straight lines intersecting at $i = -K_{ic}$; (b, d, f) plots of a/v against i at each substrate concentration yield a set of straight lines intersecting at $i = -K_{iu}$. Plots (a) and (b) show the patterns expected for competitive inhibition (K_{iu} infinite); (c) and (d) show the patterns expected for mixed inhibition (both inhibition constants finite); and (e) and (f) show the patterns expected for uncompetitive inhibition (K_{ic} infinite).

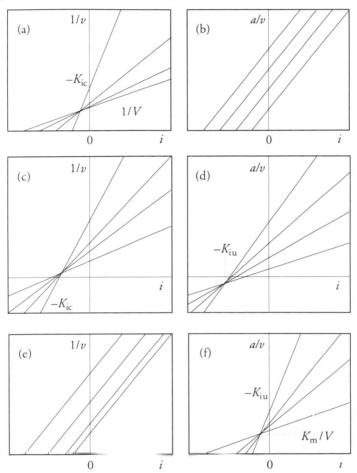

For purely descriptive purposes it is often convenient to express data that deviate from Michaelis–Menten kinetics in terms of the Hill equation (38), which may be written for a substrate-concentration dependence as $v = Va^h/(K_{0.5}^h + a^h)$, and for an inhibitor-concentration dependence as $v = v_0/[1 + (i/K_{0.5})^h]$: in the first case V is the limiting rate and $K_{0.5}$ is the substrate concentration for half-saturation, and in the second case v_0 is the uninhibited rate and $K_{0.5}$ is the inhibitor concentration that gives $v = 0.5v_0$. In both cases h, the Hill coefficient, is a measure of the degree of departure from simple kinetics: it has a value greater than 1 for positive cooperativity and a value of less than 1 for negative cooperativity (and for the heterogeneous systems considered in the previous paragraph). As the Hill equation is not derived from a physical model we should not expect it to fit experimental data exactly, but in positively cooperative systems it often fits remarkably well over a wide range of concentrations. When this is so we expect to

Figure 16. Hunter–Downs plot. For any linear type of inhibition this plot places observations from all combinations of substrate and inhibitor concentrations on a single line: a curve for mixed inhibition, a straight line with a positive slope for competitive inhibition or a horizontal straight line for uncompetitive inhibition.

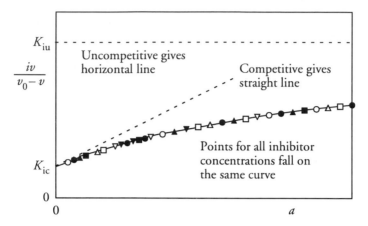

Figure 17. Primary and secondary plots for a two-substrate reaction. Any kind of primary plot, such as (a) a/v against a (cf. *Figure 7*) or (b) $1/v$ against $1/a$ (cf. *Figure 6*) can be used to obtain apparent Michaelis–Menten parameters, from which the dependences on the other substrate concentration of the apparent values of (c) K_m/V and (d) $1/V$ yield the parameters of the full rate equation. For a substituted-enzyme mechanism the apparent value of K_m/V is independent of the concentration of the second substrate.

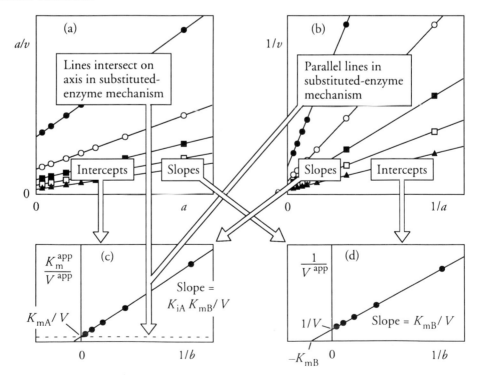

Figure 18. Appearance of the common Michaelis–Menten plots when the reaction is due to a mixture of catalytic activities. In plots of (a) v against a, (b) $1/v$ against $1/a$ and (c) a/v against a, the deviations from the behavior expected for a pure enzyme are not easily visible, and so none of these plots is suitable for recognizing if a mixture of activities is present. By contrast, in (d) the plot of v against v/a is noticeably curved over the whole range. Note that with this plot the component activities can be added along lines through the origin, and that the resultant curve is typically quite far from any of the lines for the components. (e) Although the departure from Michaelis–Menten behavior is not very obvious in the plot of v against log a, this plot is valuable for showing in a clear way how the individual components contribute to the total.

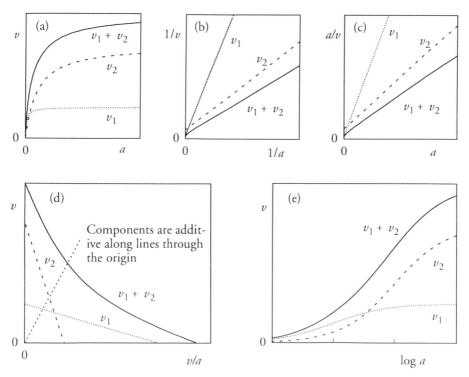

observe a good straight line in plots of $\log[v/(V-v)]$ against log a or $\log[v/(v_0-v)]$ against log i, as illustrated in *Figure 19*. There are always deviations from a straight line (except in the trivial case with $h=1$), at both ends of the former plot, and at the low end of the latter.

18. PLOTS FOR TESTING ENZYME HYPOTHESES

Table 10 lists miscellaneous plots for testing specific hypotheses about enzyme mechanisms and behavior. In Dixon's analysis of pH behavior (*Figure 20*), the logarithm of a Michaelis–Menten parameter is plotted against pH, and the resulting curve is interpreted as if it consisted of straight-line segments with slopes of 1, 0 or -1 (or, in principle, other integers), and each 'corner' corresponds to the pK for an ionization. Ionization of the free enzyme or free substrate affects V/K_m (*Figure 20b*), whereas ionization of the enzyme–substrate complex affects V (*Figure 20a*). As K_m is affected by both kinds of ionization, the plot is more complicated (*Figure 20c*), but the interpretation is clear enough as long as one remembers that $\log K_m = \log V - \log(V/K_m)$: then each decrease in slope with increasing pH corresponds to an ionization of the enzyme–

Figure 19. Hill plots for (a) substrate dependence and (b) inhibition. In each case the insets show the corresponding plots of rate against concentration, both slightly sigmoid for the weakly positive cooperativity shown. Hill plots for substrate dependence normally tend towards unit slope (represented by the broken lines) at both extremes, those for inhibition at the low end only.

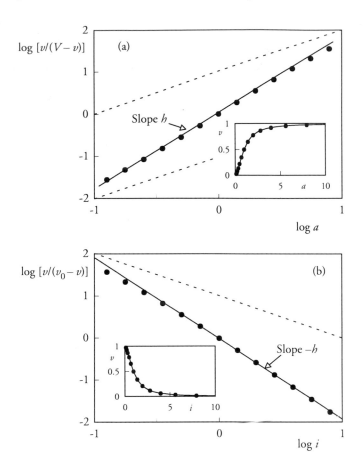

substrate complex, and each increase in slope to an ionization of free enzyme or free substrate. (As none of the three parameters is dimensionless, and in fact they have three different dimensions, the calculation may seem at worst meaningless and at best arbitrarily dependent on the units of measurement. However, although this complication is real, it affects only the vertical locations of the 'corners', which are not used in the analysis; it has no effect on their positions along the pH axis.) (See also detailed treatment of pH effects in Chapter 5, Part A.)

Table 10 also includes three other plots that are less widely used than those in *Figure 20*, but which are useful for testing particular hypotheses. Examples and more detailed descriptions may be found elsewhere (36).

Table 10. Plots for testing enzyme hypotheses

Hypothesis and test	Description	Name (reference)
There is an essential residue with a given pK. A change of slope at pH $=$ pK confirms the hypothesis	A plot of log (*parameter*) against pH is interpreted as a series of straight-line segments interrupted by changes of slope (normally $+1$ or -1). Plots of V/K_m and V as parameter give information about the free enzyme and enzyme–substrate complex, respectively (see *Figure 20*)	Dixon plot[a] (32)
Chemical modification affects μ essential residues. A straight line confirms the hypothesis	Fraction of remaining catalytic activity is raised to the power $1/\mu$ and plotted against the number of residues modified per molecule of enzyme. Normally plotted for several values of μ (typically 1, 2 and 3) superimposed on the same plot	Tsou plot, Tsou Chen-Lu plot (36, 39)
Enzyme is stable during the assay. Superimposable curves at different enzyme concentrations confirm the hypothesis	Amount of product is plotted against $e_0 t$, where e_0 is the total enzyme concentration and t is time. The plot should be made at at least two enzyme concentrations differing by at least a factor of 2	Selwyn test (36, 40)
Two substrates react at the same site. A horizontal line confirms the hypothesis	Starting concentrations $a = a_0$ and $b = b_0$ are found by experiment that give the same rate v_0 when only one substrate is present. Mixtures of the two substrates are then prepared with concentrations $a = (1-p)a_0$, $b = pb_0$, for $p = b/b_0$ in the range 0 to 1, and the total rate, v, is measured for each mixture and plotted against p	Competition plot (36, 41)

[a]Not the same as the Dixon plot for competitive inhibition (31).

Figure 20. Dixon's analysis of pH dependence. Plots of the logarithm of any Michaelis–Menten parameter against pH approximate to straight-line segments interrupted by unit changes in slope. (a) Ionizations of the enzyme–substrate complex produce slope changes in the plot of log V at the corresponding pK value; (b) ionizations of the free enzyme or free substrate produce slope changes in the plot of log(V/K_m); (c) in the plot of log K_m each increase in slope corresponds to an ionization of the free enzyme or free substrate, whereas each decrease in slope corresponds to an ionization of the enzyme–substrate complex.

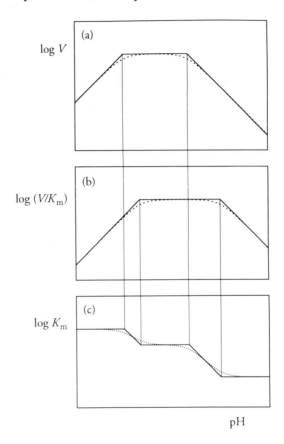

19. REFERENCES

1. Dalziel, K. (1957) *Acta Chem. Scand.,* **11,** 1706.

2. Michaelis, L. and Menten, M.L. (1913) *Biochem. Zeitschr.,* **49,** 333.

3. Lineweaver, H. and Burk, D. (1934) *J. Am. Chem. Soc.,* **56,** 658.

4. Alberty, R.A. (1953) *J. Am. Chem., Soc.,* **75,** 1928.

5. Cleland, W.W. (1963) *Biochim. Biophys. Acta.,* **67,** 104.

6. Dalziel, K. (1969) *Biochem. J.,* **114,** 547.

7. Dalziel, K. (1958) *Trans. Faraday Soc.,* **54,** 1247.

8. Pettersson, G. (1972) *Biochim. Biophys. Acta,* **276,** 1.

9. Mahler, H.R., Baker, R.H. and Shiner, V.J. (1962) *Biochemistry,* **1,** 47.

10. Dalziel, K. (1963) *J. Biol. Chem.,* **238,** 2850.

11. Hart, G.J. and Dickinson, F.M. (1982) *Biochem. J.,* **203,** 617.

12. McGibbon, A.K.H., Buckley, P.D. and Black-well, L.F. (1977) *Biochem. J.,* **165,** 455.

13. Dickinson, F.M. (1985) *Biochem. J.,* **225,** 159.

14. Boyer, P.D. (1959) *Arch. Biochem. Biophys.*, **82,** 387.

15. Silverstein, E. and Boyer, P.D. (1964) *J. Biol. Chem.*, **239,** 3901.

16. Silverstein, E. and Boyer, P.D. (1964) *J. Biol. Chem.*, **239,** 3908.

17. Dalziel, K. and Dickinson, F.M. (1966) *Biochem. J.*, **100,** 34.

18. Cornish-Bowden, A. (1995) *Analysis of Enzyme Kinetic Data.* Oxford University Press, Oxford.

19. Cárdenas, M.L. and Cornish-Bowden, A. (1993) *Biochem. J.*, **292,** 37.

20. Airas, R.K. (1976) *Biochem. J.*, **155,** 449.

21. Haldane, J.B.S. and Stern, K.G. (1932) *Allgemeine Chemie der Enzyme.* Steinkopff, Dresden, p. 119.

22. Hanes, C.S. (1932) *Biochem. J.*, **26,** 1406.

23. Eadie, G.S. (1942) *J. Biol. Chem.*, **146,** 85.

24. Hofstee, B.H.J. (1952) *J. Biol. Chem.*, **199,** 357.

25. Augustinsson, K.-B. (1948) *Acta Physiol. Scand.* **15,** Suppl. 52.

26. Eisenthal, R. and Cornish-Bowden, A. (1974) *Biochem. J.*, **139,** 715.

27. Cornish-Bowden, A. and Eisenthal, R. (1974) *Biochem. J.*, **139,** 721.

28. Cornish-Bowden, A. and Eisenthal, R. (1978) *Biochim. Biophys. Acta*, **523,** 268.

29. Jennings, R.R. and Niemann, C. (1955) *J. Am. Chem. Soc.*, **77,** 5432.

30. Cornish-Bowden, A. (1975) *Biochem. J.*, **149,** 305.

31. Dixon, M. (1953) *Biochem. J.*, **55,** 170.

32. Dixon, M. (1953) *Biochem. J.*, **55,** 161.

33. Cornish-Bowden, A. (1974) *Biochem. J.*, **137,** 143.

34. Hunter, A. and Downs, C.E. (1945) *J. Biol. Chem.*, **157,** 427.

35. Webb, J.L. (1963) *Enzyme and Metabolic Inhibitors,* vol. 1. Academic Press, New York.

36. Cornish-Bowden, A. (1995) *Fundamentals of Enzyme Kinetics (2nd edn).* Portland Press, London.

37. Scatchard, G. (1949) *Ann. N. Y. Acad. Sci.*, **51,** 660.

38. Hill, A.V. (1910) *J. Physiol. (Lond.)*, **40,** iv.

39. Tsou, C.-L. (1962) *Sci. Sin.*, **11,** 1535.

40. Selwyn, M.J. (1965) *Biochim. Biophys. Acta,* **105,** 193.

41. Chevillard, C., Cárdenas, M.L. and Cornish-Bowden, A. (1993) *Biochem. J.*, **289,** 599.

CHAPTER 4
PATTERNS OF ENZYME INHIBITION
K.F. Tipton

1. INTRODUCTION

Enzyme inhibitors may act either reversibly or irreversibly. These show very different behavior both *in vivo* and *in vitro*. Inhibitors can be of value in studies of the mechanism of action and behavior of enzymes and in metabolic studies. Furthermore, many clinically important drugs are enzyme inhibitors. The basic differences between reversible and irreversible inhibitors are summarized in *Table 1*, although there are some exceptions to this oversimplification which will be considered later. This chapter will summarize the kinetic behavior of the different types of enzyme inhibitor; further treatments have been published elsewhere (1–3).

2. SIMPLE REVERSIBLE INHIBITION

For reversible inhibition the full inhibition is usually obtained extremely rapidly since there is no chemical reaction involved, there is simply a noncovalent interaction (*Table 1*).

There are several types of reversible inhibitor which can be distinguished in terms of their response to substrate. In the absence of any inhibitors, the dependence of the initial rate of a simple enzyme-catalysed reaction on the substrate concentration is given by the Michaelis–Menten equation (*Figure 1*).

The kinetics of behavior of simple reversible inhibitors are summarized in *Tables 2 and 3*.

Often the Michaelis–Menten equation is rearranged by turning it into double-reciprocal form, as shown in *Figure 1*. Although other graphical methods or a direct fit to a rectangular hyperbola,

Table 1. Some basic differences between reversible and irreversible inhibitors

Basic equation	Kinetic constant	Rate of inhibition	Reversibility	
			In vitro	*In vivo*
Reversible				
$E + I \underset{k_{-1}}{\overset{k_{+1}}{\rightleftharpoons}} E.I$	$K_i = \dfrac{k_{-1}}{k_{+1}}$	Usually fast	Dialysis dilution; gel filtration	Elimination of free inhibitor
Irreversible				
$E + 1 \overset{k}{\rightarrow} E{-}1$	k	Often slow	None	Synthesis of more enzyme

Figure 1. Simple Michaelis–Menten kinetics.

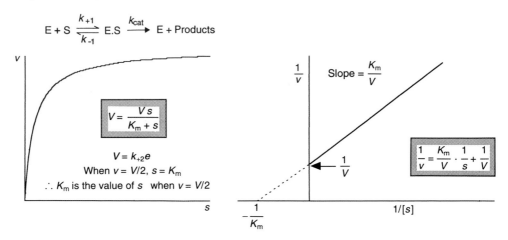

using a computer (4–7) are more accurate, the double-reciprocal plot is useful for presenting data and has most commonly been used for presenting inhibition patterns (see *Figures 2* and *3*). The patterns given by the other linear transformations are shown in *Figure 4* and those that would be seen if the data were plotted by the direct-linear procedure (8) are shown in *Figure 5*.

2.1. Competitive inhibition

The inhibitor binds to the same site on the enzyme as the substrate; substrate analogs and often the products of enzyme-catalysed reactions are competitive inhibitors. The reactions can be represented as shown in *Table 2*. The enzyme can bind either substrate *or* inhibitor; it cannot bind both at once. At very high substrate concentrations all inhibitor will be displaced from the enzyme, therefore V is unchanged but, in the presence of the inhibitor, more substrate will be required to attain V. Thus more substrate will be required to get to $V/2$. So for competitive inhibitors V is unchanged but K_m is increased, as shown mathematically in *Table 2*.

Pharmacologists frequently use i_{50} (or IC_{50}), the inhibitor concentration required to give 50% inhibition. But the measured i_{50} value depends on the substrate concentration, as shown in *Table 3*. At any fixed inhibitor concentration, the degree of inhibition will decrease as the substrate concentration (s) is increased, tending to zero as s becomes very large, as shown in *Figure 6*.

Most reversible enzyme inhibitors used as drugs are designed to look like a substrate (or product) for the enzyme and thus to be competitive inhibitors. *Table 4* lists some common reversible inhibitors. Although the types of inhibition and K_i values are shown, these will depend on assay conditions and substrate used (see also Section 2.6), the species and tissue from which the enzyme was obtained, and whether a specific isoenzyme is being studied. A quite comprehensive listing of enzyme inhibitors has been published (39).

The above discussion has considered the competitive inhibitor binding to the same site on the enzyme as the substrate, giving rise to mutually exclusive binding. It is also possible that the inhibitor might bind to a distinct site on the enzyme, causing a structural change that prevents substrate binding.

Table 2. Kinetic mechanisms of reversible inhibition

Type	Basic mechanism	Kinetic equations
Competitive	$E + S \rightleftharpoons E.S \longrightarrow E + P$ $I \updownarrow K_i$ $E.I$	$v = \dfrac{V.s}{s + K_m(1 + i/K_i)} = \dfrac{V}{1 + \dfrac{K_m}{s}\left(1 + \dfrac{i}{K_i}\right)}$
Uncompetitive	$E + S \rightleftharpoons E.S \longrightarrow E + P$ $I \updownarrow K_i$ $E.S.I$	$v = \dfrac{\dfrac{V.s}{(1+i/K_i)}}{\dfrac{K_m}{(1+i/K_i)} + s} = \dfrac{V}{1 + \dfrac{i}{K_i} + \dfrac{K_m}{s}}$
Noncompetitive	$E + S \underset{K_s}{\rightleftharpoons} E.S \longrightarrow E + P$ $I \updownarrow K_i \quad I \updownarrow K_i$ S $E.I \underset{K_s}{\rightleftharpoons} E.S.I$	$v = \dfrac{\dfrac{V.s}{(1+i/K_i)}}{K_s + s} = \dfrac{V}{\left(1 + \dfrac{K_s}{s}\right)\left(1 + \dfrac{i}{K_i}\right)}$
Mixed	$E + S \underset{K_s}{\rightleftharpoons} E.S \longrightarrow E + P$ $I \updownarrow K_i \quad I \updownarrow K'_i$ S $E.I \underset{K'_s}{\rightleftharpoons} E.S.I$	$v = \dfrac{V.s/(1+i/K'_i)}{s + K_s \cdot \dfrac{(1+i/K_i)}{(1+i/K'_i)}} = \dfrac{V}{(1+i/K'_i) + \left(1 + \dfrac{K_s}{s}\right)\left(1 + \dfrac{i}{K_i}\right)}$
Note: $\dfrac{K_i}{K'_i} = \dfrac{K_s}{K'_s}$		

s and i represent the concentrations of substrate and inhibitor respectively. K_s, K'_s, K_i and K'_i are dissociation constants for the steps indicated.
*In the noncompetitive case, omission of the $E.I \rightleftharpoons E.S.I$ step gives the same equations.

Table 3. Inhibitor and substrate concentration effects in reversible inhibition

Type	%Inhibition	i_{50}
Competitive	$\dfrac{100}{1+\dfrac{K_i}{i}\left(1+\dfrac{s}{K_m}\right)}$	$K_i\left(1+\dfrac{s}{K_m}\right)$
Uncompetitive	$\dfrac{100}{1+\dfrac{K_i}{i}\left(1+\dfrac{K_m}{s}\right)}$	$K_i\left(1+\dfrac{K_m}{s}\right)$
Noncompetitive	$\dfrac{100}{1+\dfrac{K_i}{i}}$	K_i
Mixed	$\dfrac{100}{\dfrac{(K_s+s)}{i}+\dfrac{K_s}{K_i}+\dfrac{s}{K'_i}}{\dfrac{K_s}{K_i}+\dfrac{s}{K'_i}}$	$\dfrac{s+K_s}{\dfrac{s}{K'_i}+\dfrac{K_s}{K_i}}$

2.2. Uncompetitive inhibition

This is quite rare, but it can occur, particularly in the case of enzyme reactions involving more than one substrate (see Section 2.6). The simplest model is that shown in *Table 2*. The kinetic equation in *Table 2* shows that both K_m and V are decreased by the same amount. As shown

Figure 2. Double-reciprocal plots in the presence of (a) competitive and (b) uncompetitive inhibitors.

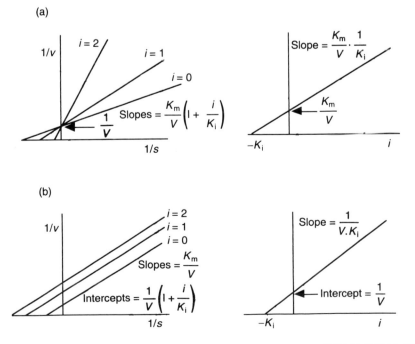

Figure 3. Double-reciprocal plots in the presence of mixed inhibitors. (a) $K'_i > K_i$; (b) $K'_i = K_i$; (c) $K'_i < K_i$.

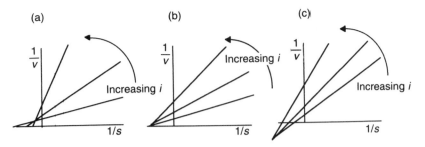

(a)   (b)   (c)

in *Figures 2–5*, competitive and uncompetitive inhibition are easily distinguished from plots of their kinetic behavior. The characteristic double-reciprocal plots resulting from these two forms of inhibition are shown in *Figure 2*. The relationship between the i_{50} and the substrate concentration is shown in *Table 3*. As s tends to ∞, $i_{50} = K_i$. Inhibition will increase as the concentration of substrate is increased. This is the exact opposite of the competitive case (*Figure 6*).

2.3. Mixed and noncompetitive inhibition
This allows inhibitor to bind both to free enzyme and to the E.S complex, as shown in *Table 2*; it is quite common. Treatment of this system by steady-state kinetics leads to an extremely

Figure 4. Effects of inhibitors on other graphical treatments of the Michaelis–Menten equation. (a) Competitive inhibition; (b) noncompetitive inhibition; (c) uncompetitive inhibition.

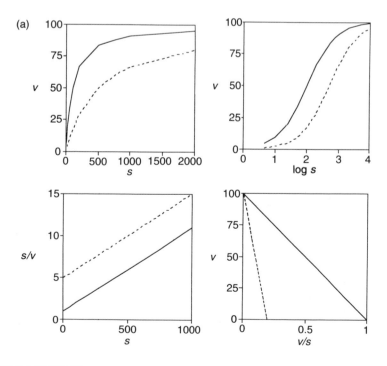

Enzyme Inhibition

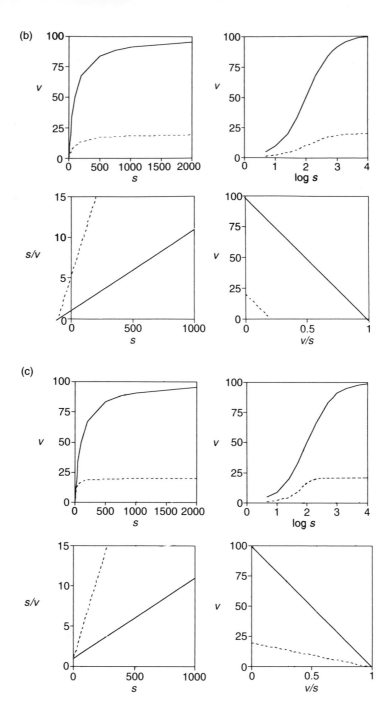

Figure 5. Inhibition patterns revealed by the direct-linear plot. (a) Competitive; (b) noncompetitive; (c) uncompetitive.

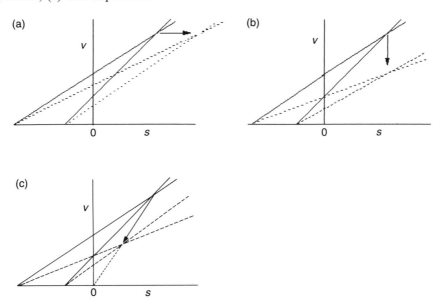

Figure 6. Dependence of the degree of inhibition resulting from a fixed inhibitor concentration upon the concentration of substrate.

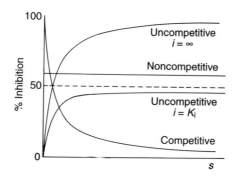

complicated equation that predicts nonhyperbolic dependence of initial velocity upon substrate or inhibitor concentration, as discussed in Section 2.8. However, if it is assumed that the rate of breakdown of the E.S complex to give products is much slower than the rates of dissociation of the complexes, all the binding steps can be regarded as equilibria. Under such conditions K_s, K'_s, K_i and K'_i will be simple dissociation constants and the kinetic equation describing the system can be written as shown in *Table 2*.

The kinetic behavior looks like a mixture of competitive and uncompetitive inhibition. Thus the double-reciprocal plot can give any of the patterns shown in *Figure 3*. The case where $K'_i = K_i$ (and therefore $K_s = K'_s$; *Figure 3b*) is often known as noncompetitive inhibition, but confusion can arise since some authors use this term to cover mixed inhibition as well.

The dependence of i_{50} on s will depend on the relative values of K_i and K'_i: it can either increase, be unaffected (noncompetitive inhibition) or decrease with s (see *Table 3*).

Table 4. Some reversible enzyme inhibitors

Enzyme	EC number	Inhibitor	$K_i(\mu M)$	Reference
Competitive				
Acetylcholinesterase	3.1.1.7	2-Hydroxybenzyltrimethyl-ammonium	0.02	9
Aconitate hydratase	4.2.1.3	Fluorocitrate	200	10
Alanine dehydrogenase	1.4.11	D-Alanine	2000	11
Aldehyde dehydrogenase	1.12.1.3	Chloralhydrate	5	12
Arginase	3.5.3.1	Adenine	14	13
Ascorbate peroxidase	1.11.1.11	Cyanide (towards H_2O_2)	8300	14
Carboxypeptidase A	3.4.17.1	Indole-3-acetic acid	78	15
Chymotrypsin	3.4.21.1	Indole	720	16
Creatine kinase	2.7.3.2	Tripolyphosphate	8	17
Dihydropteroate synthase	2.5.1.15	4-Aminobenzoylglutamate	1000	18
Isocitrate dehydrogenase (NAD)	1.1.1.41	3-Mercapto-2-oxoglutarate	0.52	19
Isocitrate dehydrogenase (NADP)	1.1.1.42	3-Mercapto-2-oxoglutarate	0.005	19
Monoamine oxidase-A	1.4.3.4	D-Amphetamine	20	20
Pepsin A	3.4.23.1	Benzamide	18	21
Succinate dehydrogenase	1.3.99.1	Malonate	40	22
Tryptophanase	4.1.99.1	Indole	50	23
Tyrosine aminotransferase	2.6.1.5	3,5,3'-Triiodothyronine	38	24
Uncompetitive				
Alanopine dehydrogenase	1.5.1.17	2-Oxoglutarate (towards NADH)	1300	25
Ascorbate peroxidase	1.11.1.11	Cyanide (towards ascorbate)	8300	14
Inositol-polyphosphate 1-phosphatase	3.1.3.57	Lithium ion	6000	26
Mixed or noncompetitive				
S-Adenosylmethionine decarboxylase	4.1.1.50	Spermine	500	27
Alanopine dehydrogenase	1.5.1.17	Succinate (towards NADH)	1000	28
Aldehyde reductase	1.1.1.21	Aldrestatin	260	29
α-Amylase	3.2.1.1	Acarbose	1500	30
Arginase	3.5.3.1	Cytidine	5	13
d-CMP deaminase	3.5.4.12	TTP	100	31
Glycine–tRNA ligase	6.1.1.14	Serine	52000	32
Isocitrate lyase	4.1.3.1	Oxalate	2000	33
Monophenol mono-oxygenase	1.14.18.1	1-Methyl-2-mercapto-imidazole	4.6	34
NAD synthase	6.3.5.1	Psicofuranine	600	35
Phenylalanine-4-mono-oxygenase	1.14.16.1	Methotrexate	38	36

Table 4. Continued

Enzyme	EC number	Inhibitor	$K_i(\mu M)$	Reference
Pyruvate dehydrogenase (lipoamide)	1.2.4.1	Isobutyryl-CoA	90	37
		Methylmalonyl-CoA	350	37
Serine acetyltransferase	2.3.1.30	L-Cysteine	0.6	38

Products of the reactions, whose inhibition type will depend on the kinetic mechanism followed (see *Tables 8–11*), are not included.

K_i values will depend on the assay conditions and perhaps also on the source of the enzyme.

In the case of the mixed inhibitors the K_i values shown are those determined from the slopes of double-reciprocal plots.

Table 5. Competition between substrates for an enzyme

Reaction scheme	Kinetic behavior

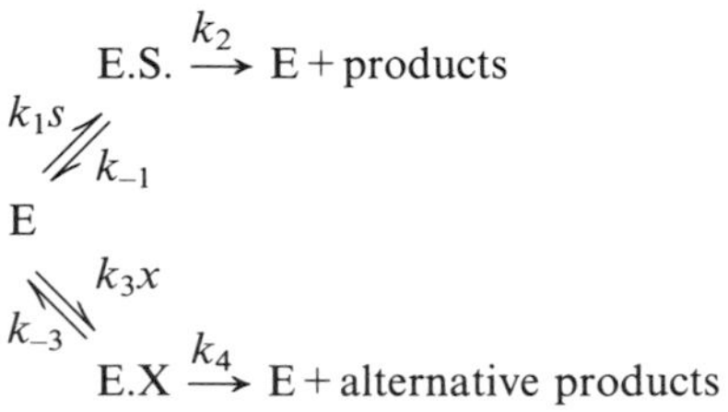

(a) As an inhibitor

$$E.S. \xrightarrow{k_2} E + products$$
$$\overset{k_1 s}{\underset{k_{-1}}{\rightleftarrows}}$$
$$E$$
$$\overset{k_3 x}{\underset{k_{-3}}{\rightleftarrows}}$$
$$E.X \xrightarrow{k_4} E + alternative\ products$$

$$v = \frac{V}{1 + \dfrac{K_m^s}{s}\left(1 + \dfrac{x}{K_m^x}\right)}$$

$$K_m^s = \frac{k_{-1} + k_2}{k_1} \quad \text{and} \quad K_m^x = \frac{k_{-3} + k_4}{k_3}$$

(b) As a competing substrate

Rate of S conversion
$$v_s = \frac{V^s s/K_m^s}{1 + \dfrac{s}{K_m^s} + \dfrac{x}{K_m^x}}$$

Rate of X conversion
$$v_x = \frac{V^x x/K_m^x}{1 + \dfrac{s}{K_m^s} + \dfrac{x}{K_m^x}}$$

Ratio of rates
$$\frac{v_s}{v_x} = \frac{V^s . s/K_m^s}{V^x . x/K_m^x} = \frac{(k_2/K_m^s) . s}{(k_4/K_m^x) . x}$$

Sum of rates
$$v_{total} = (v_s + v_x) = \frac{V^s/K_m^s . s + V^x/K_m^x . x}{1 + \dfrac{s}{K_m^s} + \dfrac{x}{K_m^x}}$$

If $\dfrac{s}{K_m^s} = \dfrac{x}{K_m^x} = 1$
$$v_{total} = \frac{V^s + V^x}{3}$$

Note: sum of rates determined separately
$$v_{sum} = \frac{V^s + V^x}{2}$$

Table 6. Some competitive substrates

Enzyme	Substrate	Competitor	Comments
Alcohol dehydrogenase (EC 1.1.1.1)	Ethanol	Allyl alcohol	Ingested allyl alcohol is metabolized by this enzyme to hepatotoxic acrolein; consumption of ethanol can protect by competition (see ref. 40)
	Retinol	Ethanol	See below (41)
Aldehyde dehydrogenase (EC 1.12.1.3)	'Biogenic aldehydes'	Acetaldehyde	Acetaldehyde derived from ethanol decreases the oxidation of the aldehyde derived from norepinephrine to its corresponding acid (42)
	Retinal	Acetaldehyde	Some of the effects of alcoholism may involve interference with the metabolism of retinol and its derivatives (41)
Dopa decarboxylase (EC 4.1.1.28)	Dopa	α-Methyldopa[a]	This compound was introduced as an enzyme inhibitor but some of its effects result from the formation of the α-methyl analogs of dopamine and norepinephrine which interfere with their functions (42)
Isoleucine-tRNA synthase (EC 6.1.1.5)	Isoleucine	Valine	Formation of incorrect aminoacyl-tRNAs by reaction of these poor competitors is prevented by 'editing' and 'proof-reading' mechanisms (43)
Valyl-tRNA synthase (EC 6.1.1.9)	Valine	Threonine	

[a] Also gives time-dependent inhibition.

Table 7. The apparent values of the Michaelis–Menten parameters in the presence of simple reversible inhibitors

Type	$K_m^{app.}$	$V^{app.}$	$(K_m/V)^{app.}$
Competitive	$K_m(1 + i/K_i)$	V	$K_m(1 + i/K_i)/V$
Uncompetitive	$K_m/(1 + i/K_i)$	$V/(1 + i/K_i)$	K_m/V
Mixed	$(1 + i/K'_i)/K_m(1 + i/K_i)$	$V/(1 + i/K'_i)$	$K_m(1 + i/K_i)/V$
Noncompetitive	K_m	$V/(1 + i/K_i)$	$K_m(1 + i/K_i)/V$

i represents the inhibitor concentration.

2.4. Competitive substrates

Some compounds that have been used as competitive inhibitors of enzymes are, in fact, alternative substrates for them. *Table 5* summarizes the behavior of a compound, X, that acts as a competitive substrate. As an inhibitor with respect to S the alternative substrate is simply competitive, except that the apparent K_i value is a Michaelis constant rather than a simple dissociation constant. Some examples of competitive substrates are shown in *Table 6*.

The situation where two substrates compete for the same enzyme is quite common in metabolic systems. In this case, equations expressed in terms of the conversion of each substrate in the presence of the other (*Table 5*) show that the discrimination between the substrates depends only on the concentrations of the competing substrates and their relative V/K_m ratios for the enzyme.

Figure 7. Replots of the slopes and intercepts of double-reciprocal plots as a function of the inhibitor concentration. (a) Competitive inhibition; (b) uncompetitive inhibition; (c) mixed; (d) noncompetitive inhibition.

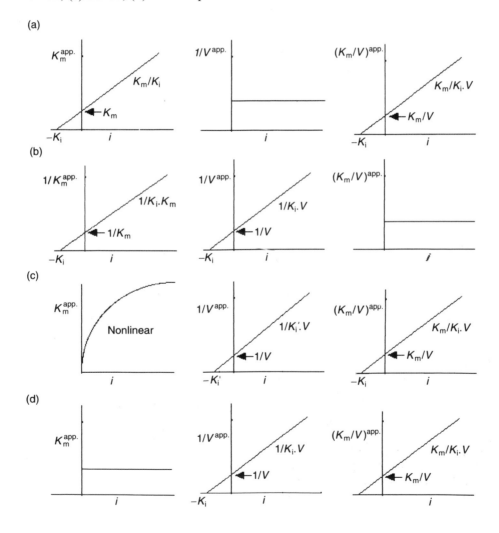

Figure 8. Dixon plots for competitive inhibition. In (a) each line is described by:

$$\frac{1}{v} = \frac{K_m}{Vs} + \frac{1}{V} + \frac{K_m}{Vs} \cdot \frac{i}{K_i}. \text{ Thus, two lines at different substrate concentrations } (s_1 \text{ and}$$

s_2) will intersect at: $\dfrac{K_m}{s_1} + 1 + \dfrac{K_m}{s_1} \cdot \dfrac{i}{K_i} = \dfrac{K_m}{s_2} + 1 + \dfrac{K_m}{s_2} \cdot \dfrac{i}{K_i}$

$$\text{or}: \frac{1}{s_1}\left(1 + \frac{i}{K_i}\right) = \frac{1}{s_2}\left(1 + \frac{i}{K_i}\right).$$

As s_1 does not equal s_2, $i = K_i$ at the intersection point. This will be true at all substrate concentrations, as shown in the broken line, for a substrate concentration sufficient to give V.

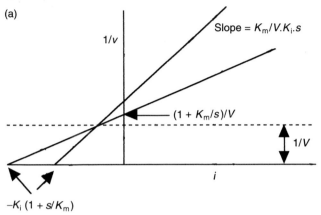

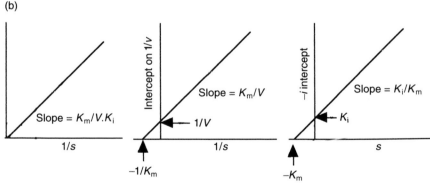

As shown in *Table 5*, the total rate of substrate conversion (S + X) will be less for the mixture than the sum of each of the activities determined separately. When each substrate is present at its K_m concentration the rate with the mixture will be two-thirds of the sum of the separate rates under the same conditions. This is a frequently used method for determining whether the same enzyme is involved in the metabolism of two different substrates.

2.5. Determination of inhibitor constants

Table 7 summarizes the relationship between the apparent values of V, K_m/V and K_m ($V^{app.}$, $(K_m/V)^{app.}$, $K_m^{app.}$, respectively) for the different types of inhibition. In the past the slopes and intercepts from double-reciprocal plots (see *Figures 2* and *3*) have most frequently been used to

Figure 9. (a) Dixon plot for uncompetitive inhibition and Cornish-Bowden plots for uncompetitive (b) and competitive (c) inhibition.

(a)

$$\frac{1}{v} = \frac{K_m}{Vs} + \frac{1}{V} + \frac{i}{K_i} \cdot \frac{1}{V}$$

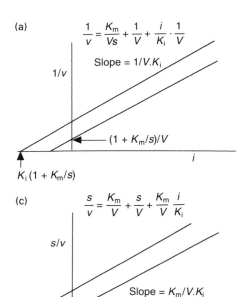

(b)

$$\frac{s}{v} = \frac{K_m}{V} + \frac{s}{V} + \frac{i}{K_i} \cdot \frac{s}{V}$$

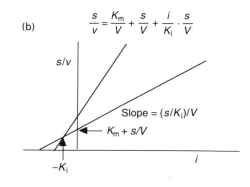

(c)

$$\frac{s}{v} = \frac{K_m}{V} + \frac{s}{V} + \frac{K_m}{V} \frac{i}{K_i}$$

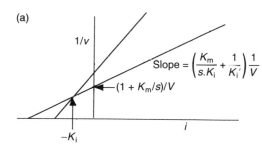

Figure 10. Dixon plots for mixed inhibition. $\dfrac{1}{v} = \dfrac{K_m}{V.s} + \dfrac{1}{V} + \dfrac{i}{V}\left(\dfrac{K_m}{s.K_i} + \dfrac{1}{K_i'}\right)$ (a) $K_i < K'_i$; (b) $K_i > K'_i$ (c) $K_i = K'_i$ (noncompetitive).

(a)

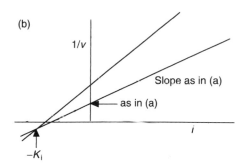

(b)

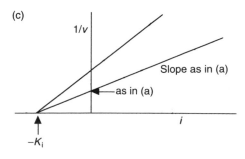

(c)

Figure 11. Cornish-Bowden plots for mixed inhibition.

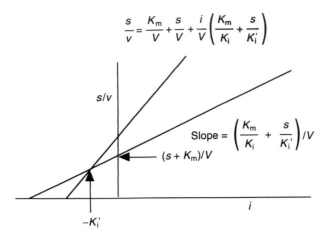

$$\frac{s}{v} = \frac{K_m}{V} + \frac{s}{V} + \frac{i}{V}\left(\frac{K_m}{K_i} + \frac{s}{K_i'}\right)$$

determine the apparent values of K_m/V and $1/V$ in the presence of inhibitor and, similarly, the extrapolated base-line intercept has been used for determining the apparent value of $-1/K_m$. The K_i values can then be obtained from plots of these values against the inhibitor concentration, as shown in *Figure 7*. However, more recently, and more accurately, direct fitting by nonlinear regression analysis is normally used to determine these values (4,6,7) and there are several commercially available programs for this as well as for determinations using the nonparametric direct-linear plot.

Dixon (44) proposed an alternative graphical procedure for determining K_i values, involving plotting reciprocal velocity against inhibitor concentration at a series of fixed substrate concentrations. For competitive inhibition the plot and underlying equations are shown in *Figure 8*. Provided it is established that an inhibitor is competitive and the maximum velocity is accurately known, it is possible, although perhaps not very accurate, to determine K_i from the intersection between one experimental line and a line drawn parallel to the inhibitor concentration axis such that its height above that axis corresponds to $1/V$. It follows from the above arguments that the lines in a Dixon plot for a competitive inhibitor must always intersect above the $-i$ axis. This has often been used as a diagnostic test for competitive inhibition, but, as discussed below, it is not valid since mixed inhibitors may also give families of lines that intersect in the same quadrant.

The plots of the slopes and intercept values from the Dixon plots against the substrate concentration or its reciprocal, shown in *Figure 8*, can be used to determine the K_m value, but there is no advantage in choosing this procedure.

In the case of uncompetitive inhibition, *Figure 9* shows that plots of $1/v$ against i at a series of fixed substrate concentrations will give a family of parallel lines from which there is no convenient procedure for obtaining the K_i value. However, Cornish-Bowden (45) has pointed out that graphs of s/v versus i effectively change the behavior of competitive and uncompetitive inhibitors, as shown in *Figure 9*.

Since mixed inhibitors give a combination of competitive and uncompetitive effects, their behavior in Dixon plots depends on the relative values of the two K_i values (K_i and K'_i), as shown in *Figure 10*. Although K_i can be determined in this way, it does not allow the easy

▶ p. 134

Table 8. Product inhibition patterns for reactions involving two substrates and two products

Mechanism	Equation	Inhibition type			
		Ax varied		B varied	
		B not saturating	B saturating	Ax saturating	Ax saturating
(a) Compulsory-order	$$v = \dfrac{V}{1 + \dfrac{K_m^B}{b} + \left(\dfrac{K_m^{Ax}}{ax} + \dfrac{K_s^{Ax}.K_m^B}{ax.b}\right)\left(1 + \dfrac{a}{K_i^A}\right)}$$	A: Competitive	Competitive	Mixed	None
	$$v = \dfrac{V}{\left(1 + \dfrac{bx}{K_i^{Bx}}\right) + \dfrac{K_m^{Ax}}{ax} + \left(\dfrac{K_m^B}{b} + \dfrac{K_s^{Ax}.K_m^B}{ax.b}\right)\left(1 + \dfrac{bx}{K_i'}\right)}$$	Bx: Mixed	Uncompetitive	Mixed	Mixed
(b) Theorell–Chance	$$v = \dfrac{V}{1 + \dfrac{K_m^B}{b} + \left(\dfrac{K_m^{Ax}}{ax} + \dfrac{K_s^{Ax}.K_m^B}{ax.b}\right)\left(1 + \dfrac{a}{K_i^A}\right)}$$	A: Competitive	Competitive	Mixed	None
	$$v = \dfrac{V}{1 + \dfrac{K_m^{Ax}}{ax} + \left(\dfrac{K_m^B}{b} + \dfrac{K_s^{Ax}.K_m^B}{ax.b}\right)\left(1 + \dfrac{bx}{K_i^{Bx}}\right)}$$	Bx: Mixed	None	Competitive	Competitive
(c) Double-displacement	$$v = \dfrac{V}{1 + \dfrac{K_m^{Ax}}{ax} + \dfrac{K_m^B}{b}\left(1 + \dfrac{a}{K_i^A}\right) + \dfrac{K_s^{Ax}.K_m^B}{ax.b}.\dfrac{a}{K_i^A}}$$ if b is varied at a fixed level of a, replace $K_s^{Ax}.K_m^B$ by $K_m^{Ax}.K_s^B$	A: Mixed	None	Competitive	Competitive
	$$v = \dfrac{V}{1 + \dfrac{K_m^{Ax}}{ax}\left(1 + \dfrac{bx}{K_i^{Bx}}\right) + \dfrac{K_m^B}{b} + \dfrac{K_m^{Ax}.K_s^B}{ax.b}.\dfrac{bx}{K_i^{Bx}}}$$ if b is varied at a fixed level of a, replace $K_m^{Ax}.K_s^B$ by $K_s^{Ax}.K_m^B$	Bx: Competitive	Competitive	Mixed	None

(a) Compulsory-order mechanism scheme:
$$Ax \;\; E.Ax \xrightarrow{\;B\;} E.Ax.B$$
$$E$$
$$A \;\; E.A \rightleftharpoons E.A.Bx \xleftarrow{\;Bx\;}$$

(b) Theorell–Chance mechanism scheme:
$$Ax \;\; E.Ax$$
$$E \;\; \overset{B}{\underset{Bx}{\rightleftharpoons}}$$
$$A \;\; E.A$$

(c) Double-displacement mechanism scheme:
$$Ax \;\; E.Ax \rightleftharpoons Ex.A \xrightarrow{\;A\;} Ex$$
$$E$$
$$Bx \;\; E.Bx \rightleftharpoons Ex.B \xrightarrow{\;B\;}$$

Table 9. Composition of the apparent K_i values for the mechanisms shown in *Table 8*

Mechanism	Product	Ax				B			
		K_i^{slope}		$K_i^{int.}$		K_i^{slope}		$K_i^{int.}$	
		Unsat. B	Sat. B	Unsat. B	Sat. B	Unsat. Ax	Sat. Ax	Unsat. Ax	Sat. Ax
(a) Compulsory-order[a]	A	K_i^A	K_i^A	—	—	$K_i^A\left(1+\dfrac{ax}{K_s^{Ax}}\right)$	None	$K_i^A\left(1+\dfrac{ax}{K_m^{Ax}}\right)$	None
	Bx	$K_i'\left(1+\dfrac{K_m^{Ax}.bx}{K_s^{Ax}.K_m^B}\right)$	K_i^{Ax}	$\dfrac{(b+K_m^B)}{\left(\dfrac{b}{K_i^{Bx}}+\dfrac{K_m^B}{K_i'}\right)}$	—	K_i'	K_i'	$K_i^{Bx}\left(1+\dfrac{K_m^{Ax}}{ax}\right)$	K_i^{Bx}
(b) Theorell–Chance	A	K_i^A	K_i^A	—	—	$K_i^A\left(1+\dfrac{ax}{K_s^{Ax}}\right)$	∞	$\left(1+\dfrac{ax}{K_m^{Ax}}\right)$	∞
	Bx	$K_i^{Bx}\left(1+\dfrac{b}{K_s^B}\right)$	∞	$K_i^{Bx}\left(1+\dfrac{b}{K_m^B}\right)$	∞	K_i^{Bx}	K_i^{Bx}	—	—
(c) Double-displacement	A	$\dfrac{K_i^A}{\left(1+\dfrac{K_s^B}{b}\right)}$	K_i^A	—	—	$\dfrac{K_i^A.K_m^B.ax}{K_s^B.K_m^{Ax}}$	∞	$K_i^A\left(1+\dfrac{ax}{K_m^{Ax}}\right)$	∞
	Bx	$\dfrac{K_i^{Bx}.K_m^A.b}{K_s^{Ax}.K_m^B}$	∞	$K_i^{Bx}\left(1+\dfrac{b}{K_m^B}\right)$	∞	$\dfrac{K_i^{Bx}}{\left(1+\dfrac{K_s^{Ax}}{ax}\right)}$	K_i^{Bx}	—	—

[a]Note: K_i' in mechanism (a) can be defined in terms of the kinetic constants for the reverse reaction as: $K_i' = \dfrac{K_s^A.K_m^{Bx}}{K_m^A}$.

Table 10. Product inhibition for two systems involving enzyme isomerization

Mechanism and equations	Inhibition type Ax varied		B varied	
	B not saturating	B saturating	Ax not saturating	Ax saturating

(a) 'Iso-compulsory-order'

$$\begin{array}{c} \text{Ax} \qquad \text{B} \\ E \rightleftharpoons E.Ax \rightleftharpoons E.Ax.B \\ \updownarrow \qquad\qquad \updownarrow \\ E' \rightleftharpoons E'.A \rightleftharpoons E'.A.Bx \\ \text{Bx} \end{array}$$

	A: Mixed	Mixed	Mixed	Uncompetitive
	Bx: Mixed	Uncompetitive	Mixed	Mixed

$$v = \frac{V}{\left(1 + \dfrac{a}{K_i'^A}\right) + \dfrac{K_m^B}{b} + \left(\dfrac{K_m^{Ax}}{ax} + \dfrac{K_s^{Ax}.K_m^B}{ax.b}\right)\left(1 + \dfrac{a}{K_i^A}\right)}$$

$$v = \frac{V}{\left(1 + \dfrac{bx}{K_i^{Bx}}\right) + \dfrac{K_m^{Ax}}{ax} + \left(\dfrac{K_m^B}{b} + \dfrac{K_s^{Ax}.K_m^B}{ax.b}\right)\left(1 + \dfrac{bx}{K_i'}\right)}$$

Note: the full equation would also contain expressions of the form $a.bx/K_i'^{ABx}$ and $a.bx/K_i^{ABx}.b$ but these may be neglected for initial-rate studies of the inhibition by a single product.

(b) 'Iso-Theorell–Chance'

$$\begin{array}{c} \text{Ax} \\ E \rightleftharpoons E.Ax \\ \updownarrow \qquad \updownarrow \diagdown^B \\ \qquad\qquad\qquad \searrow \text{Bx} \\ E' \rightleftharpoons E'.A \\ \text{A} \end{array}$$

	A: Mixed	Mixed	Mixed	Uncompetitive
	Bx: Mixed	None	Competitive	Competitive

$$v = \frac{V}{1 + \dfrac{a}{K_i'^A} + \dfrac{K_m^B}{b} + \left(\dfrac{K_m^{Ax}}{ax} + \dfrac{K_s^{Ax}.K_m^B}{ax.b}\right)\left(1 + \dfrac{a}{K_i^A}\right)}$$

$$v = \frac{V}{1 + \dfrac{a}{K_i'^A} + \dfrac{K_m^{Ax}}{ax} + \left(\dfrac{K_m^B}{b} + \dfrac{K_s^{Ax}.K_m^B}{ax.b}\right)\left(1 + \dfrac{bx}{K_i^{Bx}}\right)}$$

Note: the full equation would also contain an expression of the form and $a.bx/K_i^{ABx}.b$ but this may be neglected for initial-rate studies of the inhibition by a single product.

Table 11. Some product inhibition patterns for random-order equilibrium systems

	Inhibition type	
	Ax varied	**B varied**

Case I

Scheme:

$$K_s^{Ax}\;\;E.Ax\;\;K_m^B \qquad E.Bx\;\;K_i^{Bx}$$
$$E \qquad\qquad E.Ax.B \rightleftharpoons E.A.Bx \qquad\qquad E$$
$$K_s^B\;\;E.B\;\;K_m^{Ax} \qquad E.A\;\;K_i^A$$

$$v = \frac{V}{1 + \dfrac{K_m^{Ax}}{ax} + \dfrac{K_m^B}{b} + \dfrac{K_s^{Ax}.K_m^B}{ax.b}\left(1 + \dfrac{a}{K_i^A}\right)}$$

For inhibition by Bx replace $\dfrac{a}{K_i^A}$ by $\dfrac{bx}{K_i^{Bx}}$ in the above equation.

Note: $K_s^{Ax}.K_m^B = K_m^{Ax}.K_s^B$

By A with:	
(i) B not saturating	Ax not saturating
Competitive	Competitive
$K_i^{slope} = K_i^A\left(1 + \dfrac{b}{K_s^B}\right)$	$K_i^{slope} = K_i^A\left(1 + \dfrac{ax}{K_s^{Ax}}\right)$
(ii) B saturating	Ax saturating
None	None
By Bx with:	
(i) B not saturating	Ax not saturating
Competitive	Competitive
$K_i^{slope} = K_i^{Bx}\left(1 + \dfrac{b}{K_s^B}\right)$	$K_i^{slope} = K_i^{Bx}\left(1 + \dfrac{ax}{K_s^{Ax}}\right)$
(ii) B saturating	Ax saturating
None	None

Case II

Scheme:

$$E.Ax$$
$$E\;\;K_s^B \qquad E.Ax.B \dashrightarrow E + A + Bx$$
$$E.B$$
$$K_i^A \Updownarrow \;\; K_i^{'A} \Updownarrow$$
$$E.A \rightleftharpoons E.A.B$$
$$K_s^{'B}$$

Note: $K_s^B.K_i^{'A} = K_s^{'B}.K_i^A$

$$v = \frac{V}{1 + \dfrac{K_m^{Ax}}{ax}\left(1 + \dfrac{a}{K_i^{'A}}\right) + \dfrac{K_m^B}{b} + \dfrac{K_s^{Ax}.K_m^B}{ax.b}\left(1 + \dfrac{a}{K_i^A}\right)}$$

By A with:	
(i) B not saturating	Ax not saturating
Competitive	Mixed
$K_i^{slope} = K_i^{'A}\dfrac{\left(1 + \dfrac{K_s^B}{b}\right)}{\left(1 + \dfrac{K_s^{'B}}{b}\right)}$	$K_i^{slope} = K_i^A\left(1 + \dfrac{ax}{K_s^{Ax}}\right)$
	$K_i^{int.} = K_i^{'A}\left(1 + \dfrac{ax}{K_m^{Ax}}\right)$
(ii) B saturating	Ax saturating
Competitive	None
$K_i^{slope} = K_i^{'A}$	

For similar inhibition by Bx with the formation of E.Bx.Ax: K_m^{Ax}/ax will be unchanged, but K_m^B/b and $K_s^{Ax}.K_m^B/ax.b$ will be increased by $(1+bx/K_i^{'Bx})$ and $(1+bx/K_i^{Bx})$, respectively. Thus product inhibition by Bx will be competitive towards B at all concentrations of Ax and mixed with respect to Ax with no inhibition occurring when b is saturating. The apparent K_i values will be as defined for Ax inhibition with appropriate changes to the constants. For inhibition by A with respect to Ax, K_i^{slope} will tend to the value of K_i^A as $b \to 0$.

Case III

In the special situation of case III where the presence of enzyme-bound A has no effect on the affinity for B: $K_s^B = K_s^{'B}$ and therefore $K_i^A = K_i^{'A}$. Therefore the case II equation can be reduced to:

$$v = \frac{V}{1 + \dfrac{K_m^B}{b} + \dfrac{K_m^{Ax}}{ax}\left(1 + \dfrac{K_s^B}{b}\right)\left(1 + \dfrac{a}{K_i^A}\right)}$$

A similar simplification will apply to inhibition by Bx under such conditions.

By A with:	
(i) B not saturating	Ax not saturating
Competitive	Noncompetitive
$K_i^{slope} = K_i^A$	$K_i^{slope} = K_i^A\left(1 + \dfrac{ax}{K_s^{Ax}}\right)$
	$K_i^{int} = K_i^A\left(1 + \dfrac{ax}{K_m^{Ax}}\right)$
(ii) B saturating	Ax saturating
Competitive	None
$K_i^{slope} = K_i^A$	

Table 12. Kinetic behavior of partial inhibitors following competitive and uncompetitive mechanisms

Type	Reaction	Kinetic equation
Partially competitive	$E - S \underset{}{\overset{K_s}{\rightleftharpoons}} E.S \xrightarrow{k} E + Products$ $K_i \updownarrow + I \quad K'_i \updownarrow + I$ $E.I \underset{K'_s}{\rightleftharpoons} E.SI \xrightarrow{k} E.I + Products$	$$v = \frac{V.s}{s + K_s\left(\dfrac{1+i/K_i}{1+i/K'_i}\right)}$$ Steady-state solution: complex
Partially 'uncompetitive'	$E + S \underset{k_{-1}}{\overset{k_{+1}}{\rightleftharpoons}} E.S \xrightarrow{k_{+2}} E + Products$ $k_{-4} \updownarrow k_{+4}.I$ $E.S.I \xrightarrow{k_{+3}} E + I + Products$ $V = k_{+2}e \quad V' = k_{+3}e$ $K_s = k_{-1}/k_{+1} \quad K_i = k_{-4}/k_{+4}$ Steady-state $K_m = (k_{-1} + k_{+2}/k_{+1}) \quad K_i = (k_{-4} + k_{+3}/k_{+4})$ $K' = \dfrac{k_{+1}(k_{-4} + k_{+3})}{k_{+4} \cdot k_{+3}}$	$$v = \frac{V}{1 + \dfrac{K_s}{s} + \dfrac{i}{K_i}} + \frac{V'}{1 + \dfrac{K_i}{i} + \dfrac{K_s}{s}}$$ $$v = \frac{V}{\left(1 + \dfrac{(V'/V)i}{K_i}\right)\dfrac{1}{s}} + \frac{(1+i/K_i)}{\left(1+\dfrac{(V'/V)i}{K_i}\right)}$$ $$\frac{1}{v} = \left(\dfrac{K_s}{1 + \dfrac{(V'/V)i}{K_i}}\right)\frac{1}{Vs} + \left(\dfrac{(1+i/K_i)}{1+\dfrac{(V'/V)i}{K_i}}\right)\frac{1}{V}$$ Steady-state solution: $$\frac{1}{v} = \left(\dfrac{K_m + i/K'_i}{1+\dfrac{(V'/V)i}{K_i}}\right)\frac{1}{s} + \left(\dfrac{(1+i/K_i)}{1+\dfrac{(V'/V)i}{K_i}}\right)\frac{1}{V}$$

Note: K_s is used instead of K_m since all binding steps are assumed to be at equilibrium.

Table 13. Effects of reversible partial inhibitors on the Michaelis–Menten parameters

Type	$1/V_{app.}$	$(K_m/V)_{app.}$
Partially competitive	$1/V$	$\dfrac{K_s}{V}\left(\dfrac{1+i/K_i}{1+i/K_i'}\right)$
Partially uncompetitive (*Table 12*)	$\left(\dfrac{(1+i/K_i)}{1+\left(\dfrac{(k_{+3}/k_{+2})i}{K_i}\right)}\right)\cdot\dfrac{1}{k_{+2}\cdot e}$	$\dfrac{1}{k_{+2}\cdot e}\cdot\left(\dfrac{K_s}{1+\left(\dfrac{(k_{+3}/k_{+2})i}{K_i}\right)}\right)$
Mixed cases Partially mixed	$\dfrac{1+i/K_i'}{(1+k'.i/k.K_i')}\cdot\dfrac{1}{V}$	$\dfrac{K_s(1+i/K_i)}{(1+k'.i/k.K_i')}\cdot\dfrac{1}{V}$
Partially noncompetitive	$\dfrac{1+i/K_i}{V+(V'.i/K_i)}$	$\dfrac{K_s(1+i/K_i)}{V+(V'.i/K_i)}$
Partially uncompetitive (*Table 14*)	$\dfrac{(1+i/K_i')}{(1+i/K_i)}\cdot\dfrac{1}{V}$	$\dfrac{K_s}{V}$

measurement of K_i'. However if, s/v is plotted as a function of i, it is the latter constant that will be given by the intersection point (*Figure 11*).

2.6. Product inhibition

As discussed above, the products of enzyme-catalysed reactions are frequently inhibitors of the reaction. For a reaction involving the transformation of a single substrate to product (see *Table 2*), the product, P, would be expected to be a simple competitive inhibitor of the initial rate of the reaction in the forward direction giving a similar relationship to that shown in *Table 2*.

Most enzyme-catalysed reactions involve the transformation of two, or more, substrates into two, or more, products. In order to demonstrate the relationship between substrates and products, a two substrate–two product reaction can conveniently be written as:

$$Ax + B \rightleftharpoons A + Bx$$

and the general initial-rate equation for the reaction in the forward direction can be written as:

$$v = \dfrac{V}{1+\dfrac{K_m^{Ax}}{ax}+\dfrac{K_m^B}{b}+\dfrac{K_s^{Ax}\cdot K_m^B}{ax.b}}$$

where ax and b are the concentrations of the substrates Ax and B, respectively; V is the maximum velocity, which will be obtained when both substrates are at very high, saturating, concentrations; and K_m^{Ax} and K_m^B are the Michaelis constants for these substrates, defined as the concentrations of Ax and B which will give half-maximum velocity at saturating concentrations of the other substrate. This can easily be seen, since setting either ax or b to levels very much greater than their respective K_m and K_s values will reduce the equation to a simple, single-substrate Michaelis–Menten equation, such as that shown in *Figure 2*. The constant, K_s^{Ax}, represents the apparent dissociation constant for substrate Ax binding to the enzyme when the concentration of B is zero. In systems where either substrate can react with the free enzyme

Figure 12. Behavior of a partially competitive inhibitor.

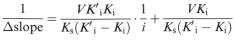

$$\Delta\text{slope} = \frac{K_s}{V}\left(\frac{1 + i/K_i}{1 + i/K'_i}\right) - \frac{K_s}{V}$$

$$\frac{1}{\Delta\text{slope}} = \frac{VK'_iK_i}{K_s(K'_i - K_i)} \cdot \frac{1}{i} + \frac{VK_i}{K_s(K'_i - K_i)}$$

(a)

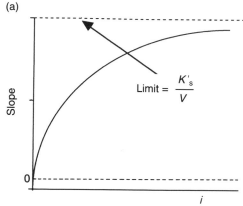

(b)

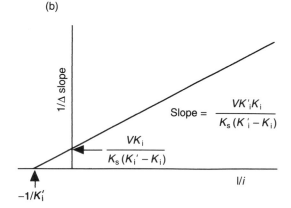

there will also be a K_s^B value, but one of these K_s values is redundant so that only one of them appears in the equation. In the double-displacement (ping-pong) mechanism (mechanism c in *Table 8*) the value of K_s^{Ax} is zero and thus the final term of the denominator in the above equation will be absent. However, that constant will have a finite value in the presence of either one of the products and so the full equation is valid for product inhibition studies.

There are several different formulations of this rate equation, each of which has its own merits (see 1, 3, 46–49 for examples). In one such formulation K_i replaces K_s, since it can be regarded as being the inhibitor constant for that substrate acting as a product inhibitor of the reverse reaction.

In such systems the types of product inhibition observed will depend on the kinetic mechanism followed. The kinetic equations and inhibition patterns given by the commonest two substrate–two product mechanisms in which the substrates interact in an ordered fashion with the enzyme

Figure 13. Behavior of an uncompetitive inhibitor that inhibits only partially.

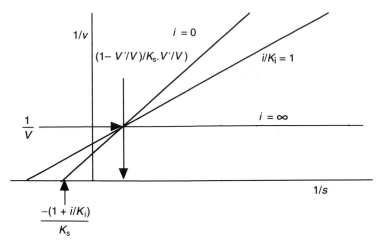

are shown in *Table 8*. The compositions of the apparent K_i values that would be obtained from replots of the slopes (K_i^{slope}) and intercepts ($K_i^{\text{int.}}$) from double-reciprocal plots against the inhibitor concentration are shown in *Table 9*.

Clearly, it is possible to envisage variations on these mechanisms and inhibitor interactions. Two such possibilities, in which the enzyme isomerizes, to a form designated E', during the reaction and reverts to the normal form (E) only after release of products are shown in *Table 10*. The initial-rate equations are unaffected in such 'iso-mechanisms' but the product inhibition patterns are altered. Slow rates of isomerization between E' and E can result in bursts or lags in the reaction progress curves (see 50–52) and appear to be the basis of the apparently cooperative behavior of some enzymes (53, 54).

A mechanism in which the two substrates can bind to the enzyme in a random order will give a complex initial-rate equation (see Section 2.8) but will be described by the simple equation if substrate dissociations can be treated as equilibria. There are several possible modes of product inhibition in such a system, as shown in *Table 11*. Hybrid systems in which one product obeys case I and the other case II are also possible.

Fuller accounts of the kinetics of enzymes involving two substrates have been published (1, 3, 55–59); see also Chapter 3. Detailed discussions of the kinetic behavior and product inhibition of reaction mechanisms involving three (3, 48, 60, 61) and four (62) substrates have also been presented.

2.7. Partial reversible inhibition

Partially competitive inhibition
It is possible that an inhibitor may not completely prevent the enzyme-catalysed reaction from occurring. For example, in the competitive case (see *Table 2*) a situation might exist where the binding of inhibitor did not prevent substrate from binding to the enzyme, but reduced the affinity of such binding without affecting the rate of product formation from the E.S or E.I.S complex, as shown in *Table 12*. In this case the inhibitor will not completely prevent the reaction from occurring, since at very high inhibitor concentrations a reaction will still occur by substrate binding to the E.I complex. Since the rates of breakdown of the E.S and

Table 14. Kinetic behavior of partially mixed inhibition

Reaction	Equation	Limiting cases
$$E + S \underset{K_s}{\rightleftharpoons} E.S \xrightarrow{k} E + Products$$ $$K_i \Updownarrow + I \quad K_i' \Updownarrow + I$$ $$+S$$ $$E.I \underset{K_s'}{\rightleftharpoons} E.S.I \xrightarrow{k'} E.I + Products$$ Relationships: $$K_i/K_i' = K_s/K_s'$$ $$V = k.e$$ $$V' = k'.e$$	$$v = \dfrac{V.s(1 + k'.i/k.K_i')\,\dfrac{(1 + i/K_i')}{(1 + i/K_i')}}{s + K_s\,\dfrac{(1 + i/K_i)}{(1 + i/K_i')}}$$	**(a) Partial noncompetitive** $$K_i = K_i' \text{ and } \therefore K_s = K_s'$$ $$v = \dfrac{V.s}{S + K_m \cdot \dfrac{\left(1 + \dfrac{i}{K_i}\right)}{\left(1 + \dfrac{i}{K_i} \cdot \dfrac{k'}{k}\right)}}$$ **(b) Partial uncompetitive** $$K_i/K_i' = k/k'$$ $$v = \dfrac{V}{\dfrac{K_s}{s} + \dfrac{(1 + i/K_i')}{(1 + k'i/k.K_i')}} = \dfrac{V}{\dfrac{K_s}{s} + \dfrac{(1 + i/K_i')}{(1 + i/K_i)}}$$

Figure 14. Replots of the slopes and intercepts of double-reciprocal plots as a function of the inhibitor concentration for a partially mixed inhibitor.

$$\Delta slope = \frac{K_s}{V} \cdot \frac{1 + i/K_i}{1 + k'.i/k.K'_i} - \frac{K_s}{V} \qquad \frac{1}{\Delta slope} = \frac{K'_i.V}{K'_s - K_s.k'/k} \cdot \frac{1}{i} + \frac{V'}{K'_s - K_s.k'/k}$$

$$\Delta intercept = \frac{1}{V} \cdot \frac{1 + i/K'_i}{1 + k'.i/k.K'_i} - \frac{1}{V} \qquad \frac{1}{\Delta intercept} = \frac{V.K'_i}{1 - k'/k} \cdot \frac{1}{i} + \frac{V'}{1 - k'/k}$$

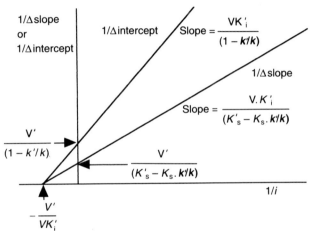

E.S.I complexes to yield products are, by definition, the same, it also follows that the maximum velocity that will be obtained at sufficiently high substrate concentrations will be unchanged by the presence of the inhibitor.

As in the case of mixed inhibition, steady-state treatment of such a system will yield a complex equation (Section 2.8). If one assumes equilibrium conditions to apply for all the binding steps, represented by K_s, K'_s, K_i and K'_i, the initial-rate equation shown in *Table 12* is obtained. Comparison of this equation with that for simple competitive inhibition (*Table 2*) indicates that double-reciprocal plots of the initial velocity against substrate concentration at a series of fixed inhibitor concentrations will be indistinguishable for the two mechanisms. However, as the inhibitor concentration rises towards infinity the velocity decreases to a finite value, given by the relationship shown in *Table 12*.

The slopes of double-reciprocal plots will, in the partial case, be given by the relationship shown in *Table 13*, which indicates that graphs of these slopes against the inhibitor concentration will be hyperbolic (see *Figure 12a*). For this reason this type of inhibition is

Figure 15. Some possible patterns of behavior of systems in which an inhibitor and substrate bind randomly under steady-state conditions.
Overall equation:

$$v = \frac{(\alpha_1 s + \alpha_2 s^2 + \alpha_3 s.i)e}{\beta_0 + \beta_1 s + \beta_2 s^2 + \beta_3 i + \beta_4 i^2 + \beta_5 s.i + \beta_6 s^2.i + \beta_7 s.i^2}.$$

At any fixed i

$$v = \frac{(\alpha'_1 s + \alpha'_2 s^2)e}{\beta'_0 + \beta'_1 s + \beta'_2 s^2}$$

The coefficients α, α', β and β' represent combinations of rate constants.

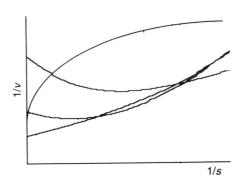

sometimes known as hyperbolic competitive inhibition. Dixon plots of $1/v$ against i will also curve downwards.

The curve of slope against i represents a hyperbola that does not pass through the origin (*Figure 12a*). Thus, if the base-line is shifted up to the point where this curve cuts the vertical axis, the data can be fitted to the equation for a rectangular hyperbola in the normal way. If the change in slope value from the axis intersection point is defined as Δslope, the relationships shown in *Figure 12b* are obtained and so a graph of $1/\Delta$slope against $1/i$ will be linear and intersect the horizontal axis at a value corresponding to $-1/K'_i$, as shown in *Figure 12b*.

Partially uncompetitive inhibition; a confusing story
The situation with uncompetitive inhibition, in which both the E.S and E.S.I complexes can react to give products, may be treated mathematically by assuming that either equilibrium or steady-state conditions apply, yielding the initial-rate equations shown in *Table 12*. Both treatments predict similar behavior. The inhibition pattern does not resemble uncompetitive inhibition at all. In fact it is unclear whether the behavior should be referred to as inhibition or activation, since i activates at low substrate concentrations and inhibits at higher ones (*Figure 13*). At very high concentrations of inhibitor ($i = \infty$) the slope of the line becomes equal to zero and the apparent value of $-1/K_s$, for the equilibrium case, becomes infinite (apparent $K_s = 0$). The slope replots against i will decrease hyperbolically.

So treatment of the uncompetitive inhibition mechanism to allow for partial inhibition does not give the characteristic parallel lines associated with uncompetitive inhibition. However, as shown in the following section, a particular case of partially mixed inhibition can, in fact, give rise to apparently uncompetitive inhibition behavior.

Table 15. Effects of an inhibitor binding twice to the same enzyme form

Mechanism	Equation	
	Equilibrium	**Steady state**

Competitive

$$
\begin{array}{c}
k_{+1}.S \\
E \rightleftharpoons E.S \rightarrow E + Products \\
k_{-1}
\end{array}
$$

$K_i \updownarrow +I$

E.I

$K'_i i \updownarrow +I$

E.I$_2$

$$v = \dfrac{V.s}{s + K_s\left(1 + \dfrac{i}{K_i} + \dfrac{i^2}{K'_i.K_i}\right)}$$

Similar but with K_m replacing K_s

Uncompetitive

$$
\begin{array}{c}
k_{+1}.S \\
E \rightleftharpoons E.S \rightarrow E + Products \\
k_{-1}
\end{array}
$$

$K_i \updownarrow +I$

E.S.I

$K'_i \updownarrow +I$

E.S.I$_2$

$$v = \dfrac{V.s}{K_s + s\left(\dfrac{i}{K_i} + \dfrac{i^2}{K'_i.K_i}\right)}$$

Similar but with K_m replacing K_s

Mixed

$$
\begin{array}{c}
K_sS \\
E \rightleftharpoons E.S \rightarrow E + Products \\
\end{array}
$$

$K_i \updownarrow +I \quad K'_i \updownarrow +I$

E.I $\rightleftharpoons$ E.I.S

$\quad\quad K'_sS$

$K''_i \updownarrow +I$

E.I$_2$

$$v = \dfrac{V.s}{s\left(1 + \dfrac{i}{K'_i}\right) + K_s\left(1 + \dfrac{i}{K_i} + \dfrac{i^2}{K_i.K''_i}\right)}$$

Complex with power terms in s and i (see *Figure 15*)

Partially mixed inhibition

Such inhibition can be described by the system shown in the equation in *Table 14*. The full kinetic equation describing the behavior of this system under equilibrium conditions is also shown in that table. It shows that at saturating substrate concentrations the maximum velocity will tend to $k.e$ (V) in the absence of inhibitor and to $k'.e$ (V') in the presence of saturating concentrations of the inhibitor. Thus the double-reciprocal plots at a series of fixed inhibitor concentrations will be indistinguishable from those for simple mixed inhibition, but plots of slopes or intercepts of these lines against the inhibitor concentration will be hyperbolic (*Table 13*), as will be Dixon plots of $1/v$ against i. Thus the plots of slopes and intercepts against i will both take the form shown in *Figure 12a*, reaching limiting values at saturating concentrations of inhibitor corresponding to K'_s/V' and $1/V'$, respectively.

Figure 16. Behavior of systems in which a complex between the enzyme and two molecules of an inhibitor is formed.

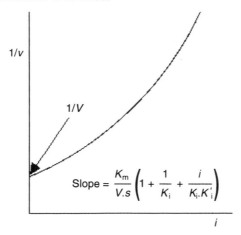

$$\text{Slope} = \frac{K_m}{V.s}\left(1 + \frac{1}{K_i} + \frac{i}{K_i.K'_i}\right)$$

As in the partially competitive case (*Figure 12*), linear plots can be obtained by plotting the reciprocal of the change in slope or intercept above the axis ($1/\Delta$slope or $1/\Delta$intercept) against the reciprocal of the inhibitor concentration, according to the relationships shown in *Figure 14*.

A limiting case of the system shown in equation of *Table 14* would be where the dissociation constants K_i and K'_i are identical (case a). Under these conditions simple noncompetitive behavior is observed in which the V is decreased by the presence of the inhibitor but K_s is unaffected. The apparent values for the slopes and intercepts of double-reciprocal plots of $1/v$ versus $1/s$ are given in *Table 13*. Dixon plots or plots of slope or intercept against the inhibitor concentration will be hyperbolic, but can be linearized if $1/\Delta$slope or $1/\Delta$intercept is plotted against $1/i$.

A second limiting case (*Table 14*, case b) where $K_i/K'_i = k/k'$ (and, as given in *Table 2*, $K_i/K'_i = K_s/K'_s$), that is the ratio of the dissociation constants for the release of S from E.S and

Table 16. Kinetics of systems in which an inhibitor can bind to two different enzyme intermediates

Reaction	Equation	Inhibition
$E \rightleftharpoons E.S \rightarrow E + \text{Products}$ $K_i \updownarrow I \quad K'_i \updownarrow I$ $E.I \qquad E.S.I$	$v = \dfrac{V.s}{s\left(1 + \dfrac{i}{K'_i}\right) + K_m\left(1 + \dfrac{i}{K_i}\right)}$	Mixed (noncompetitive if $K_i = K'_i$)
$E.I \rightleftharpoons E \rightleftharpoons E.S \rightarrow E + \text{Products}$ $\overset{K'_i}{\underset{I}{}} \quad \overset{S}{\underset{K_s}{}}$ $K_i \updownarrow I \quad K_i \updownarrow I$ $E.I \overset{S}{\underset{K_s}{\rightleftharpoons}} E.S.I$	$v = \dfrac{V.s}{s\left(1 + \dfrac{i}{K_i}\right) + K_s\left(1 + \dfrac{i}{K_i/(1 + K_i/K'_i)}\right)}$ Assuming equilibrium conditions for substrate and inhibitor dissociation	Mixed

Figure 17. Behavior of systems in which a complex between an inhibitor and substrate is the inhibitory species.

$$K_i = \frac{s_f.i_f}{si} = \frac{s_f(i - si)}{(s - s_f)} = \frac{s_f.i - s_f(s - s_f)}{(s - s_f)}$$

$E + S \rightleftharpoons E.S \rightarrow E + Products$
$K_i \updownarrow I$

S.I $\quad$ s and s_f are the total and
$\quad\quad$ *free* substrate concentrations
$\quad\quad$ i and i_f are the total and
$\quad\quad$ *free* inhibitor concentrations
$\quad\quad$ si is the concentration of
$\quad\quad$ the S.I complex

$$= \frac{s_f.i - s_f.s + s_f^2}{(s - s_f)}$$

$$K_i(s - s_f) = s_f.i - s.s_f + s_f^2 \quad \text{and}$$

$$s_f^2 + (i - s + K_i)s_f - K_i.s = 0$$

$$\boxed{s_f = \frac{[(i - s + K_i)^2 + 4K_i.s]^{\frac{1}{2}} - (i - s + K_i)}{2}}$$

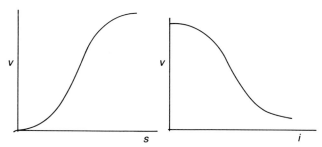

E.I.S is equal to the ratio of the rates of reaction of these two complexes to yield products, predicts 'classical' uncompetitive patterns in double-reciprocal plots, each with a slope of K_s/V. However, replots of the $1/V$ intercepts obtained against the inhibitor concentration will be hyperbolic (see *Table 13*) but a graph of $1/\Delta$intercept) against the reciprocal of the inhibitor concentration can be used to determine K_i (see *Figure 14*).

2.8. More complex inhibition behavior

Apart from partial inhibition and artifacts due to incorrect assay procedures (63), which can lead to apparently complex behavior, a number of inhibitor interactions can show departure from simple Michaelis–Menten kinetics.

Steady-state alternative pathway systems
As mentioned above (Section 2.3) steady-state kinetic treatment of systems in which there are alternative pathways leading to the enzyme complexes (such as in mixed or noncompetitive inhibition, *Table 2*, and the partial competitive and mixed inhibition cases shown in *Tables 12* and *14*) yields a complex equation, which can be written in simplified form as shown in *Figure 15*. The presence of terms in s^2 and i^2 in this equation predicts that the dependence of initial velocity on substrate or inhibitor concentration will not follow a rectangular hyperbola (nonlinear double-reciprocal plots). Only in the absence of inhibitor will the behavior reduce to that described by the simple Michaelis–Menten equation.

Table 17. Kinetics of systems where a substrate–inhibitor complex is the true inhibitor

System	Equations	Behavior and treatment
(a) General		Responses to total s and i will be sigmoid, since si, s_f and i_f do not remain constant as the total substrate concentration is varied. Thus the concentration must be calculated for each value of s (see *Figure 17*) so that the competitive kinetics can be revealed

$E + S \rightleftharpoons E.S \rightarrow E + Products$

$K_i \updownarrow I$

$E + S.I \rightleftharpoons E.S.I$

$\quad K_{si}$

$$v = \frac{V.s}{K_m(1 + si/K_{si}) + s}$$

$$= \frac{V.s_f}{K_m\left(1 + \dfrac{s_f.i_f}{K_i.K_{si}}\right) + s_f}$$

(b) Simplification

If s is not significantly reduced by inhibitor binding ($s_f \approx s$):

$$K_i = \frac{(i - si).s}{si} \quad \text{and thus} \quad si = \frac{i.s}{K_i + s}$$

$$v = \frac{V.s}{K_m\left(1 + \dfrac{i.s}{K_{si}(K_i + s)}\right) + s}$$

This system will follow simple competitive kinetics provided substrate depletion remains negligible

Such equations can give rise to a number of different types of curve describing the variation of initial velocity with substrate concentration (64–68). Some of the possibilities are shown, in the form of double-reciprocal plots, in *Figure 15*. The type of curve obtained will depend on the values of the individual rate constants and, in some cases, deviations from simple behavior may not be apparent. The use of a narrow range of substrate or inhibitor concentrations may also obscure the complexities. For example, at very high inhibitor concentrations variation of the substrate concentrations at low levels may make the terms in s^2 less important and may thus give apparent Michaelis–Menten behavior (66, 69).

Similar complex equations and initial-rate behavior can occur for an enzyme that binds two substrates in a random order to give the productive ternary complex (see, for example, ref. 65), but the case considered above differs in that the behavior would only be expected to depart from that of a normal Michaelis–Menten relationship in the presence of inhibitor.

Cooperativity
The cooperative binding of many allosteric inhibitors will, of course, lead to kinetic behavior that does not follow the simple Michaelis–Menten predictions. Treatment of such systems is beyond the scope of this chapter but is dealt with in detail elsewhere (see 1, 3, 55, 68, 70, 71).

More than one inhibitor-binding site
Systems in which more than one molecule of inhibitor can bind to the same form of the enzyme will lead to kinetic equations in which the inhibitor concentration is raised to a power. The equations given, under both equilibrium and steady-state conditions, for some simple mechanisms involving two molecules of inhibitor binding to the enzyme are shown in *Table 15*. These will yield normal

Figure 18. Inhibition by high substrate concentrations in a simple system; ▲, with substrate inhibition; ○, without substrate inhibition.

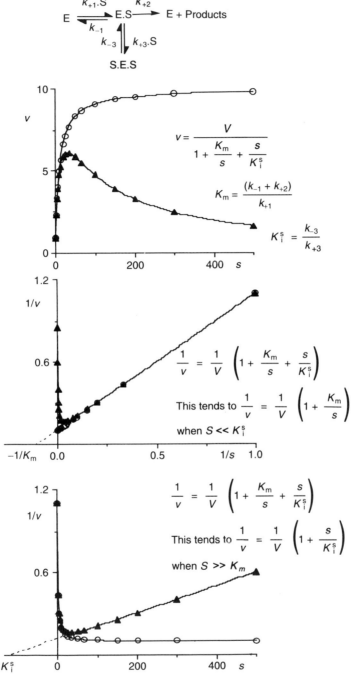

kinetic behavior with respect to substrate concentration, but curves (slope/intercept or Dixon plots) against i will be upwardly curving, or parabolic, as shown for a Dixon plot in *Figure 16*. Clearly it is possible to devise more elaborate mechanisms than those shown.

ENZYMOLOGY LABFAX

Table 18. Possible types of high-substrate inhibition in compulsory-order systems involving two substrates

Mechanism	Equation

(a) B competitive towards Ax

$$
\begin{array}{c}
\text{E.B} \rightleftharpoons \text{E} \xrightarrow[A]{Ax} \text{E.Ax} \xrightleftharpoons[\text{}]{B} \text{E.Ax.B} \\
\text{E.A} \rightleftharpoons \text{E.A.Bx} \\
Bx
\end{array}
$$

with K_i on the E.B↔E step.

$$
v = \frac{V}{1 + \dfrac{K_m^B}{b} + \left(\dfrac{K_m^{Ax}}{ax} + \dfrac{K_s^{Ax}.K_m^B}{ax.b} \right)\left(1 + \dfrac{b}{K_i} \right)}
$$

(b) Ax competitive towards B

$$
\begin{array}{c}
\text{E.Ax}_2 \\
K_i \updownarrow \\
\text{E} \xrightarrow{Ax} \text{E.Ax} \xrightleftharpoons[\text{}]{B} \text{E.Ax.B} \\
\text{E.A} \rightleftharpoons \text{E.A.Bx} \\
Bx
\end{array}
$$

$$
v = \frac{V}{1 + \dfrac{K_m^{Ax}}{ax} + \dfrac{K_m^B}{b}\left(1 + \dfrac{ax}{K_i} \right) + \dfrac{K_s^{Ax}.K_m^B}{ax.b}}
$$

(c) B binds to EA uncompetitively

$$
\begin{array}{c}
\text{E} \xrightarrow{Ax} \text{E.Ax} \xrightleftharpoons[\text{}]{B} \text{E.Ax.B} \\
\text{E.A} \rightleftharpoons \text{E.A.Bx} \\
Bx \\
K_i \updownarrow \\
\text{E.A.B}
\end{array}
$$

$$
v = \frac{V}{1 + \dfrac{b}{K_i'} + \dfrac{K_m^B}{b} + \dfrac{K_m^{Ax}}{ax} + \dfrac{K_s^{Ax}.K_m^B}{ax.b}}
$$

This is the general form of an equation that represents the behavior, K_i' is not a simple dissociation constant

It should be noted that if the inhibitor binds to more than one different enzyme intermediate, the system need not give rise to complex rate equations. For example, the systems shown in *Table 16* where there is no enzyme species with two molecules of inhibitor bound, will give rise to simple equations which correspond to a simple addition of the separate inhibitory effects.

Inhibitor binds to substrate rather than enzyme
Such a system might be represented as shown in *Figure 17*. In this system the inhibitor simply reduces the amount of substrate available to react with the enzyme. It is, of course, the free substrate (s_f) that the enzyme will bind. Thus the problem becomes that of deriving the free substrate concentration from the amounts of substrate and inhibitor added. As shown in *Figure 17* the solution to this yields an unpleasant expression for the free substrate concentration, which shows that in the presence of a fixed concentration of inhibitor the free substrate concentration will vary sigmoidally as the total substrate concentration is increased. In other words, the substrate is being titrated against the inhibitor. Therefore, although the variation of velocity with free substrate concentration would obey normal Michaelis–Menten kinetics, the

Figure 19. Initial-rate behavior of compulsory-order systems with high-substrate inhibition (see *Table 18*). (a) B competitive towards Ax; (b) Ax competitive towards B; (c) B binds to EA uncompetitively. Note: in all cases ax1 < ax2 < ax3 and b1 < b2 < b3, etc. Broken lines are used in some cases for clarity.

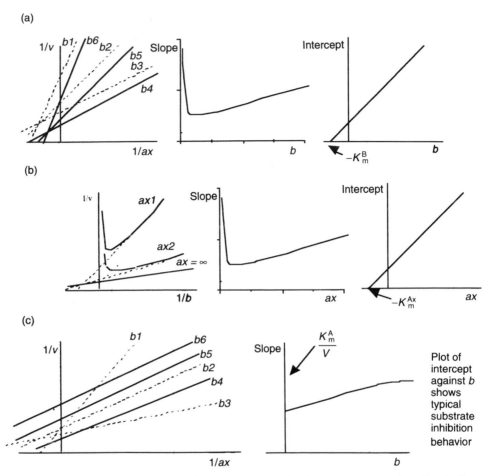

variation will describe a sigmoid curve with respect to total substrate concentration. By similar reasoning, plots of velocity against total inhibitor concentration will also be sigmoid (*Figure 17*) and plots of $1/v$ as a function of i will be parabolic.

An alternative possibility is that the S.I complex is the true inhibitor, as shown in *Table 17*. This situation can become very complicated, since the velocity can be affected both by inhibition by S.I and by depletion of free substrate. A simplifying situation will occur if the binding of inhibitor does not significantly deplete the total substrate concentration (*Table 17*).

2.9. Inhibition by high substrate concentrations

It is not uncommon for enzymes to be inhibited by high concentrations of one, or more, of their substrates, leading to kinetic plots such as those shown in *Figure 18*. Such behavior can be useful in helping to deduce the kinetic mechanism involved (see 1, 3, 49, 72) but it can restrict the range of substrate concentrations that can be used for determining K_m and V values.

Figure 20. Inhibition by high substrate concentrations in the double-displacement mechanism.

(a) By one substrate

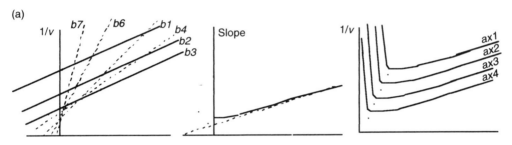

$$v = \frac{V}{1 + \dfrac{K_m^{Ax}}{ax}\left(1 + \dfrac{b}{K_i}\right) + \dfrac{K_m^{B}}{b}}.$$

(b) By both substrates

$$v = \frac{V}{1 + \dfrac{K_m^{Ax}}{ax}\left(1 + \dfrac{b}{K_i^{B}}\right) + \dfrac{K_m^{B}}{b}\left(1 + \dfrac{ax}{K_i^{Ax}}\right)}.$$

Note: in all cases ax1 < ax2 and b1 < b2 < b3, etc. Broken lines are used in some cases for clarity.

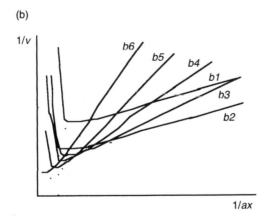

(b)

Plots of 1/v against 1/b at a series of concentrations of Ax give similar behavior

Table 18 shows some possible types of product inhibition in two substrate–two product reactions and *Figure 19* indicates the kinetic behavior that can arise from them. For a double-displacement mechanism, high substrate inhibition in which one or both of the substrates can bind to both free forms of the enzyme, as shown in *Figure 20*, can be quite characteristic.

Table 19. Effects of inhibitor contamination of the substrate

Inhibitor type and equations	Initial-rate behavior

(a) Competitive

$$v = \cfrac{V}{1 + \cfrac{K_m}{s} + \cfrac{K_m}{s} \cdot \cfrac{i}{K_i}}$$

$$= \cfrac{V}{1 + \cfrac{K_m}{s} + \cfrac{K_m}{s} \cdot \cfrac{x.s}{K_i}}$$

$$= \cfrac{V}{1 + \cfrac{K_m}{s} + \cfrac{x.K_m}{K_i}}$$

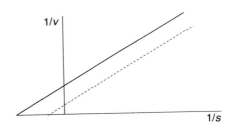

(b) Uncompetitive

$$v = \cfrac{V}{1 + \cfrac{i}{K_i} + \cfrac{K_m}{s}}$$

$$= \cfrac{V}{1 + \cfrac{K_m}{s} + \cfrac{x.s}{K_i}}$$

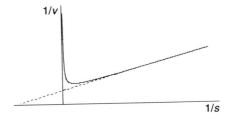

(c) Mixed

$$v = \cfrac{V}{1 + \cfrac{i}{K'_i} + \cfrac{K_m}{s} + \cfrac{K_m.i}{K_i.s}}$$

$$= \cfrac{V}{1 + \cfrac{x.s}{K'_i} + \cfrac{K_m}{s} + \cfrac{x.K_m}{K_i}}$$

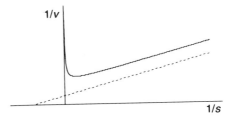

Inhibitor concentration = x(substrate concentration).
Broken lines represent behavior in the absence of inhibitor.

Figure 21. Effect of inhibitor contamination of the enzyme.

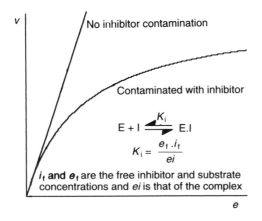

Table 20. Analysis of tight-binding inhibitors by the Dixon method

General relationships	Competitive	Noncompetitive
$e_t = e_f + es + ei$ and $i_t = i_f + ei$ subscripts f and t represent free and total concentrations, respectively		

$$v_0 = \frac{V}{1 + K_m/s} \quad \text{and} \quad V = k_{cat}.e$$

$$\boxed{x = K_m} \qquad \boxed{x = (K_m + s)}$$

$$\frac{i_t}{V - v_i(1 + K_m/s)} = \frac{K_i}{v_i}\cdot\frac{s}{x} + \frac{1}{k_{cat}} \qquad = \frac{K_i}{v_i}\cdot\frac{s}{K_m} + \frac{1}{k_{cat}} \qquad = \frac{K_i.s}{V(K_m + s)} + \frac{1}{k_{cat}}$$

v_0 and v_i are the rates in the absence and presence of inhibitor, and x is the component of the Michaelis equation affected by I

Rate at any point 'n' on a v versus i curve:

$$v_n = \frac{v_0}{n} = \frac{V}{n(1 + K_m/s)}$$

Corresponds to:

$$\frac{i'_t}{V - v_n(1 + K_m/s)} = \frac{i'_t}{V(1 - 1/n)}$$

$$= \frac{i'_t}{V}\cdot\frac{n}{n-1} = K_i\cdot\frac{s}{x}\cdot\frac{n}{v_0} + \frac{1}{k_{cat}} \qquad = K_i\cdot\frac{s}{K_m}\cdot\frac{n}{v_0} + \frac{1}{k_{cat}} \qquad = \frac{K_i.s}{(K_m + s)}\cdot\frac{n}{v} + \frac{1}{k_{cat}}$$

where i'_t is the value of i_t at point 'n'

and since a straight line through v_0 and n will, by geometry, cut the base-line at i_n where:

$$i_n = (n/n-1)i'_t$$

$$= nK_i\frac{s}{K_m}\cdot\frac{V}{v_0} + e_t$$

$$i_n = nK_i\frac{s}{x}\frac{V}{v_0} + e_t \qquad = nK_i(1 + K_m/s) + e_t \qquad = nK_i\frac{V.s}{v_0(K_m + s)} + e_t$$

Thus (K), the distance between the i_n values for a series of such lines at n values differing by unity (i.e. $v_0/2$, $v_0/3$, $v_0/4$, etc.) will be:

$$K = i_n - i_{n-1} = K_i\frac{s}{x}\frac{V}{v_0} \qquad = K_i(1 + s/K_m) \qquad = K_i$$

Although high-substrate inhibition is commonly considered as being a result of aberrant binding of substrate to the enzyme, a particular case can exist if the true substrate for the enzyme is a substrate–activator complex. This is frequently the case with metal ion-dependent enzymes; for

Figure 22. Analysis of tight-binding inhibition by the procedure of Dixon.

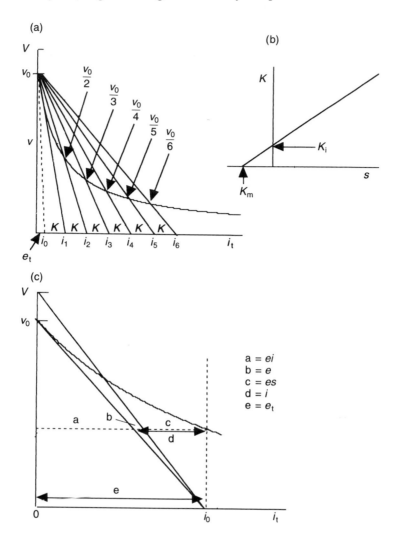

example the magnesium–ATP complex is the true substrate for many kinases. If the uncomplexed substrate can bind to the enzyme as an inhibitor, this will give rise to apparent high-substrate inhibition unless experiments are designed to ensure that the metal ion concentrations are adequate to allow the substrate to remain complexed as its concentration is increased. Some enzyme reactions involve metal ions as essential activators. In such cases, reduction of the free metal ion concentration through chelation by one of the substrates can also result in inhibition. Cases occur where a complex between the enzyme and a metal ion is the active species and the true substrate is a complex between substrate and the same metal ion, and it is necessary to calculate the concentrations of metal–substrate complex and free metal ions, from the stability constants of the metal–ligand complexes, in order to interpret the kinetic behavior (73, 74). The solution of the quadratic equations for cases where there are several metal-binding species would be a tedious procedure and computer solutions are available (see, for example, ref. 75).

Figure 23. Analysis of the species concentrations for tight-binding noncompetitive inhibition by the procedure of Dixon.

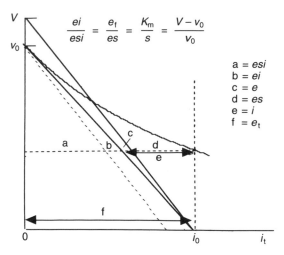

$$\frac{ei}{esi} = \frac{e_f}{es} = \frac{K_m}{s} = \frac{V - v_0}{v_0}$$

a = esi
b = ei
c = e
d = es
e = i
f = e_t

2.10. Inhibitor contamination

The presence of contaminants in the substrate will lead to incorrect estimates of the concentrations of solutions prepared by weight, but it will also lead to errors in the determination of kinetic constants if any of the impurities is inhibitory. If a substrate is contaminated by an inhibitor, the initial-rate behavior will depend on the nature of the inhibition that occurs (76–78), as shown in *Table 19*. In such cases the inhibitor is present in the substrate and so the amount present will increase as the substrate concentration is increased. Since the inhibitor concentration is a constant ratio to that of the substrate it can be represented as a constant fraction of the substrate concentration ($i = x.s$) in determining the rate equations shown in *Table 19*.

If the enzyme solution is itself contaminated with an inhibitor, this will not cause deviations from the normal behavior with respect to substrate concentration at any fixed enzyme concentration, although the determined Michaelis–Menten parameters will be altered by an extent dependent on the nature of the inhibition and the concentration of endogenous inhibitor in relation to its K_i values. However, the dependence of initial velocity upon enzyme concentration will become nonlinear since the equilibrium relationship predicts that the amount of enzyme–inhibitor complex will increase with increasing enzyme concentration (*Figure 21*).

3. TIGHT-BINDING INHIBITORS

Some inhibitors have such high affinities for the enzyme that the concentrations required for inhibition are comparable to those of the enzyme. In such cases, the binding of the inhibitor to the enzyme will significantly reduce the free inhibitor concentration and so the assumption that the total inhibitor concentration is equal to the free inhibitor concentration, which has been implicit in all the above treatments, is no longer valid and it is necessary to derive relationships in terms of the concentrations of free enzyme (e_f) and free inhibitor (i_f), as shown in *Table 20*.

Dixon (79) has devised a graphical procedure for treating tight-binding competitive and non-competitive inhibitors. The relationships involved are shown in *Table 20* and the graphical

Table 21. Analysis of tight-binding inhibitors by the Henderson procedure

Inhibition type	N_i/K_i (see below)	$\dfrac{i_t}{\left(1-\dfrac{v_i}{v_0}\right)}$	Kinetic behavior
(a) Competitive	K_m/K_i	$K_i\left(1+\dfrac{s}{K_m}\right)\dfrac{v_0}{v_i}+e_t$	
(b) Uncompetitive	s/K_i	$K_i\left(1+\dfrac{K_m}{s}\right)\dfrac{v_0}{v_i}+e_t$	
(c) Noncompetitive	$\dfrac{(s+K_m)}{K_i}$	$K_i\cdot\dfrac{v_0}{v_i}+e_t$	
Mixed	$\dfrac{s}{K'_i}+\dfrac{K_m}{K_i}$	$K_i\left(\dfrac{s+K_m}{\dfrac{K_m}{K_i}+\dfrac{s}{K'_i}}\right)\dfrac{v_0}{v_i}+e_t$	Depends on the relative values of K_i and K'_i

General tight-binding inhibitor equation:

$$\frac{i_t}{\left(1-\dfrac{v_i}{v_0}\right)}=\left(\frac{K_m+s}{\sum\left(\dfrac{N_i}{K_i}\right)}\right)\frac{v_0}{v_i}+e_t$$

N represents the inhibitor concentration and
a combination of rate constants defining
the fraction of enzyme present in the E.I
complex

Replot of slopes against
substrate concentration

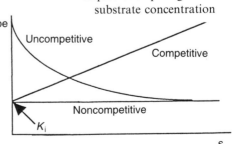

Table 22. Reaction kinetics for enzyme–inhibitor interactions

Reaction	Rate	Integrated equation	Reaction order (dimensions)	Common values for enzymes
$E + I \xrightarrow{k_{+1}} E.I$	$\dfrac{dei}{dt} = k_{+1}e.i$	$\dfrac{1}{e_0 - i_0}\ln\dfrac{i_0.e}{e_0.i} = k_{+1}t$	Second-order $(M^{-1}.sec^{-1})$	Theoretical maximum range for a diffusion-controlled interaction between a small (I) and a large (E) molecule: $10^9–10^{11}\ M^{-1}.sec^{-1}$.
If $e_0 = i_0$		$\dfrac{1}{i} - \dfrac{1}{i_0} = k_{+1}t$	Second-order	Some enzymes show rates close to these values: $(10^7–10^8\ M^{-1}.sec^{-1})$ but others show very much lower values
If $e_0 << i_0$	$\dfrac{dei}{dt} = k_{+1}e.i_0$	$\ln\dfrac{e_0}{e} = \ln ei = k_{+1}i\,t = k'.t$	*Pseudo* first-order	
$E.I \xrightarrow{k_{-1}} E + I$	$-\dfrac{dei}{dt} = k_{-1}ei$	$\ln\dfrac{ei_0}{ei} = k_{-1}.t$	First-order (sec^{-1})	Maximum values $10^9–10^{12}\ sec^{-1}$, but very much slower in many cases
$E + I \underset{k_{-1}}{\overset{k_{+1}}{\rightleftharpoons}} E.I$	$\dfrac{dei}{dt} = k_{+1}e.i - k_{-1}ei$	$\dfrac{ei_{eq}}{e_0^2 - ei_{eq}^2}\ln\dfrac{ei_{eq}(e_0^2 - ei_{eq})}{e_0^2(ei_{eq} - ei)} = k_{+1}t$ assuming $e_0 = i_0$	First-order, in forward direction	

e_0 and i_0 are the initial concentrations of E and I; ei_{eq} is the equilibrium concentration of E.I; and e, ei and i are their concentrations at any time (t) see ref. 43 for a discussion on the rates of the processes involved.

Table 23. Some tight-binding enzyme inhibitors

Enzyme	Substrate	$K_m(\mu M)$	Analog	$K_i(nM)$
Carnitine acetyltransferase	AcetylCoA	34	$(CH_3)_3\overset{+}{N}CHOCH_2SCoA$ with C=O above O-(2-(S-CoA)acetyl)carnitine	<12
	$(CH_3)_3\overset{+}{N}CH_2CHCH_2OH$ (OH above) Carnitine	120	CH_2COO^-	
Aspartate transcarbamoylase	Carbamoyl phosphate	27	$NHCOCH_2PO_3^{2-}$ $\vert$ $CHCOO^-$	
	Aspartate	11 000	CH_2COO^- $\vert$ CH_2COO^- Phosphonoacetylaspartate	27
Adenylate kinase	MgATP	100	$Ado-OPOPOPOPOP-Ado.Mg^{2+}$ P1,P5-Di(adenosine)-pentaphosphate Ado = adenosine	2.5
	AMP	625		
Lactate dehydrogenase	NADH	24	R = ribose Oxaloethyl-NADH	<1
	Pyruvate	140		

Table 23. Continued

Enzyme	Substrate	$K_m(\mu M)$	Analog	$K_i(nM)$
Cytidine deaminase	HO H HN O N–R **Putative transition state analog**	20	NH HN O N–R **Cytidine**	< 2
	3,4,5,6-Tetrahydrouridine			
Lysozyme	CH₂OH O R O X **Putative transition state analog**	9	CH₂OH O R O X O CH₂OH O OH X **(NAG)₅**	83
	(NAG)₃—NAG lactone		NAG = N-acetylglucosamine X = —NHCOCH₃ R = (NAG)₃	
Monoamine oxidase-A	CH₃O N N–H H CH₃ **Harmaline**	~ 100	HO N–H NH₂ **5-Hydroxytryptamine**	~ 6

For further details see refs 83–89.

Table 24. Behavior of nonspecific irreversible inhibitors

Reaction and notes	Kinetic equations	Kinetic behavior
$E + I \rightarrow E.I$	$\dfrac{dei}{dt} = k.i$	
e_0 and e are the enzyme concentrations at times zero and t, respectively. A_0 and A are the enzyme activities at times zero and t, respectively.	$\ln ei = \ln \dfrac{e_0}{e} = k.i.t$ $\ln \dfrac{A}{A_0} = -k.i.t$ $\qquad = k'.t$ $\log_{10} \dfrac{A}{A_0} = \dfrac{-t}{2.303}k.i$	
It is assumed that: $i \gg e$	$\qquad = \dfrac{-t}{2.303}k'$	
	where $k' = k.i$	

The half-life for inhibition ($t_{1/2}$) is:
$$t_{1/2} = \frac{\ln 2}{k} = \frac{0.693}{k}$$

Table 25. Some commonly used nonspecific enzyme inhibitors

Reagent/treatment	Amino acid groups affected
Acetic anhydride	Acetylates-NH_2 groups and also -OH and -SH
N-Acetylimidazole	Acetylates tyrosine -OH and more slowly with -NH_2
2,3-Butanedione	Reacts with arginine residues
1,2-Cyclohexanedione	Arginine residues; more slowly with -NH_2
Diazonium-1-H-tetrazole	Histidine and tyrosine; may also react with -NH_2
Diethylpyrocarbonate	Histidine; may also react with tyrosine and -SH
2,4-Dinitrofluorobenzene	-NH_2, -SH, tyrosine, -OH and imidazolc
5,5'-Dithiobis-(2-nitrobenzoate)	-SH
p-Hydroxymercuribenzoate (p-chloromercuribenzoate)	-SH
2-Hydroxy-5-nitrobenzyl bromide	Reacts with the indole of tryptophan and more slowly with -SH
Hydrogen peroxide	Oxidizes methioninc and also -SH
Iodination	Tyrosine and more slowly histidine
Iodoacetate/iodoacetamide	-SH, reacts more slowly with tyrosine, -OH, -NH_2, methionine and histidine at lower pH
Methylacetimidate	Amino groups
O-Methylisourea	Amino groups
Photo-oxidation	Cys, CysS–SCys, His, Met, Trp, Tyr
Tetranitromethane	Tyrosine
Trinitrobenzene sulfonic acid	Amino groups and also -SH

Note: some of these reagents are potentially hazardous.
Further details may be found in refs 91–93.

Figure 24. Irreversible reaction of amino acid residues in an enzyme that results in inhibition. (a) Mechanism; (b) first-order rates of reaction of residues X and Y, assuming $i \gg e$, $k' = k.i$; (c) loss of enzyme activity; (d) kinetics of inhibition: if both residues X and Y are essential for activity, $k_A = k'_1 + k'_2$; if only one were essential, e.g. X, $k_A = k'_1$; any residue(s) reacting faster than the rate of activity loss, e.g. a residue Z with rate constant k'_z, cannot be essential for activity.

(a)

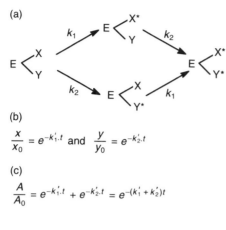

(b)

$$\frac{x}{x_0} = e^{-k'_1.t} \text{ and } \frac{y}{y_0} = e^{-k'_2.t}$$

(c)

$$\frac{A}{A_0} = e^{-k'_1.t} + e^{-k'_2.t} = e^{-(k'_1 + k'_2)t}$$

(d)

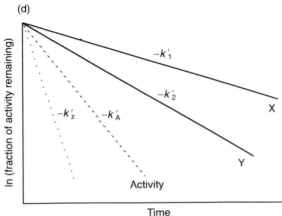

analysis is given in *Figure 22*. Dixon (79) has extended this approach to show that straight lines drawn from v_0 and V to cut the base-line at the calculated value of e_t (i_0) will allow the concentrations of the species present to be determined directly for competitive inhibition (*Figure 22c*) and a slightly more elaborate procedure will enable the same to be done for noncompetitive inhibition (*Figure 23*).

Henderson (80) proposed a procedure for determining K_i and e_t, according to the relationships shown in *Table 21*, where the ratios of the velocities over a range of inhibitor concentrations (v_i) to that in the absence of inhibitor (v_0) are determined at fixed enzyme concentrations and at a series of fixed substrate concentrations. A computer program for this procedure that also incorporates statistical analysis has been presented (81).

Frequently the rate of development of inhibition for tight-binding inhibitors is relatively slow (see ref. 82, for a more detailed treatment). If the enzyme and inhibitor concentrations

Table 26. Behavior of specific irreversible inhibitors

Reaction and notes	Kinetic equations	Kinetic behavior
$E + I \underset{k_{-1}}{\overset{k_{+1}}{\rightleftharpoons}} E.I \xrightarrow{k_{+2}} E - I$ (a) If $k_{+2} \ll k_{-1}$ $\quad K_i = k_{-1}/k_{+1}$ $\quad = \dfrac{(e-ei)i}{ei}$ (b) If k_{+2} is not $\ll k_{-1}$ $\quad K_i = \dfrac{(k_{-1}+k_{+2})}{k_{+1}}$ (c) If $k_{+2} > k_{-1}$ neither of the above treatments is valid and the behavior would deviate from first-order kinetics e_0 is the initial enzyme concentration and e is that after time t. $e-i$ is the concentration of the irreversible complex. A_0 and A are the enzyme activities at times zero and t, respectively It is assumed that: $i \gg e$	$\dfrac{de-i}{dt} = k_{+2}.ei$ $es = \dfrac{e_0}{(1 + K_i/i)}$ $\ln e-i = \ln \dfrac{e_0}{e} = k'.t$ $\ln \dfrac{A}{A_0} = k'.t$ $\log_{10} \dfrac{A}{A_0} = \dfrac{-t}{2.303} k'$ where: $K' = \dfrac{k_{+2}}{(1 + K_i/i)}$ $\dfrac{1}{k'} = \dfrac{1}{k_{+2}} + \dfrac{K_i}{k_{+2}.i}$ The half-life for inhibition $(t_{1/2})$ is: $t_{1/2} = \dfrac{\ln 2}{k_{+2}} = \dfrac{0.693}{k_{+2}}$	

are comparable and both relatively low, the rate of E.I formation will follow second-order kinetics, as shown in *Table 22*. That table also shows values for maximum rates for the association and dissociation processes. In practice these can be very much lower than the theoretical maxima. For example, the rate may be limited by a conformational change, and values as low as $10–10^4$ sec^{-1} have been reported for such processes in several enzymes (see ref. 43 for a discussion). Substitution of low reactant concentrations in the second-order rate equation (*Table 22*) at different possible values of k_{+1} will show that it is possible to have rates of inhibition that are relatively slow. Thus the rate constant for association can be determined directly from these measurements, provided that either e_0 and i_0 are known or $e_0 = i_0$.

Table 27. Specific irreversible inhibition: a fuller case

Reaction and notes	Kinetic equations	Kinetic behavior
$$E + I \underset{k_{-1}}{\overset{k_{+1}}{\rightleftharpoons}} E.I \xrightarrow{k_{+2}} E.I^* \xrightarrow{k_{+3}} E\text{–}I$$ E.I* represents an intermediate activated species	As in *Table 26* except that: $$k' = \frac{\dfrac{k_{+2}k_{+3}}{(k_{+2}+k_{+3})}}{1 + K_i/i}$$ $$K_i = \frac{k_{-1}k_{+3}+k_{+2}k_{+3}}{k_{+1}(k_{+2}+k_{+3})}$$	Identical to that shown in *Table 26* in all cases
If $k_{+2} \ll k_{+3}$	$$k' = \frac{k_{+2}}{(1+K_i/i)}$$ (as in *Table 26*) $$K_i = \frac{(k_{-1}+k_{+2})}{k_{+1}}$$ (as in *Table 26*, case b)	

Table 28. Substrate protection from specific irreversible inhibition

Mechanism and equations	Kinetic behavior
$$E + I \underset{k_{-1}}{\overset{k_{+1}}{\rightleftharpoons}} E.I \xrightarrow{k_{+2}} E\text{–}I$$ $$k_{-3} \Big\Uparrow k_{+3}S$$ $$E.S \xrightarrow{k_{+4}} E + \text{Products}$$ $$\ln \frac{A}{A_0} - k't$$ $$k' = \frac{k_{+2}}{1 + K_i(1 + s/K_m)/i}$$ $$\frac{1}{k'} = \left(\frac{K_i}{i} + \frac{K_i}{i}\cdot\frac{s}{K_m} + 1\right)\frac{1}{k_{+2}}$$	

Definitions of terms and other relationships are as shown in *Table 26.*

By definition, tight-binding inhibitors will dissociate from the enzyme rather slowly and thus it should be possible to determine k_{-1} directly from the first-order regain of activity after removal of the excess inhibitor, for example by dilution or rapid gel filtration.

Tight-binding inhibitors may, at least in some cases, resemble the structure of the reaction transition state. Some examples are given in *Table 23.*

Table 29. Treatment of reaction time courses in the presence of a nonspecific irreversible inhibitor

Mechanism and equations	Kinetic behavior

$$E + I \underset{k_{-2}}{\overset{k_{+1}}{\rightleftharpoons}} E - I$$

$$k_{-2} \Big\Vert k_{+2}S$$

$$E.S \xrightarrow{k_{+3}} E + Products$$

$$P_t = P_\infty (1 - \exp^{-k' \cdot t})$$

$$\ln(P_\infty - P_t) = \ln P_\infty - k't$$

or

$$\log_{10}(P_\infty - P_t) = \log_{10} P_\infty - k't/2.303$$

$$k' = \frac{k_{+1} \cdot i}{1 + \dfrac{s}{K_m}}$$

$$\frac{i}{k'} = \frac{1}{k_{+1}} + \frac{s}{K_m k_{+1}}$$

Product vs Time plot, labeled "Additional enzyme", "Additional substrate (at complete inhibition this will not restart reaction)"

$\ln(P_\infty - P_t)$ vs Time plot, with $\ln P_\infty$ intercept and Slope $= -k'$

$\dfrac{i}{k'}$ vs i plot

P_∞ is the amount of product formed at complete inhibition and P_t is that formed at any time (t).

4. IRREVERSIBLE INHIBITION

Inhibition is not reversed by dialysis or dilution because a chemical reaction between the enzyme and inhibitor occurs (see *Table 1*). The inhibition usually takes time to develop because a chemical reaction is involved. In the simplest type of irreversible inhibition the inhibitor reacts directly with a group, or groups on the enzyme to cause inactivation.

4.1. Nonspecific irreversible inhibition

This type of inhibition involves the direct reaction between inhibitor and a group, or groups, in the enzyme without the prior formation of a noncovalent enzyme–inhibitor complex and gives the kinetic behavior shown in *Table 24* (see also ref. 90). The inhibitory compounds are usually not specific for any particular enzyme but they will react with different functional groups in many enzymes and frequently with several different residues in the same enzyme. Some examples are shown in *Table 25*.

Although this lack of specificity limits their value for many purposes, they can help to identify groups in an enzyme that are necessary for catalysis (94). An example of the approach is shown

Table 30. Treatment of reaction time courses in the presence of a specific irreversible inhibitor

Mechanism	Equations	Analytical treatments
(a) Graphical Determine k' and P_∞ as described in *Table 29* at varying i and s $$E + I \underset{k_{-1}}{\overset{k_{+1}}{\rightleftharpoons}} E.I \xrightarrow{k_{+2}} E\text{-}I$$ $$k_{-3} \Big\Vert k_{+3}S$$ $$E.S \xrightarrow{k_{+4}} E + \text{Products}$$	$$k' = \cfrac{k_{+2}}{1 + \cfrac{K_i}{i}\left(1 + \cfrac{s}{K_m}\right)}$$ $$\frac{1}{k'} = \frac{1}{k_{+2}} + \frac{K_i}{k_{+2}.i}\left(1 + \frac{s}{K_m}\right)$$ $$\frac{i}{k'} = \frac{i}{k_{+2}} + \frac{K_i}{k_{+2}}\left(1 + \frac{s}{K_m}\right)$$	
(b) Direct curve-fitting Directly fit the progress curve in order to obtain k' and P_∞. Calculate K_i and k_{+2} if e is known and K_m and V are determined in separate experiments. Alternatively, use graphical procedures as above	$$P_t = P_\infty\left(1 - e^{-\left\{\frac{k_{+4}e}{VP_\infty}\left[\frac{K_m k_{+2}i/t}{K_i s}\right]\right\}}\right)$$ $$k' = (\text{exponential term})/t$$ $$P_\infty = \frac{k_{+4}.s.e}{k_{+2}.K_m.i}$$ $$k'.P_\infty = \cfrac{k_{+4}.e}{1 + \cfrac{K_m}{s}\left(1 + \cfrac{i}{K_i}\right)}$$	

Table 31. Inhibition of the two monoamine oxidase (MAO) enzymes by acetylenic amine derivatives

Compound	MAO-A		MAO-B	
	$K_i(\mu M)$	$k_{+2}(\text{min}^{-1})$	$K_i(\mu M)$	$k_{+2}(\text{min}^{-1})$
Pargyline	13	0.2	0.5	0.2
Clorgyline	0.05	>0.76	50	0.06
(−)-Deprenyl	25	0.14	0.97	>0.99
R = H	0.24	0.1	3.7	0.3
R = OH	4.0	0.28	8.9	0.29
R = OCH$_3$	0.0008	0.006	0.004	0.7

K_i and k_{+2} are defined in terms of the model (see *Table 26*): $E + I \underset{K_i}{\rightleftharpoons} E.I \xrightarrow{k_{+2}} E\text{–I}$.
Data from refs 106, 107.

in *Figure 24* for an enzyme with two residues, X and Y, which both react with a reagent at different rates and where modification of either leads to loss of activity. Thus, following the rates of loss of enzyme activity and of the reaction of individual amino acid residues in the enzyme can indicate the numbers and types of amino acid side-chains involved. This approach has been extended to more complex reactions that can lead to enzyme inhibition (94, 95). Although this procedure has been to a large extent superseded by site-directed mutagenesis, which offers greater specificity, selectivity and controllability, it retains the advantage of allowing rates of modification to be correlated with rates of inhibition, thus revealing situations where activity loss may result from a slower conformational change following alteration of the residue(s).

4.2. SPECIFIC IRREVERSIBLE INHIBITION

This involves the formation of an initial noncovalent enzyme–inhibitor complex (E.I) which subsequently reacts to give the irreversibly modified species (E–I). The kinetic behavior of such systems (see refs 96–99) is summarized in *Table 26*. *Table 27* shows that a, perhaps more realistic, case in which the E.I complex is transformed into an activated species (EI*) which then reacts to produce the irreversibly inhibited species, gives identical kinetic behavior. In some

▶ p. 165

Table 32. Inhibition of chymotrypsin and trypsin by substrate-derived acylating agents

Compound	Reaction					
	Chymotrypsin	Chymotryp-sinogen	Denatured chymotrypsin	Trypsin	Trypsinogen	Denatured trypsin
$\begin{array}{c}COOC_2H_5\\ \mid\\ CH_2CHNH-tosyl\end{array}$ N-Tosylphenylalanine ethyl ester	Substrate	Not substrate	Not substrate	Not substrate	Not substrate	Not substrate
$\begin{array}{c}COCH_2Cl\\ \mid\\ CH_2CHNH-tosyl\end{array}$ N-Tosylphenylalanine chloromethylketone (TPCK) $$COOC_2H_5 \overset{E}{\rightleftharpoons} E.S \longrightarrow E + Acid + Alcohol$$	Inhibitor (reacts with only one histidine; out of two in the enzyme)	No reaction	No reaction	No reaction	No reaction	No reaction
$H_2NCH_2CH_2CH_2CH_2CHNHtosyl$ N-Tosyllysine ethyl ester $$COCH_2Cl \overset{E}{\rightleftharpoons} E.I \longrightarrow E\text{-}I + NCl$$	Not substrate	Not substrate	Not substrate	Substrate	Not substrate	Not substrate
$H_2NCH_2CH_2CH_2CH_2CHNHtosyl$ N-Tosyllysine chloromethylketone (TLCK)	No reaction	No reaction	No reaction	Inhibitor	No reaction	No reaction

Reaction (row 1): $E \rightleftharpoons E.S \longrightarrow E + acid + alcohol$

Reaction (row 2): $E \rightleftharpoons E.I \longrightarrow E\text{-}I + HCl$

Enzyme Inhibition

Figure 25. Formation of a Michael acceptor in mechanism-based inhibition. (a) The Michael addition reaction

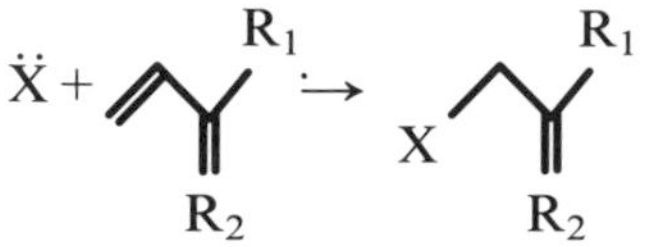

(b) Reaction of an enzyme containing a nucleophilic group with a substrate

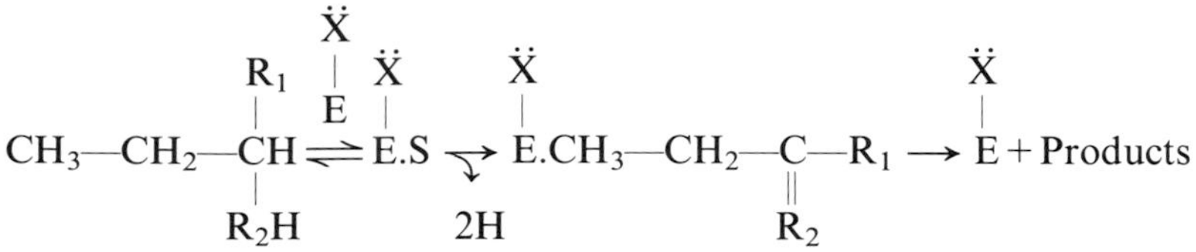

(c) Reaction of the enzyme with a mechanism-based inhibitor

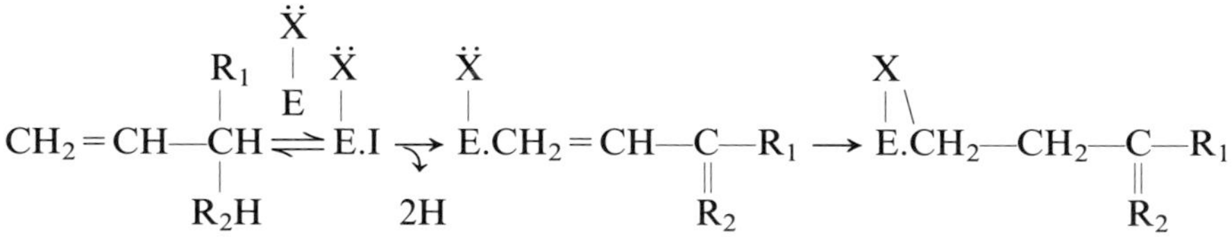

Figure 26. Substrate oxidation and mechanism-based inhibitor interaction of monoamine oxidase.

$RCH_2NR'R'' + E{-}FAD$
Amine
$\updownarrow$

$E{-}FAD{\cdot}RCH_2NR'R''$
$\downarrow$

$E{-}FADH_2{\cdot}RCH = \overset{+}{N}R'R''$
$\quad \overset{\frown}{\rvert} H_2O$
$\downarrow$

$RCHO + NHR'R'' + E{-}FADH_2$

The FADH is subsequently reoxidized with oxygen to form hydrogen peroxide

$CH \equiv C{-}CH_2{-}N(CH_3)_2 + E{-}FAD$
$\quad\quad\quad\quad\quad\quad$ Dimethylpropargylamine
$\updownarrow$

$E{-}FAD{\cdot}CH \equiv C{-}CH_2{-}N(CH_3)_2$
$\downarrow$

$E{-}FADH_2{\cdot}CH \equiv C{-}CH = \overset{+}{N}(CH_3)_2$
$\downarrow$

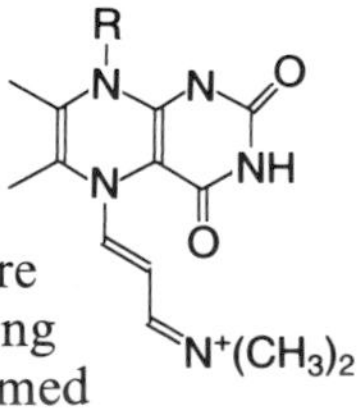

$E{-}FADH_2{-}CH = C{-}CH = \overset{+}{N}(CH_3)_2$

Partial structure
of FAD showing
the adduct formed
with dimethylpropargylamine

Figure 27. Mechanism-based inhibition of some pyridoxal-phosphate enzymes. (a) Vinylglycine as an inhibitor of aspartate aminotransferase. (b) Gabaculine and GABA (γ-aminobutyric acid) transaminase.

(a)

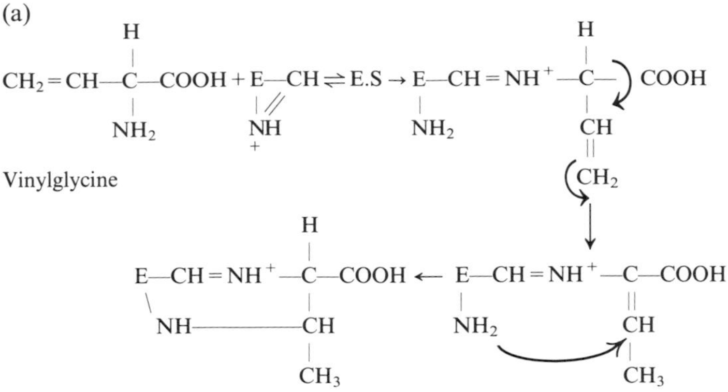

$$CH_2 = CH-\underset{\underset{NH_2}{|}}{C}-COOH + E-\underset{\underset{+}{\overset{NH}{\|}}}{CH} \rightleftharpoons E.S \rightarrow E-\underset{\underset{NH_2}{|}}{CH}=NH^+-\underset{\underset{CH}{|}}{\overset{H}{\underset{\|}{C}}}-COOH$$

Vinylglycine

$$E-CH=NH^+-\underset{\underset{CH}{|}}{\overset{H}{\underset{|}{C}}}-COOH \leftarrow E-\underset{\underset{NH_2}{|}}{CH}=NH^+-\underset{\underset{CH}{\|}}{C}-COOH$$

NH————CH   CH₃
        |
        CH₃

Note: in this enzyme the pyridoxal phosphate does not exist as the free aldehyde but forms an internal Schiff base with a lysine residue which is written as: $E-\underset{\overset{\|}{\underset{+}{NH}}}{CH}$.

Gabaculine section structures

H_3N^+ Gabaculine $+$ E–CHO →

E–CH=NH⁺

E–CH₂–NH

E–CH₂–NH⁺

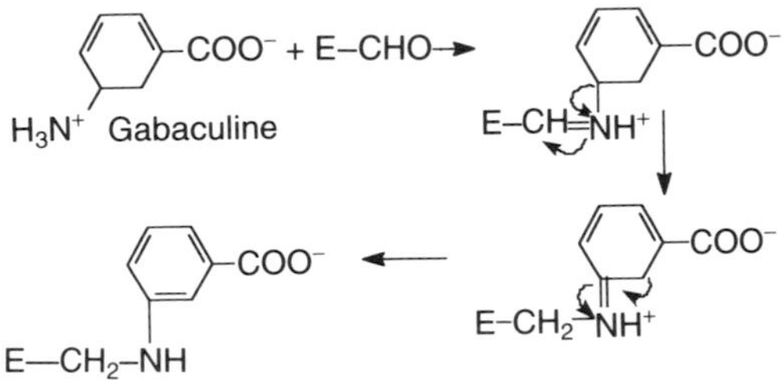

cases this species may not react to form a covalent bond with the enzyme but functions as a tight-binding inhibitor (100).

4.3. Effects of substrate concentration

If the inhibitor binds to the active site of the enzyme, it will compete with the substrate, giving rise to the behavior shown in *Table 28*. Although the analysis shown in *Tables 26* and *27* has been most widely used for determining the kinetic parameters for specific irreversible inhibition, they can, often more conveniently, be determined from the progress curve for substrate conversion in the presence of fixed amounts of the inhibitor (101–105). An example of such progress curve analysis as applied to nonspecific inhibition where the substrate protects competitively is shown in *Table 29*. There is, of course no *a priori* reason to suppose that the substrate will protect against nonspecific inhibition, and, indeed, it may not do so or may even enhance inhibition in some cases. In the case of specific inhibition, it is likely that the initial noncovalent binding will occur at the active site, in which case protection by at least one of the substrates will probably be competitive. The treatment of such systems is shown in *Table 30*. Similar analytical procedures have been presented for noncompetitive and uncompetitive substrate effects (102).

Figure 28. Recovery of enzyme activity *in vivo* after treatment with irreversible inhibitor. (a) The shaded area represents the time during which the irreversible enzyme inhibition occurs and the excess inhibitor is eliminated. Data obtained during this period would be difficult to interpret. After excess inhibitor has been eliminated enzyme activity will rise to a steady-state level. The rate of recovery can be described by the relationship:

$$\text{Precursors} \xrightarrow{\;k_{\text{synthesis}}\;} \text{Enzyme} \xrightarrow{\;k_{\text{degradation}}\;} \text{Breakdown products}$$

or: $\dfrac{de}{dt} = k_{\text{synthesis}} - k_{\text{degradation}} \cdot e$

where e is the concentration of enzyme at any time, t. After a sufficient period the recovery of activity will be complete and the level will remain constant at that occurring before the addition of inhibitor. This steady-state concentration of enzyme, e_{ss}, will be maintained because the rate of synthesis is exactly balanced by the rate of degradation. Therefore:

$$k_{\text{synthesis}} = k_{\text{degradation}} \cdot e_{ss}$$

$$\frac{de}{dt} = k_{\text{degradation}}(e_{ss} - e).$$

This is a first-order relationship which integrates to give:

$$\ln(e_{ss} - e) = -k_{\text{degradation}} \cdot t + \text{constant};$$

or in terms of recovery of activity (A) to the steady state level (A_{ss}):

$$\ln(A_{ss} - A) = -k.t + \text{constant}$$

as shown in (b).

The half-life for enzyme turnover will be given by:

$$t_{1/2} = \frac{\ln 2}{k_{\text{degradation}}} = \frac{0.693}{k_{\text{degradation}}}.$$

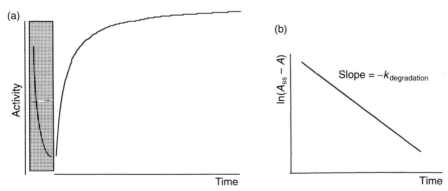

The effectiveness of specific irreversible inhibitors will depend on the values of the two constants that define the system. This is illustrated in *Table 31* in terms of the abilities of some acetylenic amine derivatives to show selectivity and potency as inhibitors of the two monoamine oxidases (A and B) which coexist in many tissues.

Table 33. The rates of turnover of some enzymes in rat liver

Enzyme	Half-life (hours)
Ornithine decarboxylase	0.2
5-Aminolaevulinate synthase	1.2
RNA nucleotidyltransferase I	1.3
Tyrosine aminotransferase	1.5
Tryptophan 2,3-dioxygenase	2.0
Hydroxymethylglutaryl-CoA reductase	4.0
Phosphoenolpyruvate carboxykinase	12.0
RNA nucleotidyltransferase II	13.0
Glucose 6-phosphate dehydrogenase	24
Glucokinase	30
Catalase	33
Acetyl-CoA carboxylase	50
Monoamine oxidase A and B	72
Glyceraldehyde 3-phosphate dehydrogenase	74
Pyruvate kinase	84
Arginase	108
Fructose 1,6-bisphosphate aldolase	117
Lactate dehydrogenase	144
6-Phosphofructokinase	168

Data taken from refs 118–122.

Table 34. The rates of turnover of monoamine oxidase (MAO) A and B

Tissue	Half-life (days)	
	MAO-A	MAO-B
Rat liver	~3	~3
Rat brain	~13	~13
Rat intestine	2.2	7.5
Rat heart	Increases with age	—
Human brain	~30	~30

See ref. 122 for further details.

4.4. Types of specific irreversible inhibitor

Specific irreversible inhibitors are recognized by the enzyme as substrates and then react within the noncovalent complex initially formed. The simplest type contain reactive groups that irreversibly inhibit the enzyme at the high local concentrations within the noncovalent complex (see refs 108–110 for discussion). Archetypal examples are the chloromethyl ketone derivatives of amino acid esters as inhibitors of proteolytic enzymes (111, 112), as shown in *Table 32*. The value of such compounds as drugs and for metabolic studies can be limited by their intrinsic reactivity.

Figure 29. The interactions of acetylcholinesterase with neostigmine and acetylcholine. The initial noncovalent enzyme–substrate complex formed with each compound has been omitted.

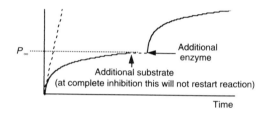

$$(CH_3)_3N^+CH_2CH_2OCCH_3$$
Acetylcholine | E–OH

$$(CH_3)_3N^+CH_2CH_2OH + E–OCCH_3$$
Acetyl-enzyme | H_2O

$$E–OH + CH_3COH$$

Neostigmine $(CH_3)_3N^+$ —◯— $OCN(CH_3)_2$ | E–OH

$(CH_3)_3N^+$ —◯— OH + E–OCN(CH_3)_2
Carbamoyl-enzyme | H_2O very slow

$$E–OH + (CH_3)_2NCOH$$

Table 35. Determination of the partition ratio for a suicide substrate

$$E + I \underset{k_{-1}}{\overset{k_{+1}}{\rightleftharpoons}} E.I \xrightarrow{k_{+2}} E.I^* \xrightarrow{k_{+4}} E–I$$
$$\downarrow k_{+3}$$
$$E + Products$$

Partition ratio $(r) = k_{+3}/k_{+4}$. The number of mol of inhibitor required to inactivate 1 mol of enzyme will be $(1+r)$.
When i is sufficiently large to give complete inhibition, i.e. when $i > (1+r)e$, $P_\infty = re$.
When $i < (1+r)e$, $P_\infty = re_i$, where e_i is the concentration of inhibited enzyme. Thus the amount of active enzyme remaining (e_r) will be given by:

$$e_r = \frac{e - i}{(1+r)}$$

and therefore:

$$\frac{e_r}{e} = 1 - \frac{i}{(1+r)e}$$

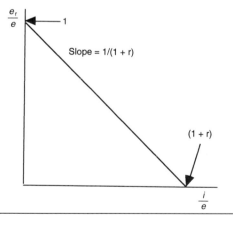

Mechanism-based (also known as suicide, enzyme-activated or k_{cat}) irreversible inhibitors have greater potential and specificity because they rely on part of the reaction catalysed by the enzyme to generate the reactive species as an intermediate in the reaction (see 100, 113–114, 135). In many cases these have used an enzyme to insert a second unsaturated bond into a compound already containing a double or triple bond, thus creating a reactive conjugated system, also known as a Michael acceptor (*Figure 25*). Oxidases and dehydrogenases frequently form an unsaturated bond in oxidizing their substrates, and an example of a mechanism-based

Table 36. Determination of the reaction kinetics for a suicide substrate

(a) For activity loss
$t = N.\ln(1 - M.z) - N'.\ln(1-z)$

Plot $i.t_{1/2}$ against i keeping the ratio
i/e constant.

$$i.t_{1/2} = \left(\frac{\ln(2 - M)}{1 - M}\right)\frac{K'}{k_{in}} + \frac{\ln 2}{k_{in}}.i$$

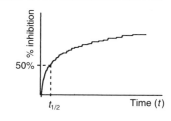

(b) For product formation
$t = N.\ln u - N'.\ln\{(u + M - 1)/M\}$

$$\left(\frac{\ln(2 - M)}{1 - M}\right)\frac{K'}{k_{in}} \rightarrow$$ Slope $= \ln 2/k_{in}$

Plot $i.t_{1/2}$ against i keeping the ratio
i/e constant.

$$i.t_{1/2} = \left(\frac{\ln\{M/(2M - 1)\}}{1 - M}\right)\frac{K'}{k_{in}} + \frac{\ln\{M/(M - 0.5)\}}{k_{in}}.i$$

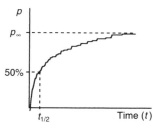

$$\frac{\ln\{M/(2M - 1)\}}{1 - M}\right)\frac{K'}{K_{in}} \rightarrow$$ Slope $= \dfrac{\ln\{M/(M - 0.5)\}}{k_{in}}$

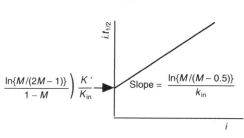

Other definitions and relationships are given in *Tables 35* and *37*. The above equations are also repeated in *Table 37* for clarity.

inhibitor of the enzyme monoamine oxidase (see refs 115, 116) that operates in this way is shown in *Figure 26*. The reaction of pyridoxal-phosphate enzymes with amino groups forms a mobile double bond in the Schiff base intermediate and the actions of two specific mechanism-based inhibitors of such enzymes (see refs 114, 117) are given in *Figure 27*.

4.5. Recovery from irreversible inhibition
For a reversible inhibitor the rate of recovery of enzyme activity will depend on the rate of removal of the inhibitor from the tissues by metabolism and elimination. For an irreversible inhibitor the recovery requires synthesis of new enzyme to replace the inhibited enzyme. This can allow the first-order rate constant and half-life of turnover to be calculated, as shown in *Figure 28*. The rates of turnover of different enzymes vary widely, as shown for some rat liver

Table 37. Kinetic behavior of a suicide substrate

Equations	Parameters	Dimensions	Relationships and definitions
(a) For activity loss			
$t = N.\ln(1 - M.z) - N'.\ln(1-z)$	$K' = \left(\dfrac{k_{-1} + k_{+2}}{k_{+1}}\right)\left(\dfrac{k_{+3} + k_{+4}}{k_{+2} + k_{+3} + k_{+4}}\right)$	Concentration	$z = \dfrac{1-u}{M}$
			$u = 1 - M.z$
$i.t_{1/2} = \left(\dfrac{\ln(2-M)}{1-M}\right)\dfrac{K'}{k_{in}} + \dfrac{\ln 2}{k_{in}}.i$	$k_{in} = \dfrac{k_{+2}k_{+4}}{k_{+2} + k_{+3} + k_{+4}}$	1/Time	$p = r.z.e$
			$t_{1/2} = $ time to 50% inhibition
(b) For product formation	$r = \dfrac{k_{+3}}{k_{+4}}$	None	$e, i, s, p = $ enzyme, inhibitor, substrate and product concentrations
$z = p/r.e$			
	$M = (1+r).\dfrac{e}{i}$	None	$e_i = $ concn of inhibited enzyme
$t = N.\ln u - N'.\ln\{(u + M - 1)/M\}$	$z = \dfrac{e_i}{e}$	None	$i_t = $ inhibitor concn at time t
$i.t_{1/2} = \left(\dfrac{\ln\{M/(2M-1)\}}{1-M}\right)\dfrac{K'}{k_{in}} + \dfrac{\ln\{M/(M-0.5)\}}{k_{in}}.i$	$u = \dfrac{i_t}{i}$	None	If $M < 1$: $z_\infty = 1$ and $u_\infty = 1 - M$
		Time	If $M > 1$: $z_\infty = 1/M$ and $u_\infty = 0$
(c) For loss of activity with added substrate	$N = \dfrac{K'}{k_{in}(1-M).i}$		
As for case (a) *except* that the value of K' obtained K'_{app} will be:	$N' = \dfrac{K'}{k_{in}(1-M).i} + \dfrac{1}{k_{in}}$	Time	where z_∞ and u_∞ are the values of these ratios at time $= \infty$
$K'_{app} = K'\left(1 + \dfrac{s}{K_m}\right)$			

Other definitions and relationships are given in *Tables 35 and 36.*

enzymes in *Table 33*. The rates of turnover of an enzyme can differ greatly between tissues and species, as shown for monoamine oxidases A and B in *Table 34*. This can have important implications for therapy with irreversible inhibitors. For comparison, in the human and in experimental animals the recovery of monoamine oxidase (MAO) from a single dose of a reversible inhibitor administered by mouth or intraperitoneally is often about 12 hours.

4.6. Irreversible inhibitors as substrates and substrates as irreversible inhibitors

Some irreversible inhibitors are in fact extremely poor substrates. Examples include neostigmine as an inhibitor of acetylcholinesterase (see refs 123, 124). *Figure 29* compares the interactions of that enzyme with acetylcholine and neostigmine. In this case the rate of hydrolysis of the carbamoyl enzyme intermediate is so slow that the inhibition can be regarded as being, for all practical purposes, irreversible. In some other cases the rate of inhibitor transformation might be such that a significant regain of activity might be seen on prolonged enzyme–inhibitor incubation. In such cases the behavior can be analysed by steady-state or transient kinetic procedures (1, 125). In practice, the definition of irreversible inhibition is defined by the time scale of the experiments.

Since mechanism-based inhibitors behave like substrates in binding to the active site of the enzyme and being converted to the reactive species through a process resembling the normal catalytic process of the enzyme, it is perhaps not surprising that it is possible in some cases for the reactive species to break down to form product. Examples include the reaction of clavulanate and β-iodopenicillanate with β-lactamase (126, 127) and the reaction of monoamine oxidase with MD 70236, 1-methyl-4-phenyl-1,2,3,6-tetrahydropyridine and phenelzine (128–130). In such cases the formation of product and the mechanism-based inhibition of the enzyme will be competing reactions according to the general scheme shown in *Table 35*. Compounds which behave in this way are sometimes referred to as 'suicide substrates'.

An important parameter is the partition ratio, which corresponds to the number of mol of product formed per mol of enzyme at complete inhibition, and procedures for calculating this (131–133) are shown in *Table 34*. Kinetic analysis of the time-course of product formation, inhibitor loss or enzyme inactivation will allow the first-order rate constant for inactivation (k_{in}) and the apparent Michaelis constant for the process (K') to be calculated (132–134), as shown in *Tables 36* and *37*.

ACKNOWLEDGMENTS

I am grateful to Dr Robert Eisenthal for helpful criticism and comments and to Bio-Research Ireland for support.

5. REFERENCES

1. Dixon, M. and Webb, E.C. (1979) *Enzymes (3rd edn)*. Longman, London.

2. Sandler, M. and Smith, H.J. (eds) (1989) *Design of Enzyme Inhibitors as Drugs*. Oxford University Press, Oxford, p. 70.

3. Segel, I.H. (1975) *Enzyme Kinetics*. Wiley-Interscience, New York.

4. Cleland, W.W. (1979) *Meth. Enzymol.*, **63**, 103.

5. Wilkinson, G.N. (1961) *Biochem. J.*, **89**, 324.

6. Leatherbarrow, R.J.(1990) *TIBS*, **15**, 445.

7. Henderson, P.J.F. (1992) in *Enzyme Assays: a Practical Approach* (R. Eisenthal and M. Danson, eds). IRL Press, Oxford, p. 277.

8. Eisenthal, R.E. and Cornish-Bowden, A. (1974) *Biochem. J.*, **139**, 715.

9. Wilson, I.B and Quan, C. (1958) *Arch. Biochem. Biophys.*, **73**, 131.

10. Peters, R.A. and Wilson, T.H. (1952) *Biochim. Biophys. Acta*, **9**, 310.

11. Yoshida, A. and Freese, E. (1970) *Meth. Enzymol.*, **17A**, 176.

12. Weiner, H., Freytag, S., Fox, J.M. and Hu, J.H. (1982) *Progr. Clin. Biol. Res.*, **114**, 91.

13. Rosenfield, J.L., Dutta, S.P., Chednag, B. and Tritsch, G.L. (1975) *Biochim. Biophys. Acta*, **410**, 164.

14. Shigeoka, S., Nakano, Y. and Kitaoka, S. (1980) *Arch. Biochem. Biophys.*, **201**, 121.

15. Smith, E.L., Lumry, R. and Polglase, W.J. (1951) *J. Phys. Colloid Chem.*, **55**, 125.

16. Cunningham, L. (1965) *Comp. Biochem.* **16**, 85.

17. Noda, L., Nihei, T. and Morales, M.F. (1960) *J. Biol. Chem.*, **235**, 2830.

18. Richey, D.P. and Brown, G.M. (1971) *Meth. Enzymol.*, **18B**, 771.

19. Plaut, G.W.E., Aogaichi, T. and Gabriel, J.L. (1986) *Arch. Biochem. Biophys.*, **245**, 114.

20. Mantle, T.J., Tipton, K.F. and Garrett, N.J. (1976) *Biochem. Pharmacol.*, **25**, 2073.

21. Mares-Guia, M. and Shaw, E. (1965) *J. Biol. Chem.*, **240**, 1579.

22. Kearney, E.B. (1957) *J. Biol. Chem.*, **229**, 363.

23. Newton, W.A., Morina, Y. and Snell, E.E. (1965) *J. Biol. Chem.*, **240**, 1211.

24. Diamondstone, T.I. and Litwack, G. (1963) *J. Biol. Chem.*, **238**, 3859.

25. Fields, J.H.A and Hochachka, P.W. (1981) *Eur. J. Biochem.*, **114**, 615.

26. Inhorn, R.C. and Majerus, P.W. (1987) *J. Biol. Chem.*, **262**, 15956.

27. Sakai, T., Hori, C., Kano, K. and Oka, T. (1979) *Biochemistry*, **18**, 5541.

28. Rubenstein, J.A., Collins, M.A. and Tabakoff, B. (1975) *Experientia*, **31**, 414.

29. Davidson, W.S. and Murphy, D.G. (1985) *Progr. Clin. Biol. Res.*, **174**, 251.

30. De Mot, R. and Verachtert, H. (1987) *Eur. J. Biochem.*, **164**, 643.

31. Maley, G.F. and Maley, F. (1964) *J. Biol. Chem.*, **239**, 1168.

32. Boyko, J. and Fraser, M.J. (1964) *Can. J. Biochem.*, **42**, 1677.

33. Smith, R.A. and Gunsalus, I.C. (1958) *J. Biol. Chem.*, **229**, 305.

34. Andrawis, A. and Khan, V. (1986) *Biochem. J.*, **236**, 91.

35. Spencer, R.L. and Priess, J. (1967) *J. Biol. Chem.*, **242**, 385.

36. Craine, J.E., Hall, E.S. and Kauffman, S. (1972) *J. Biol. Chem.*, **247**, 6082.

37. Martin-Requero, A., Corkey, B.E. and Cerdan, S. (1983) *J. Biol. Chem.*, **258**, 15338.

38. Smith, I.K. and Thompson, J.F. (1971) *Biochim. Biophys. Acta*, **227**, 288.

39. Zollner, H.Z. (1990) *Handbook of Enzyme Inhibitors*. VCH, Weinheim.

40. Stoner, H.B. and McGee, P.N. (1957) *Br. Med. Bull.*, **13**, 102.

41. Leo, M.A. and Lieber, C.S. (1983) *Alcohol. Clin. Exp. Res.*, **7**, 15.

42. Turner, A.J., Illingworth, J.A. and Tipton, K.F. (1974) *Biochem. J.*, **114**, 353.

43. Fersht, A.R. (1985) *Enzyme Structure and Mechanism*. W.H. Freeman, San Francisco.

44. Dixon, M. (1953) *Biochem. J.*, **55**, 170.

45. Cornish-Bowden, A. (1974) *Biochem. J.*, **137**, 143.

46. Alberty, R.A. (1953) *J. Am. Chem. Soc.*, **75**, 1928.

47. Bloomfield, V., Peller, L. and Alberty, R.A. (1962) *J. Am. Chem. Soc.*, **84**, 4367.

48. Cleland, W.W. (1970) in *The Enzymes* (P.D. Boyer, ed.). Academic Press, New York, p. 1.

49. Dalziel, K. (1957) *Acta Chem. Scand.*, **11**, 1706.

50. Frieden, C. (1970) *J. Biol. Chem.*, **245**, 5788.

51. Kurganov, B.I., Dorozhko, A.I., Kagan, Z.S. and Yakovlev, V.A. (1976) *J. Theor. Biol.*, **60**, 247, 271 and 287; **61**, 531.

52. Neet, K.E. and Ainslie, G.R. (1980) *Meth. Enzymol.*, **64**, 192.

53. Meunier, J.C., Buc, J., Navarro, A. and Ricard, J. (1974) *Eur. J. Biochem.*, **49**, 209.

54. Ricard, J., Meunier, J.C. and Buc, J. (1974) *Eur. J. Biochem.*, **49**, 195.

55. Cornish-Bowden A. (1979) *Fundamentals of Enzyme Kinetics*. Butterworths, London.

56. Engel, P.C. (1981) *Enzyme Kinetics. The Steady-state Approach*. Chapman & Hall, London.

57. Fromm, H.J. (1975) *Initial Rate Enzyme Kinetics*. Springer-Verlag, Heidelberg.

58. Roberts, D.V. (1977) *Enzyme Kinetics*. Cambridge University Press, Cambridge.

59. Rudolph, F., Purich, D.L. and Fromm, H.J. (1968) *J. Biol. Chem.*, **243**, 5539.

60. Dalziel, K. (1969) *Biochem. J.*, **114**, 547.

61. Rudolph, F. (1979) *Meth. Enzymol.* **63**, 467.

62. Elliott, K.R.F. and Tipton, K.F. (1974) *Biochem. J.*, **141**, 789.

63. Tipton, K.F. (1992) in *Enzyme Assays: a Practical Approach* (R. Eisenthal and M. Danson, eds). IRL Press, Oxford, p. 1.

64. Botts, J. and Morales, M. (1952) *Trans. Faraday Soc.*, **49**, 696.

65. Hearon, J.Z., Bernhard, S.A., Friess, S.L., Botts, D.J. and Morales, M.F. (1959) in *The Enzymes (2nd edn)* (P.D. Boyer, H.A. Lardy and K. Myrbäck, eds). Academic Press, New York, p. 49.

66. Frieden, C. (1964) *J. Biol. Chem.*, **239**, 3522.

67. Bardsley, W.G. and Childs, R.E. (1972) *Biochem. J.* **149**, 313.

68. Wong, J. T.-F. (1975) *Kinetics and Enzyme Mechanism*. Academic Press, New York.

69. Pettersson, G. (1970) *Acta Chem. Scand.*, **24**, 1271.

70. Tipton, K.F. (1979) in *Companion to Biochemistry* (A.T. Bull, J.R. Lagnado, J.O. Thomas and K.F. Tipton, eds). Longman, London, Vol. 2, p. 327.

71. Whitehead, E.P. (1970) *Prog. Biophys. Mol. Biol.*, **21**, 323.

72. Cleland, W.W. (1979) *Meth. Enzymol.*, **63**, 500.

73. Morrison, J.F. (1979) *Meth. Enzymol.*, **63**, 257.

74. Tipton, K.F. (1985) in *Techniques in the Life Sciences*, B1/II Supplement, BS 113 (K.F. Tipton, ed.). Elsevier, Dublin, p. 1.

75. Storer, A.C. and Cornish-Bowden, A. (1976) *Biochem. J.*, **159**, 1.

76. Tubbs, P.K. (1962) *Biochem. J.*, **82**, 36.

77. Dalziel, K. (1962) *Biochem. J.*, **83**, 28P.

78. Houslay, M.D. and Tipton K.F. (1973) *Biochem. J.*, **135**, 735.

79. Dixon, M. (1972) *Biochem. J.*, **129**, 197.

80. Henderson, P.J.F. (1972) *Biochem. J.*, **127**, 321.

81. Henderson, P.J.F. (1973) *Biochem. J.*, **135**, 101.

82. Williams, J.W. and Morrison, J.F. (1979) *Meth. Enzymol.*, **63**, 437.

83. Chase, J.F.A. and Tubbs, P.K. (1969) *Biochem. J.*, **111**, 225.

84. Collins, K.D. and Stark, G.R. (1971) *J. Biol. Chem.*, **246**, 6599.

85. Lienhard, G.E. and Secemski, I.I. (1973) *J. Biol. Chem.*, **248**, 1121.

86. Kapmeyer, H., Pfleiderer, G. and Trommer, W.E. (1976) *Biochemistry*, **15**, 5024.

87. Cohen, R.M. and Wolfenden, R. (1971) *J. Biol. Chem.*, **246**, 7561.

88. Schindler, M. and Sharon, N. (1976) *J. Biol. Chem.*, **251**, 4330.

89. Nelson, D.L., Herbert, A., Pétillot, Y., Pichat, L., Glowinsky, J. and Hamon, M. (1979) *J. Neurochem.*, **32**, 1817.

90. Tipton, K.F. (1989) in *Design of Enzyme Inhibitors as Drugs* (M. Sandler and H.J. Smith, eds). Oxford University Press, Oxford, p. 70.

91. Means, G.E. and Feeney, R.E. (1971) *Chemical Modification of Proteins*. Holden Day, San Francisco.

92. Thomas, J.O. (1974) in *Companion to Biochemistry* (A.T. Bull, J.R. Lagnado, J.O. Thomas and K.F. Tipton, eds). Longman, London, Vol. 1, p. 87.

93. Dawson, R.M.C., Elliott, D.C., Elliott, W.H. and Jones, K.M. (1986) *Data for Biochemical Research*. Oxford Science Publications, Oxford.

94. Ray, W.J. and Koshland, D.E. (1961) *J. Biol. Chem.*, **236**, 1973.

95. Rakitzis, E.T. (1984) *Biochem. J.*, **217**, 341.

96. Kitz, R. and Wilson, I.B. (1962) *J. Biol. Chem.*, **337**, 3245.

97. Malcolm, A.D.B. and Radda, G.K. (1970) *Eur. J. Biochem.*, **15**, 555.

98. Brocklehurst, K. (1979) *Biochem. J.*, **181**, 775.

99. Cornish-Bowden, A. (1979) *Eur. J. Biochem.*, **93**, 383.

100. Silverman, R.B. and Hoffman, S.J. (1984) *Med. Res. Rev.*, **44**, 415.

101. Tian, W.X. and Tsou, C.L. (1982) *Biochemistry*, **21**, 1028.

102. Tsou, C.L., Tian, W.X. and Zhao, K.Y. (1985) in *Molecular Architecture of Proteins*. Academic Press, New York, p. 15.

103. Liu, W. and Tsou, C.L. (1986) *Biochim. Biophys. Acta*, **242**, 185.

104. Forsberg, A. and Puu, G. (1984) *Eur. J. Biochem.*, **140**, 153.

105. Walker, B. and Ellmore, D.T. (1984) *Biochem. J.*, **221**, 277.

106. Fowler, C.J, Mantle, T.J. and Tipton, K.F. (1982) *Biochem. Pharmacol.*, **31**, 3555.

107. Tipton, K.F., Balsa, D. and Unzeta, M. (1993) in *Monoamine Oxidase: Basic and Clinical Aspects* (H. Yashuhara, S.H. Parvez, K. Oguchi, M. Sandler and T. Nagatsu, eds). VSP, Utrecht, p. 1.

108. Walsh, C.T. (1979) *Enzyme Reaction Mechanisms*. W.H. Freeman, San Francisco.

109. Plapp, B.V. (1983) in *Contemporary Enzyme Kinetics and Mechanism* (D.L. Purich, ed.). Academic Press, New York, p. 321.

110. Shaw, E. (1989) in *Design of Enzyme Inhibitors as Drugs* (M. Sandler and H.J. Smith, eds). Oxford University Press, Oxford, p. 49.

111. Shaw, E. (1970) *Physiol. Rev.*, **50**, 244.

112. Shaw, E. and Ruscica, J. (1971) *Arch. Biochem. Biophys.*, **145**, 484.

113. Rando, R.R. (1984) *Pharmacol. Rev.*, **36**, 111.

114. Palfreyman, M.G. (1987) *Essays Biochem.*, **23**, 28.

115. Maycock, A.L., Abeles, R.H., Salach, J.I. and Singer, T.P. (1976) *Biochemistry*, **15,** 114.

116. Dostert, P.H., Strolin Benedetti, M. and Tipton, K.F. (1989) *Med. Res. Revs,* **9,** 45.

117. Jung, M. and Danzin, C. (1989) in *Design of Enzyme Inhibitors as Drugs* (M. Sandler and H.J. Smith, eds). Oxford University Press, Oxford, p. 257.

118. Goldberg, A.L. and Dice, J.F. (1974) *Ann. Rev. Biochem.,* **43,** 835.

119. Hopgood, M.F. and Ballard, F.J. (1974) *Biochem. J.,* **144,** 371.

120. Schimke, R.T. (1973) *Adv. Enzymol.,* **37,** 137.

121. Schimke, R.T. and Katanuma, N. (1975) *Intracellular Protein Turnover.* Academic Press, New York, p. 131.

122. O'Brien, E.M. and Tipton, K.F. (1994) in *Monoamine Oxidase Inhibitors in Neurological Diseases* (A. Lieberman, C.W. Olanow, M.B.H. Youdim and K.F. Tipton, eds). Marcel Dekker, New York, p. 31.

123. O'Brien, R.D. (1968) *Mol. Pharmacol.,* **4,** 121.

124. Aldridge, W.N. (1989) in *Design of Enzyme Inhibitors as Drugs* (M. Sandler and H.J. Smith eds). Oxford University Press, Oxford, p. 294.

125. John, R.A. (1985) in *Techniques in the Life Sciences,* B1/II Supplement, BS 118 (K.F. Tipton, ed.). Elsevier, Dublin, p. 1.

126. Frère, J.-M., Dormans, C., Lenzini, V.M. and Duyckaerts, C. (1982) *Biochem. J.,* **207,** 429.

127. Frère, J.-M., Dormans, C., Duyckaerts, C. and De Graeve, J. (1982) *Biochem. J.,* **207,** 437.

128. Tipton, K.F., Fowler, C.J., McCrodden, J.M. and Strolin Benedetti, M. (1983) *Biochem. J.,* **209,** 235.

129. Tipton, K.F., McCrodden, J.M. and Youdim, M.B.H. (1986) *Biochem. J.,* **240,** 379.

130. Yu, P.H. and Tipton, K.F. (1989) *Biochem. Pharmacol.,* **38,** 4245.

131. Knight, G.C. and Waley, S.G. (1985) *Biochem. J.,* **225,** 435.

132. Tatsunami, S., Yago, M. and Hosoe, M. (1981) *Biochem. Biophys Acta,* **662,** 226.

133. Waley, S.G. (1985) *Biochem. J.,* **277,** 843.

134. Waley, S.G. (1980) *Biochem. J.,* **185,** 771.

135. Walsh, C.T. (1984) *Ann. Rev. Biochem,* **53,** 493.

CHAPTER 5
PHYSICAL FACTORS AFFECTING ENZYME ACTIVITY

A. pH-DEPENDENT KINETICS

K. Brocklehurst

1. INTRODUCTION

The status of the proton as the least sterically demanding general perturbant of protein structure, and the importance of the protonation state of specific side chains in protein (particularly enzyme) function account for the continuing interest in and development of experimental techniques relating to the pH-dependence of kinetic parameters (1,2), and theoretical treatments of intrinsic titration properties and interaction of ionizing groups (3–5). IUPAC recommends the term hydron for H^+ when a particular isotope is not specified (see ref. 5); the term proton is then reserved for $^1H^+$. Hydron is not yet in general use, however, and proton is used in its traditional general sense in this chapter, particularly as isotope effects are not dealt with. Many aspects of pH-dependent kinetic studies on enzymes have been reviewed in depth (6,7).

The principal objectives of pH-dependent kinetic studies are the determination of pK_a values approximating to those of individual ionizing groups and rate constants characteristic of reactions of specific ionic forms of reactants, and their perturbation by ligand binding, notably by additional protons and substrate or effector molecules, and by structural modification. It is important, therefore, to know when and how it is possible to obtain information pertaining to the pK_a values of individual groups and to the intrinsic reactivity characteristics of, usually, their conjugate bases. It is worth emphasizing that pH-dependent kinetic studies require greater than average care in their design and interpretation if meaningful results are to be obtained and mechanistically useful conclusions drawn.

2. POTENTIAL PITFALLS: GENERAL CONSIDERATIONS

The well-established, but still sometimes overlooked pitfalls that exist in the application of pH-dependent kinetics to the study of mechanism (2,8,9) relate mainly to the need to be aware of:

(i)   the fact that, in analysis, the kinetically influential ionizations must be confined to either the free enzyme molecule or to a single–substrate complex or intermediate.
(ii) The characteristics of reactions proceeding in parallel pathways through thermodynamic boxes.
(iii) The relationship between microscopic (group) and macroscopic (molecular or system) pK_as.
(iv) The kinetic consequence of reaction via minor ionization states.
(v)   Overlapping kinetically influential ionizations and multiple reactive protonic states.
(vi) The possibility of change in rate-determining step leading to 'mirage' pK_as.

As is now well-known, the pH-dependence of rate (v) at arbitrary constant substrate concentration may have complex significance and is usually of little value. This is because the pH-dependent rate equation for the steady state rate of an enzyme-catalysed reaction contains all of the pK_a values of the kinetic model (i.e. of the free enzyme (and possibly substrate) and of all of the complexes and intermediates of the mechanism).

By contrast, the pH-dependence of the catalytic constant (k_{cat}) can sometimes provide pK_a values of a particular enzyme–substrate complex or intermediate, and that of the specificity constant (k_{cat}/K_m) can, and usually does, provide uniquely pK_a values of the free reactant state (8, 10, 11).

Direct study by rapid reaction techniques complements the thoughtful use of steady-state kinetics in the characterization of single events. Examples of the latter are:

(i) the use of specific amide and ester substrates for serine proteinases to provide information about the pH-dependence of acylenzyme intermediate formation (k_{+2}) and hydrolysis (deacylation; k_{+3}) (see ref. 12 and references therein); in both cases the pH-dependence of the steady-state parameter, k_{cat}, is determined; in terms of the minimal three-step acylenzyme model $k_{cat} = k_{+2}k_{+3}/(k_{+2}+k_{+3})$ which can provide the values of k_{+2} or k_{+3} if one or other step is rate determining.

(ii) The use of arylester and anilide substrates for cysteine proteinases to provide information about formation and breakdown of tetrahedral intermediates (see ref. 13 and references therein).

3. PARALLEL PATHWAYS

Some of the potential difficulties with pH-dependent kinetics relate to the need to take proper account of fluxes in parallel pathways through a thermodynamic box. Two types of situation in which parallel pathways are not taken into account may be distinguished:

(i) some of the pathways are explicitly excluded from the kinetic model (14) (*Figure 1a*).
(ii) The kinetic model shows the parallel pathways (*Figure 1b*) but they are not taken into account in the analysis.

Direct interconversion in only one ionization state (*Figure 1a*) is implausible in the general case (15, 16) and the lack of recognition of the parallel pathways shown in *Figure 1b* in the derivation of rate equations leads to serious errors (ref. 17). The complexity that can exist in the pH-dependence of k_{cat}/K_m when due recognition is given to parallel pathways is revealed by analysis of the model in *Figure 1b* in which the 'vertical' proton-binding reactions (ionizations) are regarded as unperturbed equilibria with steady states around EH_2S, EHS^- and ES^{2-} for the 'horizontal' reactions (10, 11).

4. THE QUASI-EQUILIBRIUM CONDITION

Many enzyme reactions have been considered to exhibit reversible dependence of some measure of catalytic activity on pH that has the same form as that of the concentration of the

Figure 1. Kinetic models for product formation from a central ionization state of an enzyme–substrate complex. Model (a) implies the somewhat implausible assumption that substrate (S) binds only to EH^-, whereas in model (b) binding to all three ionization states of the enzyme is recognized.

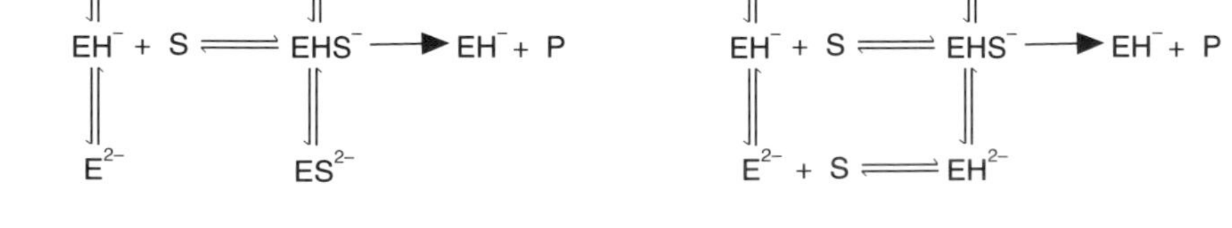

monoprotonated species of a two-site acid (6, 18). In such cases, where in addition the same step is rate-determining at all pH values (19–21) the complexities that can arise from parallel pathways within the thermodynamic box disappear when it is permissible to make the quasi-equilibrium assumption not only for the ('vertical') ionizations but also for the ('horizontal') steps in which free substrate progresses through one or more intermediate complexes or compounds to products (10, 11). The conventional criterion for this assumption of quasi-equilibrium, in terms of the simplest (two-step) model

$$EH + S \underset{k_{-1}}{\overset{k_{+1}}{\rightleftharpoons}} EHS \xrightarrow{k_{+2}} EH + P, \text{ i.e. } k_{+2} \ll k_{-1}$$

may be expressed also in the practically more useful form $k_{cat}/K_m \ll k_{+1}$ (see ref. 9).

Thus a useful transform of the steady-state expression, *Equation 1*, is *Equation 2*

$$k_{cat}/K_m = \frac{k_{+1}k_{+2}}{k_{-1} + k_{+2}} \qquad\qquad \text{Equation 1}$$

$$k_{+2}/k_{-1} = \frac{k_{cat}/k_{+1}K_m}{1 - (k_{cat}/k_{+1}K_m)} \qquad\qquad \text{Equation 2}$$

from which it follows that the conventional condition for quasi-equilibrium is a necessary consequence of the condition $k_{cat}/K_m \ll k_{+1}$.

Another indication that the quasi-equilibrium condition obtains is the observation under appropriate circumstances that K_m does not change with change in pH whereas k_{cat} does (22).

5. SIX TYPES OF pK_a

The six types of pK_a encountered in kinetic analysis – microscopic (group), macroscopic (molecular and system), titration, transition state and 'mirage' – are defined and their significance indicated in *Table 1*.

6. KINETIC CONSEQUENCES OF REACTION VIA A MINOR IONIZATION STATE

This may be illustrated by consideration of *Figure 3* and *Equations 7* and *8* but the concept applies equally to analogous equations relating to molecular pK_as and their constituent group pK_as (see ref. 44) for a discussion of 'crossed over' intrinsic ionizations).

(i)   The experimentally determined pK_a values are mixed constants and the pH-independent rate constant ($\tilde{k}'$) is the 'true' rate constant ($\tilde{k}$) multiplied by an assembly of the two acid dissociation constants of *Figure 3 (Equation 7)*.
(ii) If pK_e and pK_r differ by more than 1, the two apparent pK_a values will provide good approximations of K_e and K_r, respectively. If the reaction proceeds via a major ionization state (i.e. $K_e \gg K_r$), $\tilde{k}' \simeq \tilde{k}$.
(iii) If, however, the reactive forms of the two reactants constitute a minor ionization state (i.e. $K_e \ll K_r$) the experimentally determined value, $\tilde{k}'$, provides a low estimate of the 'true' rate constant and needs to be multiplied by $(K_e + K_r)/K_e$ (which approximates to K_r/K_e) to provide the value of $\tilde{k}$.
(iv) The possibility of reaction via either a major or minor ionization state (e.g. one or other of the constituents of the XH state of *Figure 2b*) and the consequent uncertainty illustrates the well-known principle that pH-dependent kinetics alone define proton stoichiometries in reactions but not their positions.

▶ p. 181

Physical Factors

Table 1. The six most commonly encountered types of acid dissociation constant and their significance

Type of constant	Significance	Ref.
Microscopic or group	The value of the pK_a of an individual group (-AH or -B$^+$H) (*Figure 2*) is of interest in connection with structural and mechanistic studies. Unfortunately, group pK_a values cannot be measured experimentally without making assumptions, although by careful choice of the pH-dependent property to be measured and by consideration of the pH-dependence of several different properties, useful estimates can sometimes be obtained. The problem derives from the fact that the pK_a value of an individual group may depend upon the state of ionization of other groups in the molecule. To take a two-site acid (HA-B$^+$H) as a simple case, the value of the pK_a of the group -B$^+$H may be different in HA-B$^+$H than in $^-$A-B$^+$H; each of these is a microscopic value. The difference may arise from change in direct electrostatic/hydrogen bonding interaction or from conformational change promoted by the ionization of -AH. Good estimates of group pK_as may sometimes be made from experimentally determined macroscopic values (see below)	23
Macroscopic (molecular and system) Molecular	The molecular acid dissociation constants of a two-site acid (K_I and K_{II}) and their relationships to the group dissociation constants (K_A, K_B, K_A' and K_B') are defined by *Equations 3* and *4* and *Figure 2*. $$K_I = K_A + K_B \qquad \text{Equation 3}$$ $$K_{II} = K_A' \cdot K_B'/(K_A' + K_B') \qquad \text{Equation 4}$$ If the XH state of *Figure 2b* is reactive (or catalytically active) and both X$^+$H$_2$ and X$^-$ are unreactive (or inactive), the pH-dependence of the reactivity reflects the bell-shaped pH-dependence of [XH]. For the special case in which the two acidic groups are identical (i.e. the two-site acid is H-Z-Z-H) and, additionally, their protonic dissociations are independent of each other, $K_A = K_B = K_A' = K_B' = K$ and *Equations 3* and *4* simplify to *Equations 5* and *6*, respectively. $$K_I = 2K \qquad \text{Equation 5}$$ $$K_{II} = K/2 \qquad \text{Equation 6}$$ Although it is molecular pK_a values (or in some cases system pK_a values, see below) that are obtained experimentally, these values can approximate to the mechanistically useful group constants. This situation obtains when the value of K_A is substantially different from that of K_B (and/or that of K_A' from K_B'). With interactive groups it is unreasonable to assume that $K_A \simeq K_A'$ or $K_B \simeq K_B'$, and K_A and K_B will often differ substantially from each other as will K_A' and K_B'. For example if $pK_A \ll pK_B$, pK_I will approximate closely to pK_A and pK_{II} to pK_B'	6, 24–26

System

There is an additional layer of complexity when the substrate or time-dependent inhibitor (I) also undergoes kinetically influential ionization. In *Figure 3* the nucleophilic form of an enzyme (E⁻) reacts with the electrophilic form of a time-dependent inhibitor (I⁺H). The acid dissociation constant for the provision of (E⁻) from EH, K_e, is a molecular constant which might or might not approximate to a group constant and the dissociation constant for deprotonation of I⁺H, K_r, in most cases is a group constant; $\tilde{k}'$, is the pH-independent rate constant determined by analysis of the data for the bell-shaped pH-dependence of the experimentally observed pH-dependent rate constant, k. The pH-dependence of k is described by *Equation 7* and the relationship of $\tilde{k}$, the 'true' rate constant for the reaction of the specific ionic forms E⁻ and I⁺H, to $\tilde{k}'$ is given by *Equation 8*. Analogous equations relate pH-dependent and pH-independent values of k_{cat} and of k_{cat}/K_m. [27, 28]

$$k = \frac{\tilde{k} \cdot \dfrac{K_e}{K_e + K_r}}{1 + \dfrac{[H^+]}{(K_e + K_r)} + \dfrac{K_e K_r/(K_e + K_r)}{[H^+]}}$$

Equation 7

$$\tilde{k}' = \tilde{k}\,\frac{K_e}{K_e + K_r}$$

Equation 8

The system acid dissociation constants, $K_e + K_r$ and $K_e K_r/(K_e + K_r)$, provide good approximations to K_e and K_r if pK_e and pK_r differ by more than 1 (see also Section 6 of the text)

Titration

The fact that the titration curve of a multi-site acid is identical to that of a mixture of hypothetical one-site acids, each of the same concentration as the multi-site acid, permits the use of the dissociation constants of these hypothetical acids which are referred to as the titration constants (G_1, G_2, etc.) of the real multi-site acid. For a two-site acid the titration constants are related in a simple way to the molecular constants by *Equations 9* and *10* although when there are more than two acidic groups the equations are more complex. [5, 18, 29–34]

$$K_I = G_1 + G_2$$

Equation 9

$$K_I K_{II} = G_1 G_2$$

Equation 10

When $pK_{II} - pK_I = 0.6$ (the limit for noncooperative systems), $pG_2 - pG_1 = 0$. As the difference $pK_{II} - pK_I$ increases, the consequent increase in $pG_2 - pG_1$, is greater and the titration constants approximate closely to the molecular constants when $pK_{II} - pK_I > c. 2$. An important conclusion is that the pH-dependence of any property of the multi-site acid may be represented by the sum of one-site titration curves characterized by the titration constants, and a fraction of each group is characterized by each of the pG values (i.e. the

Physical Factors

Table 1. Continued

Type of constant	Significance	Ref.
	group partitions among them). In terms of these ideas, 68% of the thiol (-SH) group of cysteine titrates with $pG = 8.36$ and 32% with $pG = 10.53$; and, conversely, 32% of the ammonio ($-N^+H_3$) group titrates with $pG = 8.36$ and 68% with $pG = 10.53$. The valuable conclusion that interacting groups titrate in fractions, each fraction following a simple one-site titration curve, applies approximately also to the analogous treatment in terms of molecular constants	
Transition state	The transition state acid dissociation constant, $K_a^\ddagger$ may be obtained from the observed rate constants $\tilde{k}_1$ and $\tilde{k}_2$ of two similar reactions (*Equations 11* and *12*), differing only in that a reactant in one case is the conjugate acid (B^+H) of the reactant (B) in the other. $$B + R \overset{\tilde{k}_1}{\rightleftharpoons} [B]^\ddagger \longrightarrow P \qquad \text{Equation 11}$$ $$B^+H + R \overset{\tilde{k}_2}{\rightleftharpoons} [B^+H]^\ddagger \longrightarrow P \qquad \text{Equation 12}$$ The virtual equilibrium constant, $K_a^\ddagger$ defined in the usual way in terms of the concentrations $[B]^\ddagger$, $[B^+H]^\ddagger$ and $[H^+]$, may be related to observable quantities via a thermodynamic cycle. The acidic hydrogen atom of the transition state serves as a reporter for factors affecting its acidity. It is possible, by using an analogous approach, to investigate transition state solvation. Transition state acid dissociation constants have been used in the study of enzyme mechanism. For example, the larger the value of $pK_a^\ddagger - pK_a$, the greater the sensitivity of a rate constant to the ionization in question and the greater the microenvironmental perturbation of the group during the activation process	35–41
'Mirage'	The treatments described above assume that the same step is rate-determining at all pH values. If this is not the case, conventional analysis of, say, a bell-shaped pH-profile may appear to provide evidence for an ionizing group that does not actually exist in the enzyme and indeed, a (spurious) numerical value for its (nonexistent) pK_a. In multistep enzyme reactions change in rate-determining step may sometimes need to be considered as an alternative to assignment of the two pK_a values of a bell-shaped curve to two different ionizing groups	9, 19

Figure 2. Protonic dissociation from a two-site acid. (a) Group acid dissociation constants (K_A, K_B, K_A', K_B'). (b) Molecular acid dissociation constants (K_I and K_{II}). The choice of one cationic acid ($-B^+H$) and one uncharged acid (A–H) is arbitrary. The relationships between the molecular constants and the group constants are given by *Equations 3* and *4 (Table 1)*.

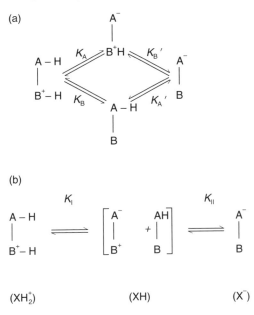

Figure 3. Reaction of the nucleophilic form of an enzyme (E^-) with the electrophilic form of a time-dependent inhibitor (I^+H). K_e, K_r and k' are defined in *Table 1*.

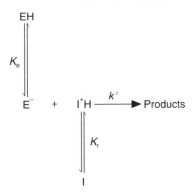

(v)   The location of protons in transition states relies on a knowledge of the chemistry of the reaction and in some cases on the constraint deriving from the diffusion-controlled limit on the value of the rate constant. If the diffusion limit would be exceeded for reaction via a particular minor ionization state, that option may be excluded.

Physical Factors

7. FORMS OF BELL-SHAPED COMPONENTS OF pH–RATE CONSTANT PROFILES

Statements apply equally to k_{cat}/K_m, k_{cat}, and a rate constant for an individual step of a catalytic mechanism or for reaction with a site-specific time-dependent inhibitor; for simplicity, all are generalized here as k. A detailed discussion of bell-shaped curves of pH-dependence is given in ref. 19.

The shape of a pH–k curve is the same for reaction via a minor ionization state as for reaction via the analogous major ionization state.

Values of the molecular pK_as, pK_I and pK_{II}, approximate closely to the pH values corresponding to half of the maximum height of a bell-shaped curve only when $pK_{II} - pK_I \geq c.$ 3.5. When they are not so widely separated, the pK_a values equate with pH values corresponding to values of k closer to the maximum of the curve. As a result the width at half maximum height becomes progressively greater than the value of $pK_{II} - pK_I$.

For the special case in which the two acidic groups are identical and their protonic dissociations are independent of each other, *Equations 5* and *6* (*Table 1*) apply. Then $pK_{II} - pK_I = 0.6$ (i.e. $K_I/K_{II} = 4$) which provides a width at half maximum height of 1.53. This provides for the narrowest possible bell-shaped curve unless the protonations responsible for generation and loss of reactivity are positively cooperative.

With positive cooperativity of protonic binding and dissociation the pH–k profile becomes narrower; when $pK_I = pK_{II}$ the width is 1.36 and the ultimate limit is 1.14 (19).

The above statements apply also to the situation where the bell-shaped curve is due to change in rate-determining step, except that in this circumstance $pK_{II} - pK_I$ is necessarily ≥ 0.6 (19).

8. pH-DEPENDENT RATE EQUATIONS AND THEIR PARAMETERS

Rate equations may be written in terms of either a pH-dependent rate constant, k (which may be a rate constant for an individual step or k_{cat} or k_{cat}/K_m) or a pH-independent rate constant, $\tilde{k}$.

For the simplest model comprising one reactive and one unreactive protonic state written in terms of a generalized reactant state, $XH \rightleftharpoons X^- + H^+$, where $[X]_T = [XH] + [X^-]$, with acid dissociation constant, K_a, and pH-dependent rate constant for product formation, k:

$$\text{rate} = k[X]_T. \qquad \qquad \text{Equation 13}$$

If X^- is the reactive state

$$\text{rate} = \tilde{k}[X^-]. \qquad \qquad \text{Equation 14}$$

Combination of *Equations 13–15* provides *Equation 16* which relates k to $\tilde{k}$ and K_a.

$$K_a = [X^-][H^+]/[XH] \qquad \qquad \text{Equation 15}$$

$$k = \tilde{k}/(1 + [H^+]/K_a). \qquad \qquad \text{Equation 16}$$

If XH is the reactive state, combination of *Equations 13, 15* and *17* provides *Equation 18*.

$$\text{Rate} = \tilde{k}[XH]. \qquad \qquad \text{Equation 17}$$

$$k = \tilde{k}/(1 + K_a/[H^+]). \qquad \qquad \text{Equation 18}$$

Equations 16 and *18* correspond to sigmoidal pH–k profiles, k increasing with increase in pH in the case of *Equation 16* and k decreasing with increase in pH in the case of *Equation 18*.

For the model comprising one reactive protonic state, XH^-, bounded by two unreactive states, XH_2 and X^{2-}, $XH_2 \rightleftharpoons XH^- \rightleftharpoons X^{2-}$, with molecular acid dissociation constants K_I and K_{II}, pH-dependent rate constant for product formation, k, and a pH-independent rate constant for reaction of XH^-, $\tilde{k}$, an analogous treatment provides *Equation 19* which corresponds to a bell-shaped pH–k profile.

$$k = \tilde{k}/(1 + [H^+]/K_I + K_{II}/[H^+]).$$

Equation 19

Although it has long been conventional to consider that the reversible dependence of enzyme activity on pH will have the same form as the concentration of the monoprotonated form of a two-site acid, careful analysis of pH–rate constant data is revealing more complex forms of pH-dependence (for example, refs 2, 44).

For a reaction with n reactive protonic states there are n pH-independent rate constants ($\tilde{k}$) and n or $(n+1)$ macroscopic pK_a values. For reactions of catalytic site nucleophiles with time-dependent inhibitors there will normally be n pK_a values because the most proton-deficient state will normally be nucleophilic. This contrasts with the situation for catalysis where there will normally be $n+1$ pK_a values because the analogous state will normally be inactive.

Rate equations for reactions in any number of protonic states may be written down or generated within a computer without the need for extensive algebraic manipulation, see Section 9. For example, the pH-dependent rate equation for reaction in four reactive protonic states $XH_3 - X$ (relative ionic charges not shown) characterized by rate constants $\tilde{k}_{XH_3} - \tilde{k}_X$ with acid dissociation constants $K_{XH_4} - \tilde{K}_{XH}$ is *Equation 20*.

$$
k = \cfrac{\tilde{k}_{XH_3}}{1 + \cfrac{[H^+]}{K_{XH_4}} + \cfrac{K_{XH_3}}{[H^+]} + \cfrac{K_{XH_3}K_{XH_2}}{[H^+]^2} + \cfrac{K_{XH_3}K_{XH_2}K_{XH}}{[H^+]^3}} + \cfrac{\tilde{k}_{XH_2}}{1 + \cfrac{[H^+]^2}{K_{XH_4}K_{XH_3}} + \cfrac{[H^+]}{K_{XH_3}} + \cfrac{K_{XH_2}}{[H^+]} + \cfrac{K_{XH_2}K_{XH}}{[H^+]^2}}
$$

$$
+ \cfrac{\tilde{k}_{XH}}{1 + \cfrac{[H^+]^3}{K_{XH_4}K_{XH_3}K_{XH_2}} + \cfrac{[H^+]^2}{K_{XH_3}K_{XH_2}} + \cfrac{[H^+]}{K_{XH_2}} + \cfrac{K_{XH}}{[H^+]}} + \cfrac{\tilde{k}_X}{1 + \cfrac{[H^+]^4}{K_{XH_4}K_{XH_3}K_{XH_2}K_{XH}} + \cfrac{[H^+]^3}{K_{XH_3}K_{XH_2}K_{XH}} + \cfrac{[H^+]^2}{K_{XH_2}K_{XH}} + \cfrac{[H^+]}{K_{XH}}}
$$

Equation 20

When catalysis is described in terms of the conventional model shown in *Figure 1b* with the quasi-equilibrium assumption applied to both 'vertical' and 'horizontal' steps, the pH-dependence of k_{cat} provides molecular pK_a values of the enzyme–substrate complex, pK_{esI} and pK_{esII} (*Equation 21*) and the pH-dependence of k_{cat}/K_m provides the molecular pK_a values of the free enzyme (for nonionizing substrates) pK_{eI} and pK_{eII} (*Equation 22*). Both pH dependences are of simple bell-shaped form.

$$k_{cat} = \tilde{k}_{cat}/(1 + [H^+]/K_{esI} + K_{esII}/[H^+])$$

Equation 21

$$k_{cat}/K_m = (\tilde{k}_{cat}/\tilde{K}_m)/(1 + [H^+]/K_{eI} + K_{eII}/[H^+])$$

Equation 22

More complex equations and hence more complex forms of pH-dependence of k_{cat}/K_m apply when the steady state rather than the quasi-equilibrium assumption is made for the 'horizontal' steps of the model of *Figure 1b* (10, 11).

It follows from the Michaelis–Menten equation $v = k_{cat}[E]_T[S]_o/(K_m + [S]_o)$ that approximate forms of pH-dependence of k_{cat} and k_{cat}/K_m are provided by pH-dependences of $v/[E]_T$ with $[S]_o \gg K_m$, and $[S]_o \ll K_m$, respectively.

9. GENERATION OF pH-DEPENDENT RATE EQUATIONS

For simple kinetic models, pH-dependent rate equations such as *Equations 16* and *18* may be derived by algebraic manipulation of rate equations written in terms of total concentrations of reactants and pH-dependent rate constants, concentrations of specific reactive forms and pH-independent rate constants, conservation equations relating the sum of the concentrations of the various forms of a given reactant to its total concentration, and equilibrium expressions for the various acid dissociation constants (see, for example, ref. 10 for this type of treatment of a more complex model).

For models in which there is multiplicity of reactive protonic states, each of which is considered to provide products in a single-step reaction (44) (*Figure 4*), rate equations for reactions in any number of protonic states may be written down or generated within the computer without the need for extensive algebraic manipulation.

This is achieved by using a simple general expression, *Equation 23*, and the two information matrices described in refs 45 and 46.

$$k = \sum_{i=1}^{n} \frac{A_i}{\left(1 + \sum_{j=1}^{n} B_{ij}\right)}$$

Equation 23

Equation 23 is a general expression that relates an experimentally determined pH-dependent rate constant k, and macroscopic acid dissociation constants K (see below), where $n =$ the number of reactive protonic states.

The numerator, $A_i(\equiv \tilde{k}_{XH_{i-1}})$ is a generalized pH-independent rate constant, where i (which may take values from 1 to n) specifies the particular protonic state such that for the X-state, $i = 1$ and thus $\tilde{k}_{X_0} \equiv \tilde{k}_X$.

Figure 4. Kinetic model for a reaction involving four reactive protonic states. The reactive protonic states of a generalized reactant state are designated X–XH$_3$ to denote relative stoichiometries in protons, and XH$_4$ is unreactive. Relative ionic charges are not shown. The model is characterized by four macroscopic acid dissociation constants K_{XH_4}–K_{XH} and by four pH-independent rate constants, $\tilde{k}_{XH_3}$–$\tilde{k}_X$. The pH-dependent rate equation corresponding to this model is *Equation 20* (see text).

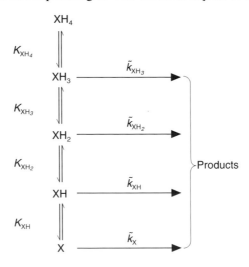

The denominator:

$$1 + \sum_{j=1}^{n} B_{ij}$$

provides for the pH-dependent variation in each of the contributions to k from reaction in a particular protonic state, i.e. reaction associated with an individual pH-independent rate constant, $\tilde{k}$ (it provides the (extended) Michaelis pH function associated with reactivity in a given protonic state).

Each term generated by j in the denominator comprises a product of $[H^+]^m$ where m is an integer with $-n \leq m < +n$, and either an acid dissociation constant, K, or a product of two or more such constants.

The construction of specific kinetic equations for particular kinetic models from *Equation 23* involves determining expressions for B_{ij} by using two information matrices. One of these (matrix 1) provides values of m (i.e. powers of $[H^+]$) and the other (matrix 2) the associated assemblies of macroscopic acid dissociation constants, each number p in a matrix element representing K_{XH_p}, with a minus sign indicating a reciprocal K. For example, for $i=3$ and $j=1$, matrix 1 specifies a term in $1/[H^+]^2$ and matrix 2 a term containing $K_{XH_2}K_{XH}$. Thus this particular term in the denominator of the expression specified by j and associated with $\tilde{k}_{XH_2}$ is $K_{XH_2}K_{XH}/[H^+]^2$. Similarly for $i=2$ and $j=3$, matrix 1 specifies a term containing $[H^+]^2$ and matrix 2 a term containing $1/K_{XH_3}K_{XH_2}$. For a reaction in four reactive protonic states, therefore, (i.e. the model shown in *Figure 4*) the right-hand side of *Equation 23* is written as the sum of four terms, each containing a numerator $\tilde{k}_{XH_3}-\tilde{k}_X$, and a denominator constructed by using matrices 1 and 2 to produce *Equation 20*.

For multistep intermediate complex models and those with multiple substrates, rate equations are most reliably written down by using the method of Cha (47) (see also refs 48–50 for a discussion of the value of this method in avoiding errors in the derivation of complex rate equations).

Cha's method is a modification of the full King–Altman schematic method of analysis (51) in which simplification is achieved by treating some of the steps of the model as quasi-equilibria.

It is particularly useful in the analysis of pH-dependence (10, 11) because rapid-equilibrium assumptions may be justified by the exceptionally high rates of proton-transfer steps in buffered aqueous media.

Computer implementations of Cha's method are available (52–54).

10. PRACTICAL PROCEDURES FOR CHARACTERIZATION OF THE pH-DEPENDENCE OF KINETIC PARAMETERS

These are exemplified below in terms of a pH-dependent rate constant for a single-step reaction, k, but identical procedures apply to the characterization of the pH dependences of k_{cat} and k_{cat}/K_m.

(i) Check the stability of both enzyme and substrate or time-dependent inhibitor at the extremes of the pH range of the proposed study by pre-incubation experiments (6). Procedures are available for kinetic analysis of reactions involving unstable inhibitors (see ref. 55 and references therein).

(ii) Select an appropriate set of constant ionic strength buffers for the pH range to be used (for example, see refs 56, 57 and other articles cited in ref. 6) and overlap different buffers in particular pH ranges to check for specific buffer effects, including pH-dependent chelation of metals (58, 59).

(iii) Allow for effects of temperature (6) and organic cosolvents (6, 60) on the pK_a values of buffer components.

(iv) The pK_a values of uncharged buffer acids (e.g. $-CO_2H$) are sensitive to changes in both organic solvent content and ionic strength because charges exist on only one side of the equilibrium ($-CO_2^-$ and H^+), whereas those of cationic acids are little affected (e.g. $-N^+H_3 \rightleftharpoons -NH_2 + H^+$).

(v) Note that this distinction does not pertain when solvent environments of ionizing groups in proteins are considered, because only the ionic species of the protein (e.g. $-N^+H_3$) is specifically solvated by the microenvironment of the protein; the charged species on the other side of the equilibrium, H^+, equilibrates with bulk solvent.

(vi) Measure the pH of the reaction mixtures rather than that of the added buffer.

(vii) Plot and inspect the form of the raw data (e.g. k versus pH) and of the logarithmic form (log k versus pH). The latter (61, 62) is particularly useful for identification of multiple ionizations and/or multiple reactive protonic states (*Table 2*).

(viii) Estimate the number of reactive protonic states and the number of kinetically influential ionizations and thus select an appropriate kinetic model; derive the pH-dependent rate equation for the chosen model.

(ix) Determine values of the characterizing parameters (pK_a values and pH-independent rate constants, $\tilde{k}$) associated with each protonic state (see *Table 3* for a list of methods that can be used to progress from provisional estimates to best-fit values with associated standard errors).

(x) Display the pH–k data superimposed on a fitted line computed from the best estimates of the parameters for the pH-dependent rate equation derived for the chosen model.

(xi) If values of k do not conform to the theoretical line, extend or change the chosen model and associated rate equation until acceptable correspondence of data and theoretical line is achieved in all pH regions.

11. AIDS TO THE ASSIGNMENT OF pK_a VALUES

(i) Note the typical pK_a values for ionizing side-chains in proteins in the absence of marked environmental perturbation: Asp/Glu, 4; His, 6; Cys, 8.5; Lys, 10; Tyr, 10.

(ii) Note the perturbing effects of various microenvironments: hydrophobic environments raise the pK_a of an uncharged acid (e.g. $-CO_2H$) and lower the pK_a of a cationic acid (e.g. $-Im^+H$); effects of proximity of charged groups to each other and/or effects of hydrogen bonding; an example of a large shift occurs in the cysteine proteinase papain where the formation of the ion-pair state containing (Cys25)-S$^-$/(His159)-Im$^+$H has pK_a 3.3–3.4 (2) rather than a value around 8.5 typical of cysteinyl thiol groups; other examples of pK_a perturbation include the catalytic site Lys of acetoacetate decarboxylase (15) and a carboxyl group in lysozyme (66).

(iii) Note that although study of the thermodynamics of ionization by determination of apparent pK_a values at various temperatures may sometimes provide confirmatory evidence of assignment (6), the problem of large compensating changes in ΔH_o and ΔS_o detracts from the value of this technique (8).

(iv) Consider using the experimentally laborious but information-rich competitive labeling method of Kaplan *et al.* (67) developed for the simultaneous modification of all accessible amino groups by acetic anhydride but with more general applicability (68).

(v) Consider using difference potentiometric titration between a protein with one group blocked to prevent protonic dissociation and the unmodified protein, or between enzyme–substrate or enzyme–inhibitor complex and native enzyme (66, 69, 70).

▶ p. 190

Table 2. Illustration of the practical value of plots of $\log k$ versus pH in identification of multiple ionizations and/or multiple reactive protonic states by consideration of the characteristics of the simple model $XH \rightleftharpoons X^- + H^+$ (acid dissociation constant K_a) with only X^- reactive (rate constant $\tilde{k}$) and deviations from it (k, a rate constant for an individual reaction or component step may be replaced by k_{cat} or by k_{cat}/K_m)

pH range	Rate equation	Equation	Behavior
From pH $<<$ pK_a to pH $>>$ pK_a	$k = \tilde{k}/(1+[H^+]/K_a)$ $\log k = \log \tilde{k}\text{-}\log(1+[H^+]/K_a)$	24	k versus pH is sigmoidal, k increasing with increase in pH; $\log k$ versus pH is linear with slope $+1$ at low pH where pH $<<$ pK_a, becomes curvilinear where pH $\simeq$ pK_a and approaches a line of slope zero where pH $>>$ pK_a (see below). If XH rather than X^- is the reactive protonic state k versus pH is sigmoidal with k increasing with decrease in pH, and the linear portion of the log plot has slope -1
pH $<<$ pK_a	$\log k = \log \tilde{k} + \text{pH} - \text{p}K_a$	25	$\log k$ versus pH is linear with slope $+1$; if both XH and X^- are reactive (XH of lower reactivity than X^-) the data deviate from the line of slope $+1$ at low pH and tend to another line of slope zero. The presence of an additional kinetically influential ionization with a pK_a value in the same pH region results in a line of slope greater than $+1$, approaching a line of slope $+2$
pH $>>$ pK_a	$\log k = \log \tilde{k}$	26	$\log k$ versus pH is a line of zero slope
pH $=$ pK_a	*Equation 24* becomes *Equation 25*		Extrapolation of the line of slope $+1$ intersects the line of slope zero when pH $=$ pK_a; this construction provides a provisional estimate of the pK_a value
	Equation 23 becomes *Equation 26* $\log k = \log \tilde{k} - \log 2$	27	The plotted value of $\log k$ falls below the value of $\log k$ at the intersection point of the lines of slope $+1$ and zero by 0.3 (i.e. $\log 2$). This behavior is characteristic of a single kinetically influential ionization

Table 3. Methods of determination of values of pK_a and $\tilde{k}$ from pH–k data (k, a rate constant for an individual reaction or component step, may be replaced by k_{cat} or by k_{cat}/K_m)

Method	Basis	Comment
(i) Inspection of a plot of k versus pH (sigmoidal curve, including each individual limb of a bell-shaped curve when the width at $\tilde{k}/2 \geq 3.5$)	Estimate $\tilde{k}$ by eye and pK_a as pH when $k = \tilde{k}/2$	Permits provisional estimates when values of k at pH $\gg pK_a$ and k at pH $< pK_a$ are available for an isolated sigmoidal component
(ii) Inspection of a plot of $\log k$ versus pH	Intersection of point of extrapolated lines of zero and unit slope (+1 or −1) is where pH = pK_a; line of zero slope has $k = \tilde{k}$ (see *Table 2*).	Permits better provisional estimates for individual sigmoidal components; the criterion that the intersection must be 0.3 above the data on the log k axis (*Equation 27, Table 2*) facilitates location of the line of unit slope
(iii) Inspection of a plot of k versus pH (bell-shaped curve)	Read the pH values, pH_1 and pH_2, that correspond to the 50% of the interpolated maximal measured value of k (not $\tilde{k}/2$) and their mean value, pK^*, and calculate $2 \log q$ as $pH_2 - pH_1$; calculate pK_I and pK_{II} as $pK^* \pm \log(q - 4 + 1/q)$ (see ref. 21)	Permits estimation of pK_I and pK_{II} even when $pK_{II} - pK_I < 3.5$, the condition that prevents the constituent limbs of the pH–k profile being treated independently as sigmoidal curves
(iv) Analysis of linear transforms of equations for sigmoidal and bell-shaped curves	Sigmoidal curves The reciprocal forms of *Equations 16* and 17 are *Equations 28* and 29, respectively: $1/k = 1/\tilde{k} + [H^+]/\tilde{k}K_a$ [28] $1/k = 1/\tilde{k} + (K_a/\tilde{k}).1/[H^+]$ [29] Bell-shaped curves *Equations 30* and 31 arise from the work of Alberty and Massey (63) and Friedenwald and Maengwyn-Davies (64) (ref. 65): $[H+]_{max} = \sqrt{K_I K_{II}}$ [30]	For sigmoidal pH–k curves, regression of $1/k$ on either $[H^+]$ (when X^- is reactive, *Equation 14*) or $1/[H^+]$ (when XH is reactive, *Equation 17*) provides values of $1/\tilde{k}$ and K_a For a bell-shaped curve, regression of $1/k$ on $([H^+] + [H^+]^2_{max}/[H^+])$ provides $1/\tilde{k}$ as ordinate intercept and $-K_I$ as intercept on the abscissa; K_{II} is calculated from *Equation 30*

Table 3. Continued

Method	Basis	Comment
	$$\frac{1}{k} = \frac{1}{\tilde{k}} + \frac{1}{\tilde{k}K_I} \left(\frac{[H^+]^2_{\text{Jmax}}}{[H^+]} + [H^+] \right) \qquad [31]$$ where $[H^+]_{\text{Jmax}}$ is $[H^+]$ at which k is maximal	
(v) Curve sketching and weighted nonlinear regression	Rate equations for reactions in any number of protonic states are generated within the computer by using the general expression, *Equation 23* (see Section 9); estimation of values of $\tilde{k}$ and pK_a for each protonic state in the generic set of equations for various models is by interactive manipulation of calculated curves or by weighted nonlinear regression	A computer program that has been used for this approach is the multitasking application program SKETCHER (45, 46) written in ANSI C by S.M. Brocklehurst (Cambridge Centre for Molecular Recognition, Department of Biochemistry, University of Cambridge, UK) running under RISCOS on an ACORN Archimedes microcomputer. Values of the parameters thus obtained are generally close to the final mean values obtained by weighted nonlinear regression performed, for example, by using the AR program from the statistical software package BMDP. For complex pH–k profiles, determination of one or more pK_a values from independent experiments helps to avoid inappropriate local minima in the nonlinear regression

(vi) Consider using direct observation of an ionizing group by using spectroscopic methods, for example electronic absorption spectroscopy for tyrosyl phenolic groups (71, 72) and cysteinyl thiol groups (73), infra-red (IR) spectroscopy for Asp and Glu carboxy groups in ^2H_2O (74), and ^1H and ^{13}C NMR spectroscopy, particularly for His imidazole groups (75–77) and Tyr phenolic groups (78).

(vii) Note that there is a serious problem with attempts to assign exact intrinsic pK_a values to groups that interact from one single experiment (23). When the degree of ligation of one group greatly affects the dissociation constant of the other, then an assumption that the property being investigated (e.g. spectroscopic signal) is completely unaffected by ligation of the second group is rendered implausible. Good approximations in pK_a assignments, however, can sometimes be made by consideration of the pH dependence of different properties and comparing the nature of the properties that provide a common result (9, 23).

12. CONCLUDING COMMENT

Perhaps the most important points to emphasize in connection with the use of pH-dependent kinetics in the study of mechanism are:

(i) to arrange to study single-step events whenever possible; and

(ii) to investigate pH-dependent characteristics by studying a multiplicity of phenomena, a point made in Section 11 in connection with attempts to assign pK_a values.

An illustration of the combined use of several experimental approaches (kinetic studies with two types of time-dependent inhibitor differing in susceptibility to activation by protonation or hydrogen bonding, kinetic studies with substrates where additional hydrogen bonding interactions are predicted to be influential in catalytic site chemistry, and potentiometric difference titrations) is provided by some recent work on cysteine proteinases (1, 2, 9, 44, 79).

ACKNOWLEDGMENTS

I thank Hasu Patel for searching the literature, Joy Smith for rapid production of the typescript and SERC, MRC and AFRC for project grants and Earmarked Studentships.

B. THE EFFECT OF TEMPERATURE ON ENZYME-CATALYSED REACTIONS

A. Walmsley

In general, an increase in temperature causes a corresponding enhancement in the rate of a reaction, whether this be purely chemical or enzyme-catalysed. The temperature dependence of a reaction can provide important information on its thermodynamics, and in situations where direct measurement by microcalorimetry may not be possible. However, care must be taken when interpreting the temperature dependence of enzyme-catalysed reactions because increasing temperature eventually leads to denaturation of the enzyme.

Consider the simple enzyme-catalysed reaction below:

$$S + E \underset{k_{off}}{\overset{k_{on}}{\rightleftharpoons}} ES \xrightarrow{k_{cat}} E + P$$

Substrate, product and enzyme are denoted S, P, and E, respectively. The substrate association and dissociation rate constants are denoted k_{on} and k_{off}, respectively. The rate constant for product formation, k_{cat} is rate-limiting for the reaction. If k_{on} and k_{off} are much faster than k_{cat}, then the binding process rapidly reaches equilibrium prior to product formation. This reaction follows Michaelis–Menten kinetics, as defined by the following equation:

$$v = \frac{[S][E]k_{cat}}{[S] + K_m}$$

Equation 32

where v is the initial rate of product formation and K_m is the half-saturation constant, which in this case is a true measure of the dissociation constant for the substrate ($K_d = k_{off}/k_{on}$). The initial rate of the reaction increases in a hyperbolic manner to a maximal level set by $[E]k_{cat}$. For this simple mechanism k_{cat} and K_m can be defined in terms of single identifiable processes. However, this is not always the case for more complex mechanisms (see Chapter 3, Part B).

This discussion considers how temperature affects rate and equilibrium constants, such as k_{cat} and K_m respectively, and illustrates the thermodynamic information that can be obtained. However, such an analysis can also be applied to more complex enzyme-catalysed reactions when the rate or equilibrium constants for individual processes can be determined by transient kinetics.

13. VARIATION OF EQUILIBRIUM CONSTANTS WITH TEMPERATURE

The equilibrium constant for a reaction increases with temperature according to the van't Hoff equation:

$$\frac{d \ln K}{d T} = \frac{\Delta H^\circ}{RT^2}$$

Equation 33

where K is the equilibrium constant, T is the absolute temperature (Kelvin), ΔH° is the standard enthalpy change for the reaction (kJ) and R is the gas constant. Thus, the equilibrium constant increases exponentially with the temperature. The integration of *Equation 33* gives the following equation:

$$\ln K = \ln A - (\Delta H^\circ / RT)$$

Equation 34

where $\ln A$ is a constant of integration. The enthalpy change (ΔH°) can be obtained conveniently from the slope of a linear plot of $\ln K$ versus $1000/T$: slope $= -\Delta H^\circ$.

Knowledge of the enthalpy change allows further thermodynamic information to be obtained, by the application of standard relationships. The Gibbs free energy (ΔG°) of a reaction (kJ mol^{-1}), which is a measure of how likely a reaction is to occur, is given by the following equations:

$$\Delta G^\circ = -RT \ln K = \Delta H^\circ - T\Delta S^\circ$$

Equation 35

where ΔS° is the standard entropy change (kJ mol^{-1}). Essentially, ΔH° is the heat required, or generated, in the course of a reaction. When heat is generated (ΔH° is $-$ve), the process is termed exothermic; and when heat is required to drive a reaction (ΔH° is $+$ve), it is termed endothermic. The simplest definition of the entropy (ΔS°) of a reaction is the change in disorder of the system. When ΔG° is negative, the reaction will have a tendency to occur spontaneously. Thus, a reaction is likely to be spontaneous if it is exothermic and involves an increase in the disorder of the system. For example, the binding of the sugar, D-glucose, to the protein responsible for the transport of D-glucose across the erythrocyte membrane, involves little change in the enthalpy but a large entropy change. The reaction is entropically driven. This behavior was postulated to be due to the exchange of hydrogen-bonds between the sugar and water for those with the transporter, leading to little change in the enthalpy. However, the release of ordered water from around the sugar and from the sugar-binding site of the protein leads to an increase in disorder that is not compensated for by the condensation of the sugar and protein. A similar situation is thought to prevail for the binding of sugars to hexokinase. Indeed, many biological interactions appear to be entropically driven, and for endothermic reactions the unfavorable enthalpy term (ΔH° +ve) is balanced by a numerically larger favorable entropy term.

14. VARIATION OF RATE CONSTANTS WITH TEMPERATURE

The rate constant for a reaction increases with temperature according to the Arrhenius equation:

$$\frac{d \ln k}{d T} = \frac{E_a}{RT^2} \qquad\qquad \text{Equation 36}$$

where k is the rate constant (sec^{-1}), T is the absolute temperature (Kelvin), E_a is the activation energy for the reaction (kJ) and R is the gas constant. Thus, the rate constant increases exponentially with the temperature. The integration of *Equation 35* gives:

$$\ln k = \ln A - (E_a/RT) \qquad\qquad \text{Equation 37}$$

where $\ln A$ is a constant of integration.

Generally, temperature dependency data are presented as an Arrhenius plot of $\ln k$ versus $1000/T$ and the activation energy (E_a) is determined from the slope of the plot (slope = $-E_a/R$). This provides a convenient means of evaluating the E_a but does not provide any statistical information, such as the standard deviation of the E_a. To obtain this information the data must be fitted to the exact equation defining the temperature dependence of the rate constant, which is obtained by taking exponentials of *Equation 37*:

$$k = A \exp(-E_a/RT). \qquad\qquad \text{Equation 38}$$

The temperature dependence of k can be fitted directly to this equation using nonlinear regression programs, such as SIGMAPLOT from Jandel Scientific. Alternatively, *Equation 37* can be rearranged as follows:

$$k = k_{(273K)} \cdot \exp[(-E_a/R) \cdot (1/273 - 1/T)] \qquad\qquad \text{Equation 39}$$

where $k_{(273K)}$ is the value of the rate constant at 273 K (0°C). Thus, the temperature dependence of rate constant k can be fitted to *Equation 39*, yielding the value for k at 273 K and the activation energy E_a.

The determination of the activation energy of a particular process can provide important thermodynamic information on the transition state, as discussed in the next section.

15. THE TRANSITION STATE

The transition state for a chemical reaction is the most unstable species on the reaction pathway. As a reaction proceeds along a notional reaction coordinate, the energy of the continuum of states will increase from that of the ground state (the unreacted substrates) through a maximum (the transition state or activated complex) and thereafter decrease to the new ground state (the products) (*Figure 5*). Generally, the transition state is considered to be a species in which chemical bonds or interactions are in the process of being formed or broken. The transition state is clearly distinguished from an intermediate of the reaction, which is characterized by a metastable minimum on the reaction profile (*Figure 5*). The activation energy represents the difference in the energies of the ground and transition states.

Figure 5. Diagrammatic representation of the change in energy of different molecular states for a hypothetical chemical reaction as the reaction proceeds.

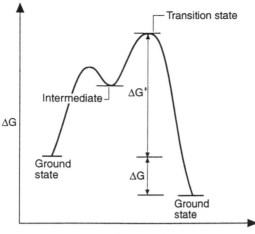

The formation of the transition state for a bimolecular reaction can be considered in terms of the following scheme:

$$A + S \rightleftharpoons AX^* \longrightarrow A + P$$

where AX* is the transition state. The concentration of the transition state is governed by the laws of thermodynamics, so that $[AX^*] = K^*[A][S]$. The thermodynamics of the formation of the transition state is given by the following standard relationships:

$$\Delta G^* = -RT \ln K^* = \Delta H^* - T\Delta S^* \qquad \text{Equation 40}$$

where ΔG^*, ΔH^* and ΔS^* are the free energy, enthalpy and entropy of formation of the transition state. The concentration of AX* is therefore given by the following relationship:

$$[AX^*] = [A][S]\exp(-\Delta G^*/RT) \qquad \text{Equation 41}$$

The rate of decomposition of AX* can be determined by the application of quantum mechanics, which states that the frequency of decomposition is the same as the vibrational frequency of the bond that is breaking:

$$v = k_b T/h \qquad \text{Equation 42}$$

where k_b is the Boltzmann constant and h the Planck constant. Using this relationship, v can be calculated as 6.25×10^{12} sec^{-1} at 300 K. The rate of decomposition of AX* is given by the following relationship:

$$-d[A][S]/dt = v[AX^*] \qquad \text{Equation 43}$$

so that the second-order rate constant for the complete reaction is given by the following relationship:

$$k = (k_b T/h)\exp(-\Delta G^*/RT) = (k_b T/h)\exp(\Delta S^*/R)\exp(\Delta H^*/RT). \qquad \text{Equation 44}$$

Taking logarithms of this function:

$$\ln k = \ln(k_b T/h) + (\Delta S^*/R) - (\Delta H^*/RT) \qquad \text{Equation 45}$$

PHYSICAL FACTORS AFFECTING ACTIVITY

and differentiating with respect to the temperature, gives the following relationship:

$$d \ln k / dT = (\Delta H^* + RT) RT^2 \qquad \text{Equation 46}$$

If this equation is compared with the Arrhenius equation (*Equation 36*), it is apparent that the activation energy is defined as follows:

$$E_a = \Delta H^* + RT. \qquad \text{Equation 47}$$

Hence, a knowledge of the activation energy E_a can be used to calculate the enthalpy, entropy and free energy of formation of the transition state, using the following relationships:

$$\Delta H^* = E_a - RT, \; \Delta S^* = (R/T).\ln(k/k_b h) + \Delta H^* \text{ and } \Delta G^* = \Delta H^* - T\Delta S^*.$$
$$\text{Equation 48}$$

These thermodynamic parameters provide valuable information on the nature of the transition state and of the reaction mechanism. For example, a large enthalpy of activation often implies that the substrate is put under considerable conformational strain in forming the transition state. This enthalpy change may be accompanied by a large decrease in entropy, indicating that the binding of the substrate to the enzyme requires the substrate to adopt a precise conformation. On the other hand, there might be little change in the enthalpy but a large increase in entropy, due to the substrate swapping H-bonds with the aqueous solvent and the enzyme and releasing ordered water molecules from around the substrate and/or displacing them from the binding site.

16. INTERPRETATION OF ARRHENIUS PLOTS

The analysis previously described is only applicable when the temperature dependence of a single rate constant is being studied and the Arrhenius plot is apparently linear. However, Arrhenius plots of the temperature dependence of V_{max} for enzymatic reactions are frequently found to be nonlinear (see *Figure 6*). There are several potential reasons for this behavior:

(i) The enzyme undergoes irreversible heat inactivation at higher temperatures, leading to a reduction in V_{max}.
(ii) A phase change in the environment of the enzyme. For example, many membrane transporters undergo a change in the apparent activation energy at a transition point in the state of the lipid membrane, as it changes from a fluid to a crystalline state at lower temperatures.
(iii) The enzyme can exist in two different conformational forms that are both active but have different activation energies.
(iv) The enzyme is reversibly inactivated, so that the enzyme exists as an equilibrium mixture of active and inactive molecules.
(v) The enzymatic reaction involves two or more processes, so that the V_{max} is governed by two rate constants, which have different activation energies.

The data obtained in the form of an Arrhenius plot are often fitted to two straight lines above and below the apparent transition temperature, but such an analysis is often unjustified. To justify such an analysis would require the data to be highly precise and usually, when the error of the individual measurements is taken into account, the data are better fitted to a continuous curve. In the author's opinion there are few conceivable circumstances that would lead to a gross change in the enzyme conformation at a precise transition temperature.

Clearly, the analysis of the temperature dependence of an enzyme reaction involves a number of difficulties and great care must be exercised in the interpretation of such data. However, there are circumstances when the nonlinearity of an Arrhenius plot can be exploited. For example, if

Figure 6. The temperature dependences of V_{max} for influx of D-$[^{14}C]$glucose into human red cells under zero trans ($\bullet$, $\circ$) and equilibrium exchange ($\blacksquare$, $\square$) conditions. These data were used in a nonlinear regression analysis of the glucose translocation cycle, as described in ref. 81, generating the rate constants for reorientation of the transporter with and without sugar.

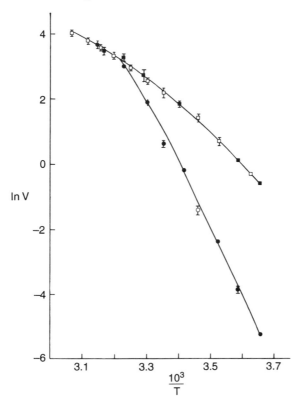

the mechanism for a reaction is known, and the V_{max} can be derived in terms of the individual rate constants, then information on the pre-steady-state kinetics of the reaction can be obtained from the temperature dependence of the steady-state kinetics. Simply, the individual rate constants in the V_{max} term are substituted by the Arrhenius equation to give an equation defining the temperature dependence of V_{max}. The V_{max} measurements, made at a series of temperatures, are then fitted to this equation to give the values for the individual rate constants, at 0°C, and their corresponding activation energies.

Consider the following reaction:

$$E + S \underset{}{\overset{K_d}{\rightleftharpoons}} \underset{\underset{P_1}{+}}{E\text{-}S} \overset{k_2}{\longrightarrow} E\text{-}I \overset{k_2}{\longrightarrow} E\text{-}P_2$$

where S is the substrate, P_1 and P_2 are sequentially released products and E-I is an intermediate formed prior to formation of the second product P_2. This type of reaction mechanism is found for the hydrolysis of esters and amides catalysed by chymotrypsin, as discussed in detail by

Fersht (80). The steady-state rate equation for this type of mechanism, with E-I reaching a steady state, can be derived by applying the steady-state treatment to E-I (see ref. 80):

$$v = [E_0][S].\frac{k_2k_3/(k_2+k_3)}{K_dk_3/(k_2+k_3)+[S]}$$

Equation 49

where E_o is the initial enzyme concentration. This is a Michaelis–Menten equation, where

$$K_m = K_d.\frac{k_3}{k_2+k_3}$$

Equation 50

and

$$k_{cat} = \frac{k_2k_3}{k_2+k_3} \text{ or } \frac{1}{k_{cat}} = \frac{1}{k_2}+\frac{1}{k_3}.$$

Equation 51

A temperature dependence equation can then be defined for *Equation 50*, by substituting each rate constant by the Arrhenius equation:

$$k_{cat}(\text{Temp}) = A.\exp(RT/E_2) + A.\exp(RT/E_3)$$

Equation 52

$$= k_2'\exp[(E_2/R).(1/T - 1/273)] + k_3'\exp[(E_3/R).(1/T - 1/273)]$$

where E_2 and E_3 are the activation energies that correspond to rate constants k_2 and k_3 respectively and k_2' and k_3' are the values for these rate constants at 0°C. Hence, values for k_2' and k_3' and their respective activation energies can be obtained from a nonlinear regression fit of the temperature dependence of k_{cat} ($k_{cat} = V_{max}/[E_0]$) to *Equation 52*. Indeed, since k_2 and k_3 are known at all temperatures over the studied range, the true K_d can be calculated and its temperature dependency determined, from *Equation 50*. This is a powerful technique for obtaining pre-steady-state information from steady-state measurements, but again great care must be exercised in its use. This procedure has been applied by the author in analysing the translocation cycle for the glucose transporter from human erythrocytes (81). However, it was also possible to measure some of the individual rate constants, by stopped-flow transient kinetic studies, at a series of temperatures, and to show that these yielded linear Arrhenius plots. Hence, the nonlinearity of the plots could not be attributed to denaturation of the transporter or membrane phase transitions.

REFERENCES

1. Mellor, G.W., Thomas, E.W., Topham, C.M. and Brocklehurst, K. (1993) *Biochem. J.*, **290**, 289.

2. Mellor, G.W., Patel, M., Thomas, E.W. and Brocklehurst, K. (1993) *Biochem. J.*, **294**, 201.

3. Bashford, D. and Karplus, M. (1990) *Biochemistry*, **29**, 10219.

4. Bashford, D. and Karplus, M. (1991) *J. Phys. Chem.*, **95**, 9556.

5. Dixon, H.B.F. (1992) *Essays Biochem.*, **27**, 161.

6. Tipton, K.F. and Dixon, H.B.F. (1979) *Methods Enzymol.*, **63**, 183.

7. Cleland, W.W. (1982) *Methods Enzymol.*, **87**, 390.

8. Knowles, J.R. (1976) *CRC Crit. Rev. Biochem.*, **4**, 165.

9. Brocklehurst, K. (1994) *Protein Engng*, **7**, 291.

10. Brocklehurst, K. and Dixon, H.B.F. (1976) *Biochem. J.*, **155**, 61.

11. Brocklehurst, K. and Dixon, H.B.F. (1977) *Biochem. J.*, **167**, 859.

12. Fink, A.L. (1987) in *Enzyme Mechanisms* (M.I. Page and A. Williams, eds). The Royal Society of Chemistry, London, p. 159.

13. Brocklehurst, K. (1987) in *Enzyme Mechanisms* (M.I. Page and A. Williams, eds). The Royal Society of Chemistry, London, p. 140.

14. Peller, L. and Alberty, R.A. (1959) *J. Am. Chem. Soc.*, **81**, 5907.

15. Schmidt, D.E. and Westheimer, F.H. (1971) *Biochemistry*, **10**, 1249.

16. Plaut, B. and Knowles, J.R. (1972) *Biochem. J.*, **129**, 311.

17. Brocklehurst, K. and Topham, C.M. (1993) *Biochem. J.*, **295**, 898.

18. Dixon, H.B.F., Brocklehurst, K. and Tipton, K.F. (1987) *Biochem. J.*, **248**, 573.

19. Dixon, H.B.F. (1973) *Biochem. J.,* **131**, 149.

20. Dixon, H.B.F. (1974) *Biochem. J.,* **137**, 443.

21. Dixon, H.B.F. (1979) *Biochem. J.,* **177**, 249.

22. Cornish-Bowden, A.J. (1976) *Biochem. J.,* **153**, 455.

23. Dixon, H.B.F. (1976) *Biochem. J.,* **153**, 627.

24. Adams, E.Q. (1916) *J. Am. Chem. Soc.,* **38**, 1503.

25. Edsall, J.T. and Wyman, J. (1958) *Biophysical Chemistry.* Academic Press, New York, Vol. 1, p. 477.

26. Michaelis, L. (1914) *Die Wasserstoffionenkonzentration.* Springer, Berlin, p. 30. (English translation (1926) Baillière, Tyndall and Cox, p. 56.)

27. Malthouse, J.P.G. and Brocklehurst, K. (1976) *Biochem. J.,* **159**, 221.

28. Brocklehurst, K. and Malthouse, J.P.G. (1980) *Biochem. J.,* **191**, 707.

29. Simms, H.S. (1926) *J. Am. Chem. Soc.,* **48**, 1239.

30. Von Muralt, A.L. (1930) *J. Am. Chem. Soc.,* **52**, 3518.

31. Dixon, H.B.F. (1975) *Biochem. J.,* **151**, 271.

32. Dixon, H.B.F. (1988) *Biochem. J.,* **253**, 911.

33. Dixon, H.B.F., Clarke, S.D., Smith, G.A. and Carne, T.K. (1991) *Biochem. J.,* **278**, 279.

34. Dixon, H.B.F. and Tipton, K.F. (1973) *Biochem. J.,* **133**, 837.

35. Kurtz, J.L. (1963) *J. Am. Chem. Soc.,* **85**, 987.

36. Kurtz, J.L. (1972) *Acc. Chem. Res.,* **5**, 1.

37. Thornton, E.K. and Thorton, E.R. (1978) in *Transition States of Biochemical Processes* (R.D. Gandour and R.S. Schowen, eds). Plenum Press, London, p. 3.

38. Parker, A.J. (1969) *Chem. Rev.,* **69**, 1.

39. Abraham, M.H. (1974) in *Progress in Physical Organic Chemistry* (A. Streitweiser and R.W. Taft, eds). Wiley Interscience, New York, Vol. 11, p. 1.

40. Critchlow, J.E. and Dunford, H.B. (1972) *J. Biol. Chem.,* **247**, 3714.

41. Dunford, H.B., Critchlow, J.E., Maguire, R.J. and Roman, R. (1974) *J. Theoret. Biol.,* **48**, 283.

42. Jencks, W.P. (1959) *J. Am. Chem. Soc.,* **81**, 475.

43. Jencks, W.P. (1969) *Catalysis in Chemistry and Enzymology.* McGraw-Hill, New York, p. 471.

44. Brocklehurst, K., Brocklehurst, S.M., Kowlessur, D., O'Driscoll, M., Patel, G., Salih, E., Templeton, W., Thomas, E., Topham, C.M. and Willenbrock, F. (1988) *Biochem. J.,* **256**, 543.

45. Brocklehurst, S.M., Topham, C.M. and Brocklehurst, K. (1990) *Biochem. Soc. Trans.,* **18**, 598.

46. Topham, C.M., Salih, E., Frazao, C., Kowlessur, D., Overington, J.P., Thomas, M., Brocklehurst, S.M., Patel, M., Thomas, E.W. and Brocklehurst, K. (1991) *Biochem. J.,* **280**, 79.

47. Cha, S. (1968) *J. Biol. Chem.,* **243**, 820.

48. Topham, C.M. and Brocklehurst, K. (1992) *Biochem. J.,* **282**, 261.

49. Selwyn, M.J. (1993) *Biochem. J.,* **295**, 897.

50. Brocklehurst, K. and Topham, C.M. (1993) *Biochem. J.,* **295**, 898.

51. King, E.L. and Altman, C. (1956) *J. Phys. Chem.,* **60**, 1375.

52. Cornish-Bowden, A. (1977) *Biochem. J.,* **165**, 55.

53. Lam, C.F. (1981) *Techniques for the Analysis and Modelling of Enzyme Kinetic Mechanisms.* Research Studies Press, Chichester, p. 1.

54. Ishikawa, H., Maeda, T., Hikita, H. and Miyatake, K. (1988) *Biochem. J.,* **251**, 175.

55. Topham, C.M. (1992) *Biochem. J.,* **287**, 334.

56. Jencks, W.P. and Regenstein, J. (1976) in *Handbook of Biochemistry and Molecular Biology, Physical and Chemical Data (3rd edn)* (G.D. Fasman, ed.). Chemical Rubber Publishing, Cleveland, OH, Vol. 1, p. 305.

57. Ellis, K.J. and Morrison, J.F. (1982) *Methods Enzymol.,* **87C**, 405.

58. Good, N.E., Winget, G.D., Winter, W., Connolly, T.N., Izawa, S. and Singh, R.M.M. (1966) *Biochemistry,* **5**, 467.

59. Sillén, L.G. and Martell, A.E. (1971) *Chem. Soc. Spec. Publ.,* Vol. 25.

60. Findlay, D., Mathias, A.P. and Rabin, B.R. (1962) *Biochem. J.,* **85**, 139.

61. Dixon, M. (1953) *Biochem. J.,* **55**, 161.

62. Dixon, M. and Webb, E.C. (assisted by Thorne, C.J.R. and Tipton, K.F.) (1979) *Enzymes (3rd edn).* Longmans, London, p. 138.

63. Alberty, R.A. and Massey V. (1955) *Biochim. Biophys. Acta,* **13**, 347.

64. Friedenwald, J.S. and Maengwyn-Davies, G.D. (1954) in *The Mechanism of Enzyme Action* (W.D. McElroy and B. Glass, eds). Johns Hopkins University Press, Baltimore, MD, p. 191.

65. Wharton, C.W. and Eisenthal, R. (1981) *Molecular Enzymology.* Blackie & Son, London.

66. Parsons, S.M. and Raftery, M.A. (1972) *Biochemistry,* **11**, 1623.

67. Kaplan, H., Stevenson, K.J. and Hartley, B.S. (1971) *Biochem. J.,* **124**, 289.

68. Cruickshank, W.H. and Kaplan, H. (1975) *Biochem. J.,* **147**, 411.

69. Fersht, A.R. and Sperling, J. (1973) *J. Mol. Biol.*, **74**, 137.

70. Lewis, S.D., Johnson, F.A. and Shafer, J.A. (1976) *Biochemistry*, **15**, 5009.

71. Timasheff, S.N. (1970) in *The Enzymes (3rd edn)* (P. Boyer, ed.). Academic Press, New York, Vol. 2, p. 371.

72. Gorbunoff, M.J. (1971) *Biochemistry*, **10**, 250.

73. Polgár, L. (1974) *FEBS Lett.*, **47**, 15.

74. Timashef, S.N. and Rupley, J.A. (1972) *Arch. Biochem. Biophys.*, **150**, 318.

75. Hunkapiller, M.W., Smallcombe, S.H., Whitaker, D.H. and Richards, J.H. (1973) *Biochemistry*, **12**, 4732.

76. Robillard, G. and Schulman, R.G. (1972) *J. Mol. Biol.*, **71**, 507.

77. Markley, J.L. (1975) *Acc. Chem. Res.*, **8**, 70.

78. Karplus, S., Snyder, G.H. and Sykes, B.D. (1973) *Biochemistry*, **12**, 1323.

79. Thomas, M.P., Topham, C.M., Kowlessur, D., Mellor, G.W., Thomas, E.W., Whitford, D. and Brocklehurst, K. (1994) *Biochem. J.*, **300**, 805.

80. Fersht, A. (1985) *Enzyme Structure and Mechanisms*. W.H. Freeman, New York.

81. Walmsley, A.R. (1988) *Trends Biochem. Sci.*, **13**, 226.

CHAPTER 6
ANALYSIS OF LIGAND BINDING BY ENZYMES

A.R. Clarke

1. INTRODUCTION TO LIGAND BINDING

All biological events, at some point, require the precise and selective binding of one molecule to another. The specific nature of such interactions and their consequences is the basis of processes as diverse as an ordered and regulated metabolism, the immune response, mobility, selective gene expression and signaling within and between cells. An understanding of these inter-molecular interactions arises from a knowledge of how they are observed and quantified. This chapter seeks to explain the techniques commonly available to acquire this knowledge.

Two things ought to be said at the outset. First, only those techniques accessible to most biological scientists will be dealt with. Secondly, to some degree the definition of enzymology will be slightly stretched, since the procedures described are most often applied to proteins which bind ligands but do not catalyse a covalent chemical transformation. In the case of enzymatic proteins, these techniques are usually performed in the following circumstances:

(i) With allosteric ligands binding in a regulatory capacity to noncatalytic sites on the enzyme.

(ii) With unreactive analogs of the true substrate which mimic its structure and bind in the same way to the active site. Good candidates for such studies are usually effective competitive inhibitors of the enzyme in question. Ideally, the binding affinity of the analog should be similar to that of the true reactive substrate. Such analogs are invaluable not only in binding studies but also in crystallizing the enzyme as a complex to provide detailed information about the contacts between the two molecules leading to selectivity and catalytic efficiency.

(iii) With enzymes for which the reaction involves two or more substrates, omission of one substrate will prevent catalysis, so providing a complex on which conventional binding studies can be performed. NB: some enzymes bind their substrates in an ordered fashion – substrate A must bind before substrate B – in which case only the enzyme : A complex is accessible (e.g. many NAD-linked dehydrogenases). Other enzymes have little or no preference in the order of binding, so that the properties of both the E : A and E : B complexes can be examined (e.g. many phosphoryl transfer enzymes) (see Chapter 3B). In other cases, the enzyme may require a cofactor which is necessary for the catalytic reaction to proceed, but is not a prerequisite for forming the enzyme–substrate complex. Typical examples are restriction endonucleases which can form complexes with DNA but only cleave the substrate once Mg^{2+} is introduced. In these enzymes the interaction of enzyme and DNA can be followed without the 'problem' of product formation.

(iv) With enzymes that catalyse reactions where the chemical step is sufficiently slow that the comparatively fast binding process can be studied at leisure prior to product formation. When choosing suitable techniques, it is important that the time scale over which data are gathered is shorter than that of the on-enzyme chemical reaction. Hence in a conventional fluorimeter or spectrophotometer a measurement can be made in about 5 sec (with practice and good stirring) giving scope to study binding steps in catalytic processes where the chemical conversion occurs on longer time scales (i.e. with rate constants of $0.05 \ sec^{-1}$ or less). If site-directed mutants of an enzyme can be made in which catalytic activity is severely diminished or abolished but binding is preserved, then all the better. An alternative means of 'catching' the binding process before any covalent chemical reaction has occurred is to use rapid mixing methods which allow us to observe stages in substrate binding which occur in a few milliseconds. This method is illustrated in Sections 2.2 and 4.

2. THEORETICAL BACKGROUND TO LIGAND BINDING

2.1. Interactions at equilibrium

Dissociation and association constants
Definition: at its simplest, the binding of a ligand (L) to a protein molecule (E) can be described by the following scheme:

$$E + L = E : L$$

The affinity of this interaction or, to put it another way, the stability of the E:L complex, is an intrinsic property of the atomic contacts made between the partners and so has a constant value irrespective of their concentrations. This value is defined by a dissociation constant (K_d), where:

$$K_d = \frac{[E]_F \cdot [L]_F}{[E : L]} \text{ (units, molar)} \qquad \text{Equation 1}$$

or an association constant (K_a) where:

$$K_a = \frac{[E : L]}{[E]_F \cdot [L]_F} \text{ (units, molar}^{-1}\text{).} \qquad \text{Equation 2}$$

For the purposes of this discussion the former convention as the standard measure of binding affinity will be used. It cannot be sufficiently stressed that $[E]_F$ is the concentration of *free* binding sites on the protein (i.e. not involved in an interaction with its ligand), and $[L]_F$ is the concentration of *free* ligand (i.e. not bound to the protein). It is also important to stress that when a system is at equilibrium, the *net* concentrations of each of the three components do not change. Although E:L is constantly dissociating to form E and L, and E and L are constantly associating to form E:L, the rates of these two processes balance. By designing experiments which measure the concentrations of E, L, and E:L, it is possible to evaluate K_d.

Basic applications
There are several reasons for determining the dissociation constant which describes a protein–ligand interaction:

(i)   It can be used to assess the extent to which a protein will be occupied by a given ligand in a biological context, if the *in vivo* concentrations can be estimated. By extension it can be used to estimate the change in occupation as the ligand concentrations fluctuate in differing cellular conditions. Such information is invaluable in establishing whether a protein : ligand interaction is a good candidate for regulating, initiating or propagating a cellular process. For instance, if it is suspected that ATP binding to a given cytoplasmic protein regulates its activity but it is found that the K_d value is 10^{-6} M, whereas the free ATP concentration never falls to within a factor of 100 of this value, then the hypothesis should be questioned because the protein will always be saturated with the nucleotide (i.e. [E : L] is always much greater than [E]).

(ii) The K_d values for several candidate ligands binding to the same site on a protein allow one to assess the specificity of the site in a quantitative way. This is well illustrated by the progress made in understanding the sequence specificity of DNA-binding proteins.

(iii) Alterations in dissociation constants for a given ligand, depending on conditions (e.g. temperature, ionic strength, pH), give insights into the physical nature of an interaction. More importantly, changes in binding affinity for one ligand in the presence of another defined molecule are often the first clues to function and mechanism (e.g. changes in the DNA binding affinity of repressor proteins in the presence of an inducer (reduced affinity) or a co-repressor (increased affinity)).

(iv) Mutational analysis of a protein binding site requires a method of evaluating changes in affinity. This technique is central to identifying binding sites in a protein and characterizing the molecular interactions which stabilize a complex.

Binding energy

It is often convenient to represent the affinity of an interaction as a free energy (ΔG) of binding between the components (at standard molar concentrations). This is represented as:

$$\Delta G = RT \ln K_d. \hspace{4cm} \text{Equation 3}$$

The most useful aspect of this relationship is that it allows quantification of the effect of altering the ligand or the protein or the conditions as a change in binding energy ($\Delta\Delta G_{A-B}$, i.e the change in the free energy of interaction in going from situation A to situation B). This is expressed as:

$$\Delta\Delta G_{(A-B)} = RT . \ln(K_{d(B)}/K_{d(A)}). \hspace{3cm} \text{Equation 4}$$

Evaluation of binding affinity and the concentration of binding sites

The crucial values to be gained from binding experiments are the affinity of the sites and the concentration of such sites in the sample. The way in which data are used in relation to *Equation 1* entirely depends on the method used to collect them. In general these can be divided into two:

Scatchard analysis. This method depends on the ability to measure the concentration of free ligand $[L]_F$ and/or bound ligand $[L]_B$ when either the total concentration of ligand $[L]_T$ (i.e. [L] + [E:L]) is changed or (less often) the total concentration of binding sites $[E]_T$ (i.e [E] + [E:L]) is changed. Methods of this type usually exploit differences between the molecular properties of the ligand and the protein : ligand complex, particularly their size. In these experiments there is a solution of a protein (either pure or crude) to which ligand is added. The total concentration of ligand added, $[L]_T$, is known and can be represented as:

$$[L]_T = [L]_F + [L]_B. \hspace{4cm} \text{Equation 5}$$

The properties of the equilibrium are being determined from a knowledge of how much of the ligand is free ($[L]_F$) and how much is bound ($[L]_B$). The concentration of free ligand $[L]_F$ is measured as $[L]_T$ is varied. The relationship between these and the two valuable parameters of K_d and concentration of binding sites in the protein sample ($[E]_T$) can then be deduced. It is worth remembering that:

(i) the number of binding sites on a protein molecule is not known initially, so even if its molar concentration is known, it is impossible to know, with certainty, the concentration of sites; and

(ii) the protein may be a crude and impure preparation of a cell extract or even with binding to receptors on the cell surface in a suspension of cells. This is not problematic because *only* a knowledge of the total ligand concentration ($[L]_T$) and the free or bound ligand concentrations ($[L]_F$ or $[L]_B$) can be used to provide this information.

For these experiments *Equation 1* can be rewritten as:

$$K_d = \frac{[E]_F \cdot [L]_F}{[L]_B}$$

where $[E]_F$ is the concentration of unoccupied or free binding *sites* on the target protein molecule and $[L]_B$ represents the concentration of occupied binding sites or, equally, the concentration of bound ligand molecules. It is known that:

$[L]_B = [L]_T - [L]_F$ and $[E]_F = [E]_T - [L]_B$

so that

$$K_d = \frac{([E]_T - [L]_B) \cdot [L]_F}{[L]_B}$$

and, rearranging:

$$[L]_B = [E]_T - K_d \cdot \frac{[L]_B}{[L]_F}.$$ Equation 6

Thus the following plot of the form $y = c - m \cdot x$ can be used to determine the relevant quantities, as shown in *Figure 1*.

Measurements of the saturation of binding sites. These usually exploit a spectroscopic difference between the free protein and the protein–ligand complex or, less frequently, a difference between the properties of the free ligand and the bound ligand. In such experiments the approach is different from that taken in Scatchard analysis. For example, if the binding of a ligand induces a fluorescence change in a protein then in this case the complex is more fluorescent than the free

Figure 1. The Scatchard plot.

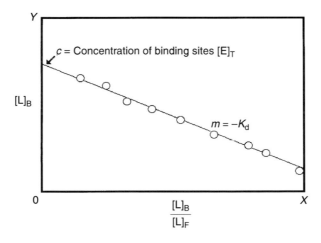

protein. Ligand is added to the solution and the fluorescence is recorded to give the data (shown in *Figure 2*).

In such experiments only the total amount of ligand added, $[L]_T$, and the degree to which the protein sites are saturated with ligand are known. The degree of saturation (α) is defined by:

$$\alpha = \frac{[E:L]}{[E] + [E:L]} = \frac{[E:L]}{[E]_T} = \frac{\Delta F}{\Delta F_M} \qquad \text{Equation 7}$$

given also that:

$$K_d = \frac{([E]_T - [E:L]) \cdot ([L]_T - [E:L])}{[E:L]} \text{ and } [E:L] = [E]_T \cdot \alpha.$$

The following expression can be derived:

$$\frac{1}{1-\alpha} = \frac{[L]_T}{\alpha} \cdot \frac{1}{K_d} - \frac{[E]_T}{K_d}. \qquad \text{Equation 8}$$

Thus the linear plot shown in *Figure 3* can be constructed from the data.

While this is a valid form of linear analysis, it has become more usual, and is statistically more accurate, to use nonlinear fitting methods in which the untransformed data are fitted directly to a nonlinear equation. In these cases $[L]_T$, the independent variable, is represented as a function of α, the dependent variable. For the above case from a quadratic solution the following equation can be derived.

$$\alpha = (([L]_T + [E]_T + K_d) - (([L]_T + [E]_T + K_d)^2 - 4 \cdot [L]_T \cdot [E]_T)^{0.5})/(2 \cdot [E]_T). \qquad \text{Equation 9}$$

This is the most important general relationship used in binding studies on proteins that bind ligands noncooperatively. It can be applied in any conditions and allows direct fitting to a set of

Figure 2. A binding curve measured by a change in the protein fluorescence intensity as the E:L complex is formed at equilibrium.

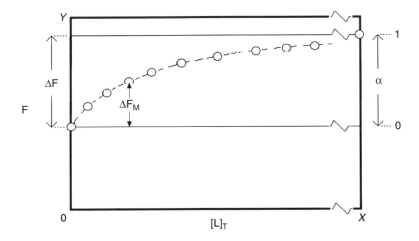

Figure 3. Linearized plot for determining sites concentration and dissociation constant from binding data at equilibrium.

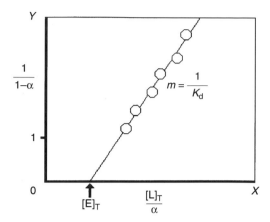

data to extract values for K_d and $[E]_T$ in standard nonlinear fitting programs. In some circumstances it is difficult to determine ΔF_M accurately from inspection of the data, making calculation of α awkward. In these circumstances ΔF_M can be fitted directly as a parameter in the equation:

$$\Delta F = \Delta F_M \cdot (([L]_T + [E]_T + K_d) - (([L]_T + [E]_T + K_d)^2 - 4 \cdot [L]_T \cdot [E]_T)^{0.5})/(2 \cdot [E]_T).$$

Equation 10

Analysis of spectroscopic data is further described in Section 3.2.

Conditions for defining K_d and the concentration of binding sites
In an experiment of the type described above where $[L]_B$ and $[L]_F$ are determined (p. 201), by definition, the amount of ligand associated with the active site must be a significant proportion (say > 5%) of the total ligand over the concentration range used. Equally, the concentration of free ligand $[L]_F$ must also be significant. These criteria set limits on the conditions in which such an experiment produces data which reliably determine K_d and $[E]_T$. In practice the best conditions for determining K_d and $[E]_T$ from a single data set are secured when $K_d - [E]_T$ and the total ligand concentration is varied from $0.2[E]_T$ to $5[E]_T$. However, the errors in determining parameters depend on both the method of measurement and on the ratio of K_d to $[E]_T$. *Figure 4* shows model examples of such experiments performed in different conditions with an assumed ±5% error in the measurements. Section 3.1 describes experimental methods for collecting Scatchard data. Some directly measure $[L]_F$ (so that $[L]_B$ must be deduced, as shown in *Figures 4a and b*) and others measure $[L]_B$ (so that $[L]_F$ must be deduced, as shown in *Figures 4c and d*). As can be seen from these examples, the type of measurement has a strong influence on the accuracy with which K_d and $[E]_T$ are determined in different conditions.

If a method is used where the saturation of protein sites is measured as a function of total ligand concentration, the criteria for accurate determination of K_d and [sites] are different. Here the K_d is best determined when $[E]_T$ is less than a tenth of K_d and the concentration of sites is best determined when $[E]_T$ is greater than K_d. Examples of each are given in *Figure 5*, assuming ±5% error in the measurement of α: curve A, $[E]_T = 100K_d$; curve B, $[E]_T = 10K_d$; curve C, $[E]_T = 0.1K_d$.

Figure 4. Conditions for collecting Scatchard binding data and their attendant errors.

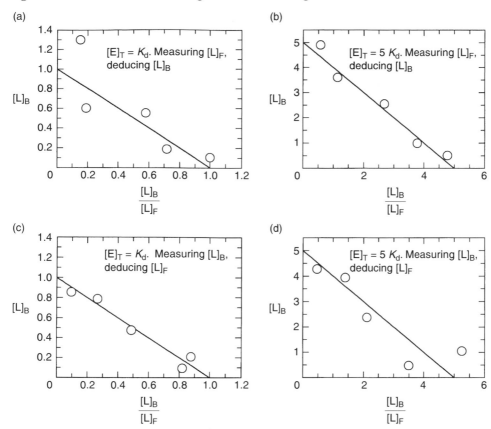

It should be noted as a crude guide to data analysis that the extrapolation of the initial part of the curve (where $[L]_T$ approaches zero) to the point of intersection of a line where α is equal to 1 gives an estimate of $K_d + [E]_T$. In general, a smooth hyperbolic curve gives a good estimation of K_d whereas an angular response, with a distinct 'corner' or break-point, gives a good estimation of $[E]_T$. These points are illustrated by the curves shown in *Figure 5*.

In the case where $[E]_T$ is much less than K_d (in curve C by a factor of 10), there is no information on the concentration of binding sites and the relevant analytic equation is analogous to the Michaelis–Menten hyperbola:

$$\alpha = \frac{[L]_T}{[L]_T + K_d} \qquad\qquad \text{Equation 11}$$

to give the linearization:

$$1\alpha = K_d \cdot 1/[L]_T + 1. \qquad\qquad \text{Equation 12}$$

The plot derived from *Equation 10* is shown in *Figure 6a*. The more usual form of this relationship is shown below (*Equation 13*) and can be used to analyse data of the type shown in *Figure 2* by nonlinear fitting, as long as $[E]_T$ is much less than K_d:

$$\Delta F = \frac{\Delta F_M \cdot [L]_T}{[L]_T + K_d}$$

Equation 13

to give the linearization:

$$1/\Delta F = K_d/\Delta F_M \cdot 1/[L]_T + 1/\Delta F_M.$$

Equation 14

This plot is shown in *Figure 6b*.

Figure 5. Binding data collected in tight binding ($[E]_T > K_d$) and weak binding ($K_d > [E]_T$) conditions.

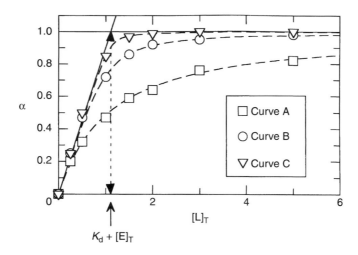

Cooperative binding

The preceding sections have dealt with simple binding processes in which all sites behave independently. In oligomeric proteins where there are several binding sites for the same ligand, the possibility of cooperative binding arises. If the binding of the first ligand affects the other sites to increase their affinity, binding is positively cooperative. If the affinity is reduced, it is negatively cooperative. The former behavior is frequently observed, whereas the latter is rare. The physical basis for positive cooperativity is most simply explained by the Monod–Wyman–Changeux (MWC) model. This proposes that the oligomeric protein exists in two, interconverting forms, the T (or 'tense') state, in which all sites bind the ligand weakly, and the R (or 'relaxed') state, in which all sites bind ligand tightly. The very fact that the ligand binds more tightly to the R state means that, as the protein begins to be saturated the equilibrium T = R moves to the right, providing more high-affinity sites. This means that, as saturation progresses, the binding affinity increases. This results in a sigmoidal shape to the binding curve (saturation versus ligand concentration) as shown in *Figure 7*. Note that this curve represents the change in an optical signal (e.g. fluorescence) as ligand is added. The signal changes from F_S (start) to F_E (end) as the protein is saturated; the degree of saturation (α) = $(F-F_S)/(F_E-F_S)$ (*Figure 7*).

Although it is possible to analyse data for cooperative binding by applying the MWC model, the analytical solution contains four constants (describing the binding affinity to the R state ($K_{d(R)}$)and the T state ($K_{d(T)}$), the T = R equilibrium in the absence of ligand ($K_{MWC}=[T]/[R]$)

Figure 6. Double reciprocal plots for the analysis of equilibrium binding data (where $K_d \gg [E]_T$).

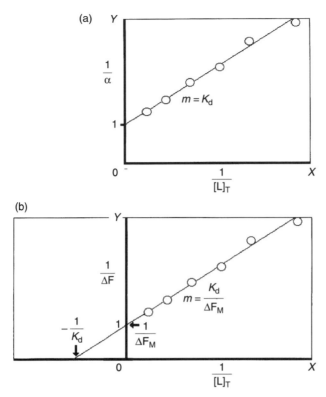

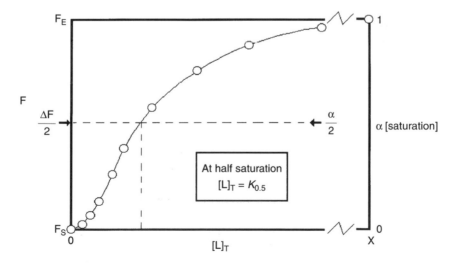

Figure 7. Equilibrium binding in a positively cooperative system.

Ligand Binding

and the number). In practice it is usually assumed that the ligand cannot appreciably bind to the T state (i.e. $K_{d(T)}$ is much larger than $K_{d(R)}$). This gives the simpler equation:

$$\alpha = [E:L]/[E]_T = \frac{A(1+A)^{n-1}}{K_{MWC} + (1+A)^n}$$

Equation 15

where, for convenience $A = [L]_T/K_{d(R)}$.

Note that this is defined in terms of the saturation of sites. A complication of this analysis lies in interpreting the signal which reports the binding process. If an optical method is used, it is difficult to say whether the signal change arises from the direct effects of binding at a single site or whether it is reporting the concerted structural change in the protein from the T state to the R state. If it is the latter, then the following equation is used:

$$\alpha = [R]/[E]_T = \frac{(1+A)^n}{K_{MWC} + (1+A)^n}$$

Equation 16

where [R] is the concentration of binding sites (subunits) in the R (relaxed) state. Both *Equations 15* and *16* apply only to conditions where the concentration of protein sites ($[E]_T$) is much smaller than $K_{0.5}$, so that $[L]_T \sim [L]_F$.

Such complexities are avoided by using the Hill analysis, which accounts for its greater popularity in the treatment of cooperative binding data. The Hill equation is not based on a realistic physical model but treats the data phenomenologically according to the following equilibrium:

$E + n.L = E:L_n$.

In such a system the multisite protein exists in only two states, unliganded or fully liganded. For this system the mass action equation is defined as:

$$K = \frac{[E]_F \cdot [L]_F{}^n}{[E:L_n]}$$

Equation 17

where n is defined as the Hill coefficient. In conditions where $[L]_F \sim [L]_T$ (i.e. where the sites concentration ($[E]_T$) is low by comparison to the binding affinity), saturation is described by:

$$\alpha = \frac{[L]_T{}^n}{[L]_T{}^n + K} \quad \text{or} \quad \Delta F = \frac{\Delta F_M \cdot [L]_T{}^n}{[L]_T{}^n + K}$$

Equation 18

where ΔF is the measured signal change (for example in fluorescence intensity) at a given ligand concentration and ΔF_M is the maximal change (at infinite ligand concentration). This equation can be used to fit data by nonlinear methods or can be linearized as:

$$\log(\alpha/(1-\alpha)) = \log(1/K) + n.\log[L] \quad (y = c + m.x).$$

Equation 19

This plot is represented in *Figure 8* and shows that data deviate from linearity at both high and low values of α. The Hill coefficient is defined by the slope of the plot when $\log(\alpha/1-\alpha)$ is 0 (or $\alpha = 0.5$). Extrapolations of the data at low and high levels of saturation define $K_{d(T)}$ and $K_{d(R)}$, respectively. The Hill coefficient is an index of the degree of cooperativity, and for a positively cooperative system has a value greater than unity with a maximum equal to the number of sites per oligomeric protein molecule. In the rarer case of negative cooperativity, the value is less than unity. In the Hill analysis two parameters are usually reported to define the binding properties of the protein, the value of n and the ligand concentration required to achieve half saturation (often abbreviated as $K_{0.5}$) (*Figure 8*).

In a Scatchard analysis of cooperative ligand binding, the data deviate from linearity, as shown in *Figure 9*. This type of plot gives different information from that derived by the Hill analysis (or the MWC model). For positive cooperativity it provides an estimate of the dissociation constant for the relaxed state ($K_{d(R)}$) but cannot, in practice, be used to determine the

Figure 8. The Hill plot.

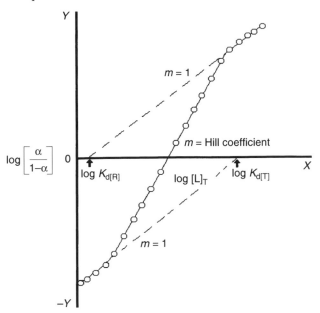

Figure 9. Scatchard plots for cooperative proteins.

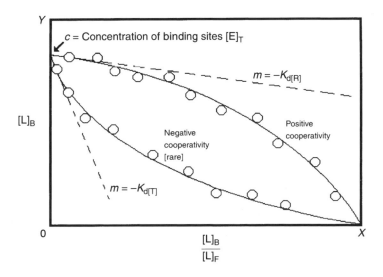

dissociation constant for the T state ($K_{d(T)}$) because of the low level of saturation required to define this parameter (i.e $[L]_B \rightarrow 0$). For negative cooperativity, the Scatchard plot produces an estimate of $K_{d(T)}$. For both it is useful in defining the number of ligand binding sites (*Figure 9*).

2.2. The kinetics of binding

In systems where there is a change in the optical properties of either the protein or the ligand upon combining in solution, equilibrium binding studies can be supplemented by measurements of the rates of steps in the process. In principle, such data can be collected by methods other than spectroscopy but, in practice, these reactions usually occur on the millisecond-to-seconds time scale and require rapid detection methods.

The simple collision/dissociation process
Take, for example, the most straightforward case of a protein–ligand interaction, where:

$$E + L \underset{k_{-1}}{\overset{k_1}{\rightleftharpoons}} E : L.$$

The rate of association is defined by k_1 which has units of M^{-1} sec^{-1} and the rate of dissociation is defined by k_{-1} which has units of sec^{-1}. The dissociation constant (K_d) for the interaction is given by the ratio k_{-1}/k_1.

If $E:L$ has a different spectroscopic signal from E, then its rate of formation can be measured. It should be stressed that it does not matter whether the formation of $E:L$ or the loss of E is measured in this reaction, the rate constant for the process is the same. In general, enough ligand has to be mixed with the protein to ensure complete conversion of E to $E:L$. The result of an experiment in which the protein is mixed with the ligand and the signal change is measured as a function of time takes the form shown in *Figure 10*.

Conditions should be used such that $[L]_T$ is at least five times greater than $[E]_T$ so that the free ligand concentration does not significantly change as the complex is produced. Such conditions are known as pseudo-first-order because the rate of the conversion depends only on the

Figure 10. Data from a single exponential reaction.

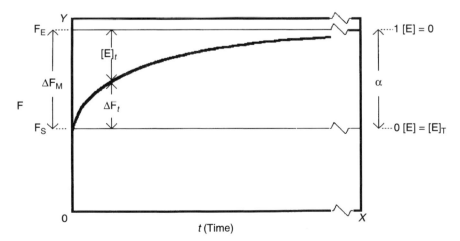

concentration of the component which is present in excess (i.e. the ligand). In such circumstances, the disappearance of E with time can be described by:

$$[E]_t = [E]_T \cdot e^{-k \cdot t}. \tag{Equation 20}$$

In this equation, $[E]_t$ is the concentration of E at any time t and k is the first-order rate constant for the process and is measured in units of sec^{-1}. Note that it is the decay in concentration of free sites with time (as they are filled with ligand) which is plotted in this analysis. Adapted forms of the above equation in which raw data can be used directly in nonlinear fitting routines are described in Section 4, *Equations 26* and *27*. Linearizing this equation for plotting gives:

$$\ln[E]_t = \ln[E]_T - k \cdot t. \tag{Equation 21}$$

This gives the plot shown in *Figure 11*.

Extracting the first-order rate constant (k) from these data, either by linearization or by nonlinear fitting to the raw data, is simplified by the fact that the units in which the signal is measured do not influence the result of a first-order analysis.

The rate constant (k) is only an observed constant (usually denoted k_{obs}) occurring at a given concentration of ligand $[L]_T$ because the formation of $E:L$ will always get quicker the higher the ligand concentration, i.e. the production of $E:L$ is collision controlled. The experiment is performed at a series of ligand concentrations and a plot of k_{obs} versus $[L]_T$ is shown in *Figure 12*.

The equation describing such a plot is:

$$k_{obs} = k_1 \cdot [L] + k_{-1}. \tag{Equation 22}$$

In principle, this allows the evaluation of the on-rate (k_1) and the off-rate (k_{-1}). In practice k_{-1} is often poorly determined in such plots and is best assessed by diluting the complex into a much larger volume of buffer. If the dilution takes the concentration of $[L]_T$ from twice the value of K_d to a tenth of it or less (i.e. a 20-fold dilution), then the observed rate of decay of the $E:L$ complex gives a first-order decay and provides a good approximation of k_{-1} according to:

$$[E:L]_t = [E:L]_T \cdot e^{-k_{-1} \cdot t}. \tag{Equation 23}$$

Figure 11. Linearization of single exponential data.

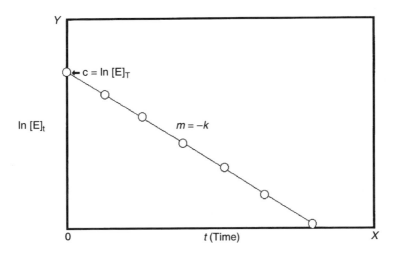

Figure 12. Relationship between observed rate constant and ligand concentration for a simple, one-step collision reaction.

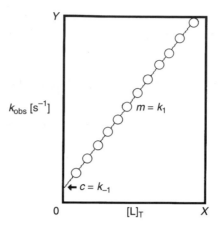

The analysis of the decay of the E:L complex is carried out exactly as described. This technique is discussed further in Section 4.

Structural rearrangements induced by ligand binding
Very often the binding of a ligand induces a structural change in the protein (i.e. there is an induced-fit mechanism). In these processes the ligand forms a relatively weak collision complex with the protein and optimal bonding contacts are only made when the binding site changes shape and, with it, the supporting protein structure. This is illustrated in *Figure 13*.

This type of binding interaction is commonly found when: (i) precise recognition is required where the enzyme must engulf the ligand to make optimal contact, and (ii) the protein must transmit the effect of binding at one site to another site in the structure (e.g. in cooperative, energy transducing or signaling proteins). In these cases the net rates of the binding and dissociation in the collision step (k_{-1} and $k_1.[L]$) are fast compared to the rearrangement rates (k_{-2} and k_2). The experiment is performed as described above and the rate of decay of the E and E:L pair to form E:L* is defined by *Equation 20*. However, the observed rate constant (k_{obs}) for the formation of E:L* (or, equivalently, the rate of decay of the E + E:L pair at differing ligand concentrations) is described by the following equation, which can be used for nonlinear curve fitting:

$$k_{obs} = \frac{k_2 \cdot [L]_T}{K_{d(1)} + [L]_T} + k_{-2}. \hspace{3cm} \text{Equation 24}$$

Figure 13. Conformational change in an induced-fit, two-step binding process.

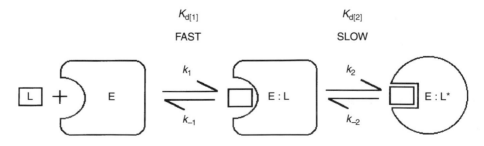

A plot of such data takes a hyperbolic form, shown in *Figure 14*. The curvature of the plot defines $K_{d(1)}$ and describes the weak binding of the ligand in the initially formed collision complex (E:L). The reason the observed rate of formation of E:L* reaches a maximum (i.e. it does not continue to get faster as the ligand concentration is increased) is that it is limited by the structural rearrangement (isomerization) of the complex which is independent of ligand concentration.

Figure 14. Relationship between observed rate constant and ligand concentration in a two-step binding process.

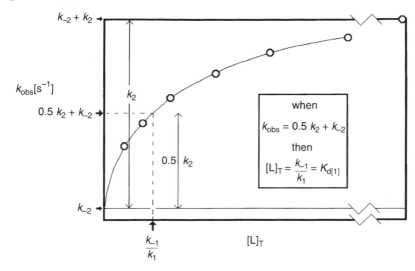

3. PRACTICAL TECHNIQUES FOR STUDYING INTERACTIONS AT EQUILIBRIUM

3.1. Methods for Scatchard analysis

The experimental techniques described in this section are designed to measure bound ligand and free ligand concentrations, thereby producing data which can be processed by the Scatchard analysis described in Section 2.1.

Equilibrium dialysis

The experiment is begun by separating equal volumes of an enzyme and a ligand solution by a dialysis membrane impermeable to protein molecules but freely permeable to the ligand. In time, the ligand will diffuse into the protein solution, allowing the formation of the complex, and equilibrium is established. The experiment is repeated at varied concentrations of ligand, and data analysed according to *Equation 6*. The principle is illustrated in *Figure 15*.

Such apparatus usually takes the form of several dialysis chambers, arranged in a circular block (illustrated in *Figure 15*) to allow several data points to be collected in a single experiment; the solutions are agitated by rotation. At equilibrium the ligand concentration in compartments 1 and 2 ($[L]_1$ and $[L]_2$) are determined to give:

$[L]_F = [L]_2$ and $[L]_B = [L]_1 - [L]_2$.

Figure 15. Equilibrium dialysis.

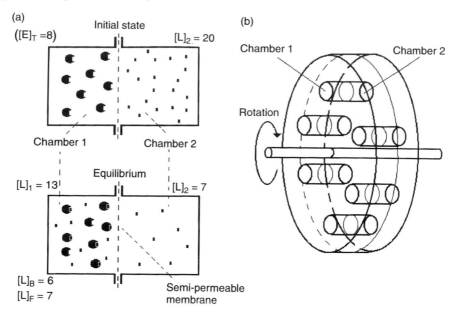

If the protein interferes with the assay for ligand concentration, then: $[L]_B = 0.5.[L]_{2(initial)} - [L]_{2(equilibrium)}$.

Ultrafiltration
A method closely related to the previous method uses filters which will exclude the protein but allow the ligand through. This method is faster to perform and has become much more convenient with the easy availability of spin filtration vessels where the force pushing the solution through the filter is provided by centrifugation in a microcentrifuge. Protein and ligand are mixed and added to the filtration chamber and the solution centrifuged for sufficient time and speed to force 10–20% of the volume into the base of the tube. In these conditions, the protein does not significantly change in concentration. The total ligand concentration ($[L]_T$) is known and that in the filtrate can be determined to establish that of the free ligand ($[L]_F$). As $[L]_B = [L]_T - [L]_F$ then values for $[L]_B$ and $[L]_F$ can be substituted into *Equation 6*. Values are redetermined at differing ligand concentrations.

When dealing with ligands binding to the surface of cells, exactly the same technique is used but in this case large-pore (0.2 µm) filters are used to exclude cells. Such cell-binding measurements are usually carried out by radioactive labeling of the ligand and forcing all the solution through to the lower chamber. The cells on the filter are then washed with a further aliquot of buffer to remove nonspecifically bound ligand. Quantitation of the radioactivity on the filter gives the amount of bound ligand and the free is calculated by subtracting this from the total radioactivity.

Gel filtration
The equilibrium dialysis and ultrafiltration techniques (discussed above) are sometimes limited by problems arising from the protein or ligand binding to the dialysis membrane or filter, or the lack of availability of a membrane which will clearly exclude the protein and not the ligand. To measure ligand binding by the 'Hummel and Dreyer' method, a gel filtration matrix is required

which will separate protein and ligand on the basis of molecular size, and a column of the matrix is equilibrated in a given concentration of the ligand. A sample of the protein is applied to the column in a buffer containing ligand at this given concentration. As fractions elute from the column, the concentrations of ligand and protein are determined in each. A typical elution pattern is shown in *Figure 16*.

These experiments are best employed to determine the concentration of binding sites in a protein sample in conditions where $[L]_F$ is equal to or greater than 10 times the value of K_d. In these circumstances the area under peak (i) [or in trough (ii)] gives the number of moles of binding site in that amount of protein material represented by the area under the protein peak (iii).

This experiment should be performed in conditions where $[E]_T = 0.5 [L]_T$ and the volume of sample is 0.05–0.1 column volumes. The nature of the technique makes a full analysis of K_d and the concentration of binding sites problematic, in that the protein dilutes as the sample migrates down the gel and so its saturation with ligand will alter and distort the data. This can be avoided by performing a series of experiments where the ligand concentration is varied over a fivefold range, say from $5[E]_T$ to perhaps $25[E]_T$ to maintain a constant saturation during the migration through the gel. However, this makes it difficult to assess the added ligand concentration ($[L]_B$ in the diagram) eluting with the protein. If it is possible to assay for ligand with an error of less than 1 or 2%, then data can be analysed according to *Equation 6*.

It should be noted that mixing protein and ligand and passing the complex down a gel filtration column which has not been equilibrated in the ligand solution gives little information on most interactions since this complex will dissociate and the components separate during the run. Only when the rate of dissociation is slow compared to the time taken to run the column is this a valid method of establishing: (i) that a stable complex is formed and (ii) the concentration of binding sites in the protein sample.

Figure 16. A protein–ligand interaction measured by gel filtration.

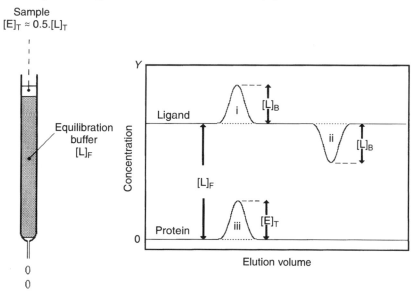

Filter binding assays (DNA–protein interactions)
Nitrocellulose filters have the property of binding proteins but not DNA. Hence when a mixture of the two components, where the DNA is radioactively labeled, is forced through such a filter the free DNA passes through whereas the protein-associated DNA does not. This allows $[L]_F$ and $[L]_B$ to be determined directly by measuring the radioactivity of the DNA on the filter and in the filtrate. The apparatus required to perform these experiments is illustrated in *Figure 17*.

The volume of protein–DNA solution used for each measurement is approximately 1 ml and, in most cases, it is advisable to wash the filter with two volumes of buffer to remove DNA molecules not associated with the protein. There are attendant problems with washing the filter in that it removes free ligand and so promotes dissociation of weak complexes if the rate of dissociation is rapid. In addition, some proteins bind only weakly to nitrocellulose filters. In view of this, it is preferable to perform the washing step as rapidly as possible.

Gel retardation (DNA–protein interactions)
An increasingly popular method of identifying and quantifying protein–DNA interactions exploits the changed electrophoretic mobility of radioactively labeled DNA when it associates with protein. The protein and DNA are mixed and applied to a well in the top of a nondenaturing polyacrylamide gel. When current is passed through the gel the free DNA migrates rapidly towards the positive pole, whereas the DNA–protein complex is retarded so that the two labeled DNA components are separated. Radioactive counting of the unretarded band establishes the amount of free DNA (from which $[L]_F$ is calculated by taking into account the volume of solution loaded on to the gel) and the radioactivity of the retarded band gives the amount bound.

Typically a 6% acrylamide mix is used (ratio of acrylamide : bis-acrylamide = 1 : 20) for casting the slab in a 0.5 × TBE buffer. In most experiments a large excess of competitor DNA or poly dIdC is included in the binding mixture to prevent the formation of nonspecific protein–DNA complexes.

Figure 17. A filter binding assay.

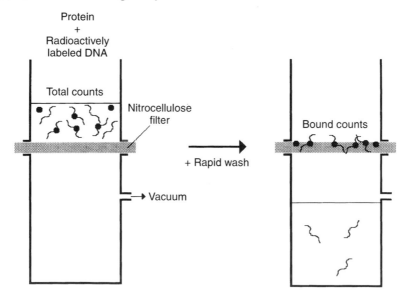

ENZYMOLOGY LABFAX

This method is applicable to short DNA molecules (25–250 base pairs), where there is a large change in mobility on formation of the complex, whereas the filter-binding technique is better suited for larger DNA molecules. One peculiarity of the retardation method is that one would imagine that once the free DNA is separated from the complex, the latter would dissociate, so smearing the band on the gel. The reason given for this not occurring (in most cases) is that once the complex passes from the well into the pores of the dense matrix, the ligand can only be released into a small, local volume. This allows rapid re-association and maintains the complex in a 'frozen' state as it migrates through the gel.

3.2. Methods for measuring saturation of binding sites

The principal optical method for directly detecting the interaction of ligands with proteins is provided by measuring changes in the fluorescence as they combine. This allows the saturation of one component with the other to be measured. Usually it is the fluorescence signal from the protein which is measured and which changes as it is saturated with the added ligand, but occasionally fluorescent ligands are used and it is the signal arising from this component which is recorded. Both the wavelength and the intensity of the emitted light from fluorophors are sensitive to their local, molecular environment and this property is widely exploited in binding studies. Most other optical signals are either not sufficiently sensitive to the local environment to act as good reporters of complex formation (for example light absorption) or require a highly specialized apparatus (for example NMR). The following fluorescence techniques are the most commonly used.

Intrinsic protein fluorescence
Most proteins contain two types of fluorescent residue, tryptophan and tyrosine. The former has both a greater efficiency of absorption and a greater quantum yield of fluorescence (i.e. a greater proportion of the absorbed photons are re-emitted as fluorescence) and so dominates the fluorescence output of tryptophan-containing proteins. The binding of a ligand close to a fluorescent residue, or the alteration in protein structure induced by binding, usually causes a change in the fluorescence spectrum, which can be exploited as a signal.

Extrinsic fluorescence of labeled proteins
If there is no change in the fluorescence of the protein itself, it is often useful to attach a fluorophor by covalent chemistry. Clearly, controls need to be performed on the labeled molecule (or dye–protein conjugate) to ensure its function has not been compromised. Changes in the extrinsic fluorescence (that from the label) on ligand binding have some advantages over intrinsic fluorescence in that the intensities are greater and the fluorophor can be chosen at will to absorb and emit at wavelengths not interfered with by other components of the system.

Ligand fluorescence
Some natural ligands are themselves fluorescent (e.g. NADH, NADPH) or can be made fluorescent without perturbing their functional properties (e.g. DNA, RNA, nucleotides, complex polysaccharides). In binding experiments with these ligands, it is preferable to add the protein to the ligand, so that the titration is done 'in reverse'. It is important to remember that, in the analysis of data from these experiments, the protein is acting as the ligand and vice versa. This means that, although K_d has the conventional sense, the sites concentration is the reciprocal of the usual convention (i.e. if a protein molecule has four binding sites, the molecular stoichiometry would be 0.25 in a standard analysis).

Resonance energy transfer
If none of the above provides a good signal, the other technique worth trying is energy transfer (see *Figure 18*). This method requires one fluorophor (either intrinsic or extrinsic) on the protein

and a different fluorophor on the ligand. The emission spectrum of one fluorophor (the donor) must overlap with the excitation spectrum of the other (the acceptor). It does not matter whether the donor is on the protein or the ligand (the same, of course applies to the acceptor). The experiment is performed by irradiating a solution of the component bearing the donor fluorophor at its excitation wavelength, adding the component-bearing acceptor fluorophor and measuring the intensity of fluorescent light at the emission wavelength of acceptor. Energy transfer occurs efficiently when donor and acceptor are within about 2.5 nm of each other, so that emitted light can only be detected when the two molecules form a complex (note that the average distance apart of noninteracting molecules in a 10^{-6} M solution is about 100 nm). The intensity of light at the emission wavelength of fluorophor B therefore can serve as a signal for the formation of the protein–ligand complex.

Practical considerations in equilibrium fluorescence spectroscopy
The fluorescence techniques described above have several advantages over other methods for monitoring ligand binding processes. The techniques can be carried out over a wide concentration range (0.05 µM upwards for most fluorophors), they measure binding in free solution with no complication introduced by the method itself, they are quick to perform and are nondestructive so that material can be reclaimed if it is valuable.

The general method for making measurements of intrinsic protein fluorescence, extrinsic protein fluorescence or ligand fluorescence is the same. A solution of the component from which the signal arises is placed in a cuvette and irradiated at its approximate excitation wavelength. The emission monochromator is then scanned towards higher wavelengths from an initial position just above the excitation wavelength. The fluorescence intensity versus wavelength plot which results is known as an emission spectrum. The emission monochromator is then fixed at a wavelength coinciding with the maximum emission intensity and the excitation monochromator is scanned upwards in wavelength to a point just below the wavelength of the emission monochromator. The resulting plot of emission intensity versus excitation wavelength is known as an excitation spectrum. The ligand is then added from a concentrated stock solution and the new emission spectrum is recorded. The emission spectra before and after formation of the complex are compared to find the wavelength at which the change in emission intensity is greatest, as shown in the example in *Figure 18a*.

In most experiments a fresh solution of the signal-bearing component is then placed in the cuvette, the excitation monochromator is kept at the peak wavelength, the emission monochromator set to the wavelength of greatest intensity change and the ligand added in small aliquots. A record of the emission intensity versus the concentration of ligand added ($[L]_T$) provides the data for analysis according to the description on pp. 202–204.

When detecting the formation of a protein–ligand complex by RET (resonance energy transfer) a solution of the donor molecule is irradiated at its excitation maximum and the intensity of the emitted fluorescent light is selected at the emission maximum of the acceptor (see *Figure 18b*). The acceptor is then added in aliquots and the increase in the emission intensity is recorded to provide the binding curve.

Fitting data
Equations 9 and *11* (simple binding) and *Equations 16* and *18* (cooperative binding) show how the type of binding curve acquired by these methods can be analysed. However, these equations are given in terms of the variation of α (proportion of sites saturated) as the total ligand concentration $[L]_T$ is increased. Values of α are measured from the binding curve by estimating the signal change at saturating ligand concentrations and representing all values as a proportion

Figure 18. A protein–ligand interaction measured by (a) a change in emission properties of the enzyme and (b) resonance energy transfer.

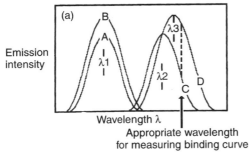

A = Excitation spectrum of free protein – emission at λ2

B = Excitation spectrum of protein–ligand complex – emission at λ3

C = Emission spectrum of free protein – excitation at λ1

D = Emission spectrum of protein–ligand complex – excitation at λ1

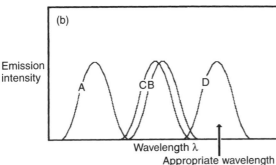

A = Excitation spectrum of donor

B = Excitation spectrum of acceptor

C = Emission spectrum of donor

D = Emission spectrum of acceptor

of this. However, it is not always easy to estimate the end point (F_E) of a curve and, in some cases where the data are 'noisy' (error prone), there is uncertainty in determining the starting signal (F_S). This problem can be overcome by loading the data into a nonlinear fitting program as a set of fluorescence intensity values versus $[L]_T$ and allowing the end point and/or start point to be fitted along with K_d and $[E]_T$. The general equation used in these circumstances is:

$$\alpha = (F - F_S)/(F_E - F_S). \qquad \text{Equation 25}$$

Therefore from *Equation 9*, F can be expressed as:

$$F = ((([L]_T + [E]_T + K_d) - (([L]_T + [E]_T + K_d)^2 - 4 \cdot [L]_T \cdot [E]_T)^{0.5})/(2 \cdot [E]_T)) \cdot (F_E - F_S) + F_S$$

where F is the measured fluorescence, or any other signal (at a given value of $[L]_T$), F_S is the starting fluorescence (before addition of ligand – this is known and should be put directly into the equation, i.e. it need not be fitted) and F_E is the fluorescence at the notional endpoint when complete saturation has been established (this is then fitted as a parameter).

In the simple case where $[E]_T$ is much smaller than K_d the binding curve produces a simple rectangular hyperbola (*Equation 11*). In these conditions the data can be analysed by the equation:

$$F = (([L]_T \cdot (F_E - F_S))/([L]_T + K_d)) + F_S.$$

For the Hill analysis (*Equation 18*) the modified equation for nonlinear fitting is:

$$F = (([L]_T^n \cdot (F_E - F_S))/([L]_T^n + K)) + F_S$$

and for the MWC analysis (*Equation 16*) where $\alpha = [R]/[E]_T$:

$$F = (((1 + A)^n \cdot (F_E - F_S))/(K_{MWC} + (1 + A)^n)) + F_S.$$

Wherever an analytic solution for a binding process can be defined in terms of saturation (α), *Equation 25* can be used to produce an equation for direct data fitting, regardless of the type of signal or the starting and finishing values.

4. PRACTICAL TECHNIQUES FOR STUDYING THE KINETICS OF LIGAND BINDING

4.1. General
In some cases the rates of association and dissociation of protein–ligand complexes and the structural rearrangements induced by these events are sufficiently slow to allow measurement by nonspecialized optical apparatus such as the conventional fluorimeter. However, these processes usually occur over short time periods and so require specialized equipment to follow the formation of complexes as a function of time. Two general techniques are employed; equilibrium perturbation and stopped-flow mixing. The former technique is too specialized to be dealt with in detail here, but is based upon the sensitivity of binding equilibria to temperature and pressure. Hence, if the equilibrium is perturbed by rapid heating or by a sudden increase in pressure, the rate at which the new equilibrium position is established can be used to analyse the kinetics of the process. These techniques can be used to measure processes occurring on the 10 µsec time scale. Stopped-flow techniques, on the other hand, are more widely used and becoming increasingly popular with the ready availability of good, commercially built machines.

4.2. Stopped-flow mixing methods
The principle behind the stopped-flow spectrometer is illustrated in *Figure 19*. The protein and ligand solutions are placed in separate syringes and rapidly mixed by forcing the syringes forwards to drive the solutions together into a quartz mixing chamber (quartz allows the use of ultraviolet wavelengths). The volume of the solutions mixed is taken up by the filling of the stopping syringe, the plunger of which is pushed out until it hits the stopping bar. The volumes of protein and ligand solution mixed are controlled by the distance between the back of the plunger and the stopping bar. For a single reaction the combined volume of the protein and ligand solutions is typically 0.1–0.2 ml, so that each time the mixing chamber is more than replenished with fresh, unassociated components. When the stopping syringe is arrested, a microswitch is depressed on the stopping bar, which activates data collection.

Figure 19. Schematic representation of a stopped-flow apparatus.

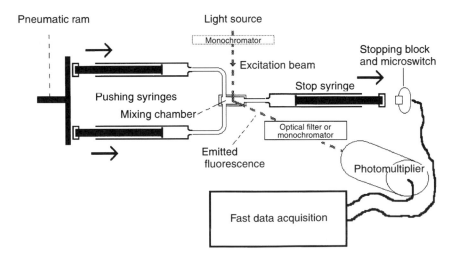

ENZYMOLOGY LABFAX

The time taken for the components to mix in the flow period is normally between 1 and 2 msec (defined as the dead time) so that any reaction occurring on faster time scales is near completion by the time reliable observations of the event are made. In practice, it is difficult to measure an observed rate constant faster than about 1000 sec^{-1}. The quantity measured during the reaction is usually the voltage output of the photomultiplier, and it is important to determine the signal arising from the components individually and from the protein–ligand complex so that the amplitude of the reaction occurring during the recording time is known relative to the overall, expected signal change. For instance, the formation of a collision complex may lead to a signal change too fast to measure at the ligand concentrations used, followed by a slow rearrangement of the complex which is easily measurable. The amplitude of the instant phase as a function of ligand concentration then serves as an additional signal in the system.

In the above type of experiment, where the binding of ligand and protein is being observed, the volumes of the two solutions mixed together are equal. As pointed out in Section 2.2, there is a further type of transient experiment which can be performed by diluting the pre-formed protein–ligand complex with 10–20 volumes of buffer to allow the complex to dissociate. In appropriate conditions (*Equation 23*) this provides a means of determining the rate constant of the slowest step in the dissociation pathway.

One of the great advantages of transient kinetic techniques carried out in pseudo-first-order conditions (as described in Section 2.2) is that it is only the form of the change in signal with time that determines k_{obs} and it is the variation of this value with ligand concentration which provides most information.

In determining values for k_{obs} from real data (of the type shown in *Figure 10*) the following equations are used for nonlinear fitting by computer programs.

For a signal (F) that decreases with time then:

$$F = \Delta F_M . \exp(-k_{obs} . t) + F_E. \qquad\qquad \text{Equation 26}$$

For a signal that increases then:

$$F = \Delta F_M - \Delta F_M . \exp(-k_{obs} . t) + F_S. \qquad\qquad \text{Equation 27}$$

Note that the equations are given with F (nominally fluorescence intensity) as the signal. In practice the signal is measured in volts in stopped-flow experiments but this makes no difference to the analysis.

Finally, and in general, it should also be remembered that an important characteristic of a tenable kinetic mechanism is that the rate constants deduced from rapid kinetics are consistent with the measured binding affinity of the ligand in equilibrium measurements.

CHAPTER 7
ENZYME COFACTORS

P.C. Engel

1. INTRODUCTION

Despite the availability of 20 different types of amino acid side chain in proteins, as produced by the ribosomal translation machinery, and of various post-translational modifications, many enzymes require the assistance of other chemical entities in order to achieve catalytic activity. These chemical adjuncts, or 'cofactors', are of two basic types chemically, metal-ion cofactors and organic cofactors. Beyond that division, however, there are further important distinctions in the way cofactors interact with enzymes (see Sections 2 and 3). Historically, the recognition of a cofactor requirement came from dialysis experiments in which a high molecular weight, usually heat-labile, protein was retained by the membrane and inactivated by the removal of a diffusible, low molecular weight and often heat-stable cofactor. Sections 2 and 3, however, also include examples of cofactors that are not removed in this way. If they are very tightly but noncovalently bound, they may be removed by denaturing the protein (e.g. precipitation with 5% trichloroacetic acid or unfolding with 6 M guanidinium HCl). A covalently attached cofactor, however, may only be isolated and identified by the methods of protein chemistry (acid hydrolysis or peptide cleavage).

2. TYPES OF METAL-ION COFACTOR (see also Chapter 8)

2.1. Structural metal ions
Such metal-ion cofactors do not participate directly in the catalytic event, but are needed to hold the protein in the active conformation. Examples: Ca^{2+} in trypsin (1); structural Zn^{2+} in horse liver alcohol dehydrogenase (2). They remain in place throughout catalysis.

2.2. Co-substrates
These cofactors are in free solution rather than tightly bound to the enzyme. They complex with a substrate to convert it to a form that is active in the enzyme-catalysed reaction. Example: Mg^{2+} complexed to ATP in most kinase and ligase reactions. They enter and leave the catalytic site once per cycle.

2.3. Essential parts of catalytic machinery
Various metal ions' capacity for valency changes, coordination chemistry and electron withdrawing properties are used as essential features of catalysis in enzymes. Examples: Cu atom in cytochrome oxidase; catalytic-site Zn^{2+} (distinct from structural Zn^{2+} in Section 2.1) in horse liver alcohol dehydrogenase (2); Mo atom in xanthine oxidase (3). They remain in place throughout catalysis; usually tightly bound.

3. TYPES OF ORGANIC COFACTOR

3.1. Coenzymes
Coenzymes are true, autonomous substrates for the enzymes they service. Example: $NAD^+/$ NADH with lactate dehydrogenase (LDH) (4).

$$\text{Pyruvate} + \text{NADH} + \text{H}^+ \xrightleftharpoons[\text{LDH}]{} \text{lactate} + \text{NAD}^+$$

The distinction between pyruvate as 'substrate' and NADH as 'coenzyme' is metabolic, in that a recyclable constant pool of NADH services the bulk throughput of metabolic substrates (e.g. glucose). Typically, although not always, the K_m values for coenzymes are low (10^{-5}–10^{-4} M) compared to those for the throughput substrate (10^{-3}–10^{-2} M), so that low pool concentrations of the coenzymes keep their enzymes close to saturation. They enter and leave the catalytic site once per cycle.

3.2. Prosthetic groups

Unlike a coenzyme, a prosthetic group is an integral part of the active enzyme. Examples: FAD/FADH$_2$ in acyl-CoA dehydrogenase (5), pyridoxal phosphate in aspartate amino transferase (6). These groups are chemically altered during the first half of the catalytic cycle, which thus produces a product and a modified enzyme. Often, although not always, such enzymes show 'ping-pong' kinetics (Chapter 3B) since the first product, once formed, may leave the enzyme before the catalytic cycle is completed. To complete catalysis, the second substrate electron transferring flavoprotein (ETF) (ETF for acyl-CoA dehydrogenase, 2-oxoglutarate for aspartate aminotransferase) must react with the prosthetic group to regenerate it in its original form and yield the second product. They remain in place throughout catalysis.

3.3. Co-substrates

Some organic cofactors work as carriers for other groups and their operation involves two enzymes. Thus, for example, coenzyme A is a coenzyme only in the vague metabolic sense that relatively small molar amounts of it are recycled in facilitating bulk metabolic processes (e.g. fatty acid β-oxidation). In a strict enzymological sense it is not a coenzyme. Typically it serves:

(i)   as a true substrate for a synthetase reaction in which, for example, a fatty acid is 'activated' by conversion to its CoA thioester.
(ii) As a co-substrate for another enzyme; for example, as part of palmitoyl-CoA it makes possible the oxidation catalysed by long-chain acyl-CoA dehydrogenase.

Such co-substrates enter and leave the catalytic site once per cycle.

4. PROPERTIES OF INDIVIDUAL COFACTORS

4.1. Adenine mononucleotide cofactors: ATP, ADP, AMP

Role

ATP is often referred to as the universal currency of the cell. Its formation is a central outcome of 'energy-yielding' (strictly energy-conserving) catabolic pathways, such as glycolysis, and of electron transport; for example, in mitochondria or photosynthesis. ATP is split in many biosynthetic processes which would not be energetically feasible without its involvement. Some of the reactions split off the γ-phosphate, yielding ADP. Examples:

$$\text{Glucose} + \text{ATP} \xrightarrow[\text{hexokinase}]{} \text{glucose 6-phosphate} + \text{ADP}$$

$$\text{Acetyl CoA} + \text{CO}_2 + \text{ATP} \xrightarrow[\text{acetyl CoA carboxylase}]{} \text{malonyl CoA} + \text{ADP} + \text{Pi}$$

Others split between the α- and β-phosphates yielding AMP and pyrophosphate. Examples:

$$\text{Fatty acid} + \text{CoASH} + \text{ATP} \xrightarrow[\text{acyl CoA synthetase}]{} \text{fatty acyl CoA} + \text{AMP} + \text{PPi}$$

Amino acid + tRNA + ATP $\longrightarrow$ amino acyl tRNA + AMP + PPi
various amino acyl tRNA synthetases

It is less common nowadays than in older textbooks to see ATP/ADP/AMP described as 'cofactors'. The term seems conceptually appropriate if, for example, ATP/ADP are viewed as low molecular weight compounds that facilitate the metabolic breakdown of glucose to pyruvate, but less appropriate if one views glucose breakdown as directed towards formation of ATP as a principal product. Nevertheless, the role of a pool of ATP and ADP in cycling phosphoryl groups is entirely analogous to that of, say, $NAD^+/NADH$ in cycling reducing equivalents.

Vitamin relationship
None. Mammals can synthesize their own purine nucleotides.

Name
Adenosine 5'-triphosphate $\equiv$ ATP
Adenosine 5'-diphosphate $\equiv$ ADP
Adenosine 5'-monophosphate $\equiv$ AMP $\equiv$ adenylic acid

Structure

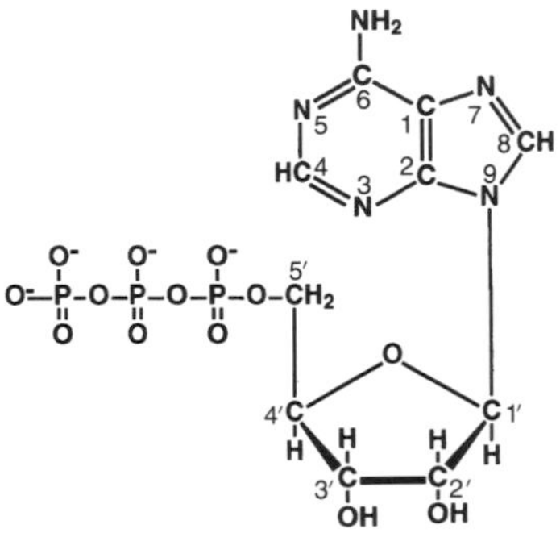

Commercially available forms

AMP

free acid	Formula weight: 347.2
Na$_2$ salt (6H$_2$O)	Formula weight water soluble: 499.2

ADP

free acid	Formula weight: 427.2
Na$_2$ salt	Formula weight: 471.2
K salt (2H$_2$O)	Formula weight: 501.3
Ba salt (insoluble)	Formula weight: 630.2
Di(monocyclohexylammonium) salt	Formula weight: 635.6
Li and Tris salts – H$_2$O content not quoted	

ATP

Na$_2$ salt	Formula weight: 551.1
Na$_2$ salt (3H$_2$O)	Formula weight: 605.2

Li, K, Mg, Ca, Ba, Tris and monoethanolammonium salts are also available; formula weights are not quoted because of variable water of crystallization.

Several of these forms of the adenine mononucleotides are available in different grades of purity for specific applications. For example, Sigma provides vanadium-free ATP and ADP for work with Na^+/K^+ ATPases.

Stability
AMP is the most stable of the three compounds and the disodium salt is stable as the solid at room temperature. ADP as the disodium salt may break down slowly to form AMP and should be stored dry at $-20°C$. As the free acid it may disproportionate to ATP + AMP, and again should be stored at $-20°C$. The crystalline potassium salt is stable at $4°C$ and the di(cyclohexylammonium) salt is also described as stable and nonhygroscopic. Solutions of ATP and ADP are most stable at weakly acidic pH values (3.5–6.8). In alkaline solution even at $0°C$ ATP splits to AMP and pyrophosphate.

Enzymatic assay
ATP may be sensitively estimated by a bioluminescence assay with luciferase (7). Alternatively, coupled enzyme assay procedures are available, for example hexokinase + glucose 6-phosphate dehydrogenase (8). Likewise, ADP may be assayed with pyruvate kinase + lactate dehydrogenase (9).

Photometric properties
The ultraviolet absorption properties of AMP, ADP and ATP are due to the adenine ring and all three have a single absorption maximum at 259 nm (1.54×10^4 mol^{-1} cm^{-1}).

Analogs
A wide variety of analogs are available commercially for studying enzymes that use ATP/ADP/AMP. These include:

(i)   etheno derivatives of AMP, ADP and ATP (10). Their absorbance and fluorescence properties are useful in studying binding.
(ii) Thiophosphate derivatives of ADP and ATP. These are useful as inhibitors and substrates. They include:
      adenosine 5'-*O*-(2-thiodiphosphate)   (ADP-β-S)
      adenosine 5'-*O*-(1-thiotriphosphate)   (ATP-α-S)
      adenosine 5'-*O*-(3-thiotriphosphate)   (ATP-γ-S).
(iv) An imido derivative in which the O between the β and γ P is replaced by NH. This is adenylimidodiphosphate, AMP-PNP and is resistant to β, γ-cleavage.
(v)   A methylene derivative in which a CH_2 replaces the O between the β and γ P. This is adenyl (β,γ-methylene) diphosphonate, AMP-PCP and again is resistant to β,γ-cleavage.

4.2. Ascorbic acid

Vitamin role
Ascorbic acid (11, 12 p. 3, 13 p. 3, 14 p. 3) is vitamin C. Although the compound was identified in the 1930s and the symptoms of scurvy resulting from its deficiency have long been recognized, its biochemical role has become clear only comparatively recently. It serves as a cofactor for prolyl hydroxylase, an iron-containing enzyme responsible for making the hydroxyproline units in collagen. Hence the connective tissue weakness characteristic of scurvy. Ascorbic acid reduces the Fe in prolyl hydroxylase. This produces dehydroascorbic acid which may be reduced once again by glutathione.

Structure

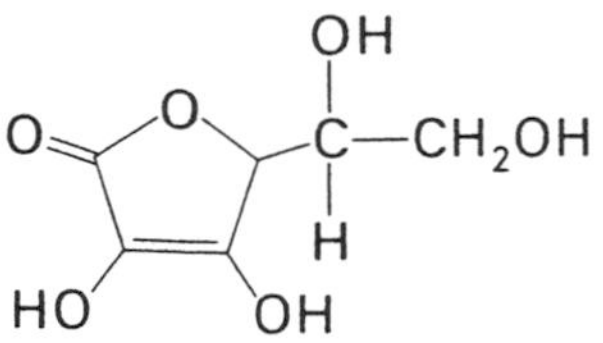

Commercially available forms

Ascorbic acid
  free acid   Formula weight 176.1
  hemicalcium salt   Formula weight 195.2
  Na salt   Formula weight 198.1

Stability
Dry solid stable. Alkaline solutions unstable. Relatively stable in solution at pH 5–6.

Photometric properties
Absorption maximum at 265 nm (7×10^3 1 mol^{-1} cm^{-1}).

4.3. Biotin

Role
Cofactor in carboxylation reactions (e.g. acetyl CoA carboxylase). ATP-dependent reaction converts biotin to carboxybiotin which serves as donor for recipient substrate. Biotin is a covalently bound prosthetic group, attached via amide linkage to a protein lysyl residue (12 p. 379, 13 p. 279, 14 p. 63, 15).

Distinct from its biological role, biotin has been given new importance in molecular biology by the exploitation of its very tight sequestration by avidin and streptavidin. Biotinylation of molecules permits sensitive detection and purification procedures based on biotin–streptavidin recognition. Streptavidin may be conjugated with enzymes (e.g. peroxidase, alkaline phosphatase), fluorophores, etc.

Vitamin relationship
Biotin itself was originally termed vitamin H.

Structure

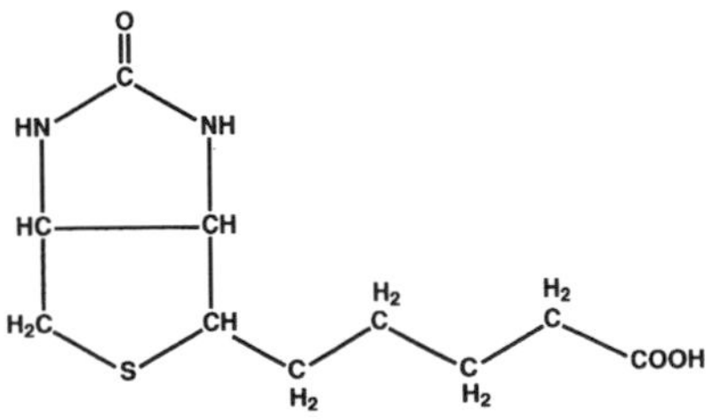

Commercially available forms

D-biotin   Formula weight 244.3
Biocytin *N*-ε-biotinyl-L-lysine   Formula weight 372.5

COFACTORS   227

Cofactors

Analogs
Diaminobiotin, 2-iminobiotin.

4.4. Cobalamin cofactors

Role
These cobalt-containing cofactors (16 p. 3, 17 p. 3, 18 p. 3) are involved in a variety of mutase and methyl transfer reactions. Examples:

3-Methyl aspartate $\longleftrightarrow$ glutamate
methylaspartate mutase

Methylmalonyl-CoA $\longleftrightarrow$ succinyl-CoA
methylmalonyl-CoA mutase

$\qquad$ 5-Methyl THF $+$ B_{12} $\longleftrightarrow$ THF $+$ methyl-B_{12}

Methyl-B_{12} $+$ homocysteine $\longleftrightarrow$ B_{12} $+$ methionine
methyl THF-homocysteine transmethylase

Vitamin relationship
The name vitamin B_{12} is given to cyanocobalamin but this arises during purification from biological material. The true biologically active species have a methyl group, a hydroxy group or a 5'-deoxy-adenosyl group replacing the cyanide coordinated to cobalt.

Names
Cyanocobalamin $\equiv$ vitamin B_{12}
5'-Deoxyadenosyl cobalamin $\equiv$ coenzyme B_{12}
Hydroxocobalamin $\equiv$ aquocobalamin
Methylcobalamin $\equiv$ methyl-B_{12}

Structure
The core of the B_{12} structure is the corrin ring system, with a cobalt atom coordinated between the nitrogen atoms of four linked pyrrole rings.

Commercially available forms

Vitamin B_{12}	Formula weight 1355.4
Coenzyme B_{12}	Formula weight 1579.6
Hydroxocobalamin	
acetate salt	Formula weight 1388.4
hydrochloride	Formula weight 1382.8
Methylcobalamin	Formula weight 1344.4

Stability
The various forms of B_{12} are reasonably stable in the dark as solids but are photolabile. Solutions are even less stable to light. Solutions of coenzyme B_{12} are most stable in the dark at pH 6–7. Vitamin B_{12} solutions are most stable in the dark at pH 4–6.

Photometric properties
B_{12} derivatives, being cobalt coordination compounds, are highly colored (red) as a result of an absorbance band at about 525 nm. There are also bands at about 350 nm and 270 nm. In the case of coenzyme B_{12}, however, the presence of an adenine ring intensifies the ultraviolet absorption and shifts the peak to 260 nm.

Absorption maxima:

Vitamin B_{12}	278 nm (1.56×10^4 l mol^{-1} cm^{-1})
	361 nm (2.77×10^4 l mol^{-1} cm^{-1})
	550 nm (8.55×10^3 l mol^{-1} cm^{-1})
Coenzyme B_{12}	260 nm (3.47×10^4 l mol^{-1} cm^{-1})
	340 nm (1.23×10^4 l mol^{-1} cm^{-1})
	375 nm (1.09×10^4 l mol^{-1} cm^{-1})
	522 nm (8.0×10^3 l mol^{-1} cm^{-1})
Hydroxocobalamin	273 nm (1.84×10^4 l mol^{-1} cm^{-1})
	351 nm (2.23×10^4 l mol^{-1} cm^{-1})
	525 nm (7.56×10^3 l mol^{-1} cm^{-1})
Methylcobalamin	268 nm (2.27×10^4 l mol^{-1} cm^{-1})
	343 nm (1.20×10^4 l mol^{-1} cm^{-1})
	525 nm (7.6×10^3 l mol^{-1} cm^{-1})

4.5. Coenzyme A

Role

Coenzyme A (12 p. 311, 13 p. 201, 14 p. 33) serves as an acceptor and donor of acyl groups and, as the thiol portion of acyl thiolester substrates, plays an essential role in activating the acyl portion for reaction, for example in fatty acid β-oxidation.

Vitamin relationship

Derived from vitamin B_5, D-pantothenic acid.

Name

The original trivial name given by Lipmann ('A' referring to acetylation of sulfonamide) has stuck. Coenzyme A is also abbreviated CoA, especially in naming its acyl derivatives, and sometimes written CoASH to emphasize the free thiol group in the underivatized form.

Structure

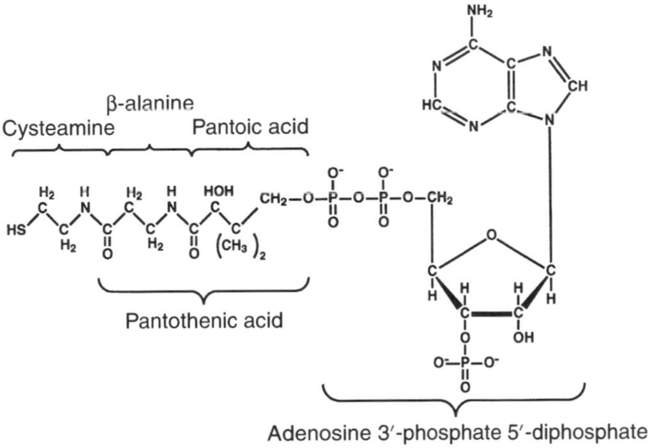

β-alanine
Cysteamine
Pantoic acid
Pantothenic acid
Adenosine 3′-phosphate 5′-diphosphate

Commercially available forms

Free acid	Formula weight 767.5
Lithium salt	Formula weight 785.4

Sodium salt   Formula weight 833.5
Oxidized form (disulfide) Li salt   Formula weight 1568.8

A wide range of acyl-CoA derivatives is also available, but these cost, on a molar basis, 5–10 times more than coenzyme A itself. For routine use it is therefore worth considering making the acyl-CoA derivatives by acylating CoA.

Stability
The dry salts are much more stable than the free acid. They should all be stored at $-20°C$ in a desiccator. Breakdown products may be potent inhibitors of some CoA-dependent enzymes, for example phosphotransacetylase. The free acid may suffer 20% decomposition in 6 months even when well stored. In solution, especially above pH 7 and in the presence of heavy-metal ions the thiol group is vulnerable to air oxidation. It may be protected by degassing, chelating agents, low pH (4–5), dithiothreitol, etc. In general, solutions should be made freshly and used promptly.

Enzymatic assay
Several assay procedures are available. Some are stoichiometric but the most sensitive ones recycle the CoA to provide a catalytic assay (19) (i.e. the CoA is used catalytically). Example:

$$\text{Acetyl phosphate} + \text{CoASH} \longrightarrow \text{acetyl CoA} + \text{phosphate}$$
$$\text{Acetyl CoA} + \text{arsenate} \longrightarrow \text{CoASH} + \text{acetyl arsenate}$$
$$\underline{\text{Acetyl arsenate} + H_2O \longrightarrow \text{acetate} + \text{phosphate}}$$
$$\text{NET: acetyl phosphate} \longrightarrow \text{acetate} + \text{phosphate.}$$

The first two reactions are catalysed by phosphotransacetylase in arsenate buffer. The third is spontaneous. Acetyl phosphate, measurable by the Lipmann and Tuttle procedure (20), is broken down at a rate that depends on the small catalytic amount of CoASH added. Alternatively acetyl CoA may be recycled by using citrate synthase and malate dehydrogenase with continuous measurement of NADH production (21).

Photometric properties
Free coenzyme A has a strong absorption band at 260 nm owing to the adenosine moiety. The acyl derivatives have an additional band at 232 nm from the thioester band.

Analogs
3'-Dephospho coenzyme A
Desulfo coenzyme A
1-N6 Etheno coenzyme A
D-Pantethine (disulfide of pantetheine)
Pantothenic acid (D or DL)
N-Acetyl cysteamine

Some CoA-'requiring' enzymes will accept fragments, for example N-acetyl cysteamine, as an alternative. For purification procedures requiring large numbers of routine assays this may offer a cheap alternative (22).

4.6. Coenzyme Q

Role
Coenzyme Q (CoQ) (16 p. 137, 17 p. 111) is one of the components of membrane-bound electron transfer systems that can readily be removed in a soluble form and equally added back. Reduction of 'electron carriers' by 'hydrogen carriers' is a crucial factor in bringing about

proton pumping across such membranes, and in this context CoQ is a hydrogen carrier, undergoing a 2H reduction from the quinone to the quinol. The generic term 'coenzyme Q' embraces a family of compounds differing in the number of repeating units in the polyisoprenyl side chain (see below). In mitochondrial electron transport CoQ mediates oxidation of cytochrome b by cytochrome c. Its reoxidation by cytochrome c is coupled to phosphorylation (site 2).

Names and abbreviations
Coenzyme Q $\equiv$ CoQ $\equiv$ ubiquinone. The length of the side chain is indicated in CoQ by the number of isoprenyl units as a subscript, for example CoQ_6 has six isoprenyl units. The same is indicated in the ubiquinone nomeclature by the number of atoms in the polyisoprenyl chain. Since each unit contributes five C-atoms, CoQ_6, for example, is ubiquinone-30. CoQ_o is 2,3-dimethoxy-5-methyl-1,4-benzoquinone.

Structure

Commercially available forms

CoQ_o	Formula weight 182.2
CoQ_6	Formula weight 590.9
CoQ_7	Formula weight 659.0
CoQ_9	Formula weight 795.2
CoQ_{10}	Formula weight 863.4

Stability
Somewhat photolabile. Reduced form slowly autoxidizable.

Photometric properties
The quinone forms have absorption maxima at 275 nm ($\sim 1.4 \times 10^4 \, l \, mol^{-1} \, cm^{-1}$) and 405 nm. In the quinol there is a single maximum at 290 nm ($\sim 4 \times 10^3 \, l \, mol^{-1} \, cm^{-1}$) (all measured in ethanol).

Analogs
Decylubiquinone, a synthetic cytochrome b_f substrate is commercially available.

4.7. Cytosine mononucleotide cofactors: CTP, CDP, CMP

Role
Apart from their role as nucleic acid precursors, these pyrimidine nucleotides have a specific role in the biosynthesis of phospholipids. Phosphatidic acid (1,2-diacyl glycerol 3-phosphate) reacts with CTP to yield CDP-diacyl glycerol, as an activated intermediate, and pyrophosphate. The CDP-DAG reacts with serine or inositol to form phospholipids, with release of CMP.

CTP is similarly involved in the activation of glycerol, ribitol, *N*-acetylneuraminic (sialic) acid, etc., in the synthesis of teichoic acid for bacterial cell walls and the cell surface sugar residues of mammalian and other cells.

COFACTORS

Names

Cytidine is the ribose mononucleoside of cytosine.
Cytidine 5'-monophosphate ≡ CMP ≡ cytidylic acid
Cytidine 5'-diphosphate ≡ CDP
Cytidine 5'-triphosphate ≡ CTP

Structure

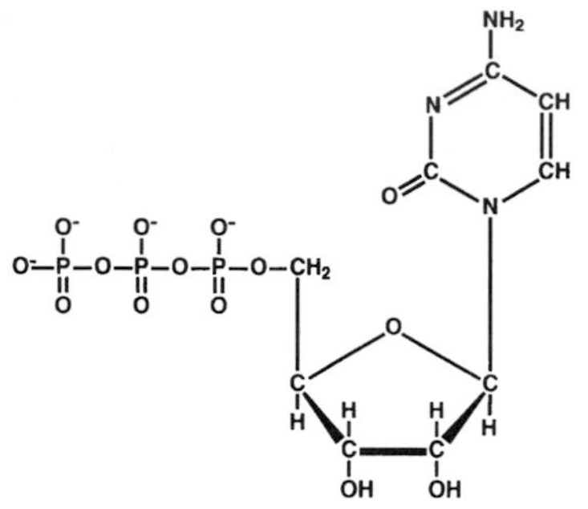

Commercially available forms

CMP
  free acid   Formula weight 323.2
  Na_2 salt   Formula weight 367.2
CDP   Formula weight 403.2
  available as Na and Tris salts
CTP   Formula weight 483.2
  available as Na_2 or Tris salt
  (solid) and as Li salt (solution)

Stability

CTP Na_2 salt should be stored dry at 4°C and may decompose 4% in 1 year. Li salt solution should be stored at −20°C, when it is stable.

Photometric properties

The ultraviolet absorption spectra of CMP, CDP and CTP are essentially identical both at pH 7 and at pH 2. At pH 7 there is an absorption maximum at 271 nm (9.2×10^3 l mol^{-1} cm^{-1}) and a shoulder at about 232 nm. At pH 2 there is a single maximum at 280 nm (1.3×10^4 l mol^{-1} cm^{-1}).

Analogs and derivatives

CDP derivatives of choline, ethanolamine and glycerol are available. Also CDP diacylglycerol (dioleoyl, dipalmitoyl or dimyristoyl).

Periodate oxidized CMP, CDP and CTP are all available (with the ribose moiety converted to a reactive dialdehyde) and in the case of CMP a borohydride reduced version is also available.

4.8. Flavin cofactors

Role

Versatile oxidoreduction cofactors (14 p. 185, 23 p. 253, 24 p. 217, 25). The isoalloxazine ring system is capable of undergoing either one-electron reduction to a semiquinone or two-electron reduction to a quinol form. This allows flavoproteins to serve as adaptors between one and two-electron oxidoreductants and also to operate over a wide range of oxidation–reduction potentials. Individual flavoproteins may cycle between fully oxidized and fully reduced, between fully oxidized and half-reduced, or between half-reduced and fully reduced. This is possible because they function as tightly bound prosthetic groups and individual proteins stabilize particular oxidation states. In the two-electron oxidation mode flavin cofactors are stronger oxidants than NAD^+. Hence their use, for example, to make $C=C$ double bonds (succinate dehydrogenase), whereas NAD^+ is used for easier oxidations (e.g. malate dehydrogenase). Most flavin cofactors are tightly noncovalently bound, but a few enzymes (e.g. succinate dehydrogenase) have covalently bound flavin.

Flavins also serve as cofactors for hydroxylation. Here they are reduced by NADPH to the quinol form which then reacts with O_2 to form a reactive hydroxylating species.

Vitamin relationship
Derived from vitamin B_2, riboflavin.

Names and abbreviations
The two commonly found flavin cofactors are riboflavin 5′-phosphate, also known as flavin mononucleotide, FMN; and flavin adenine dinucleotide, FAD, in which FMN is in phosphodiester linkage with adenosine.

Structure

Commercially available forms

FMN sodium salt
FAD disodium salt   Formula weight 829.6

Stability
Solids stable at 4°C if kept dry and dark. Solutions should be protected from light.

Photometric properties
FMN and FAD are both strongly yellow as the result of an absorbance band at 450 nm. They also have maxima at 370 and 260 nm. Excitation of the 450 nm band, either directly or

by energy-transfer from the short wavelength bands, gives green fluorescence maximal at 540 nm. This is approximately 10 times more intense for FMN than for FAD, where stacking with the adenine ring causes quenching. On binding to the protein ('apoenzyme'), flavin cofactors' absorption may become either more or less intense, and equally the fluorescence may be quenched or enhanced. The local environment frequently causes a 'resolution' of the 450 nm band producing shoulders and often a shift of the wavelength of maximum absorbance.

Reduction to the quinol form bleaches flavins. The spectrum of this 'leuco' flavin is featureless in the visible and near UV region.

Flavoproteins that yield a semiquinone may form either the blue neutral form or the red anionic form, with characteristic and distinct spectra. Some flavoproteins also, in either the oxidized or fully reduced state or both, form charge transfer complexes with strong long-wavelength absorbance bands.

Analogs
A number of analogs have been used successfully to probe the protein environment of bound flavins (26). They are not, however, commercially available.

4.9. Folic acid coenzymes

Role
Tetrahydrofolate, THF (14 p. 251, 23 p. 599, 24 p. 429), is a carrier of one-carbon fragments which are required in the synthesis of purines, pyrimidines, methionine, formylmethionine, etc. Serine is the major donor of the one-carbon units.

Vitamin relationship
The active coenzyme forms are derived from dietary folic acid (pteroylglutamic acid). In some micro-organisms *p*-aminobenzoic acid is the essential nutrient serving as the precursor for endogenous synthesis of folic acid.

Name and abbreviations
Folic acid ≡ pteroylglutamic acid ≡ vitamin M
Tetrahydrofolic acid ≡ THF ≡ H_4 Pte Glu
Dihydrofolic acid ≡ DHF ≡ H_2 Pte Glu
Folinic acid ≡ N5-formyl THF ≡ citrovorum factor ≡ leucovorin

Structure

Folic acid   Formula weight 441.4
Dihydrofolic acid   Formula weight 443.4
5-Methyltetrahydrofolic acid
     Ba salt   Formula weight 594.8
     Mg salt   Formula weight 481.8
     Na_2 salt   Formula weight 503.4
Tetrahydrofolic acid   Formula weight 445.4
Folinic acid Ca salt   Formula weight 511.5

Stability
Folic acid is photolabile. The solid is stable in the dark. Dihydrofolic acid is unstable above pH 6. It is more stable in dilute acid, without O_2 and cold. Tetrahydrofolic acid: solid oxidizes in air; store without O_2. Solutions are most stable at pH 7.4 and in the presence of a reductant, for example ascorbic acid. Unstable in acid.

Photometric properties
Folic acid, absorption maxima at 282 nm (2.7×10^4 l mol^{-1} cm^{-1}) at 350 nm and (7×10^3 l mol^{-1} cm^{-1}). Dihydrofolic acid, absorption maximum at 282 nm (2.8×10^4 l mol^{-1} cm^{-1}). Tetrahydrofolic acid, absorption maximum at 298 nm (2.8×10^4 l mol^{-1} cm^{-1}).

Analogs
Methotrexate $\equiv$ amethopterin $\equiv$ 4-amino 10-methylfolic acid
Aminopterin $\equiv$ 4-amino folic acid

4.10. Glutathione

Role
Glutathione's coenzyme functions (27) derive from the fact that, as a cysteinyl peptide, it carries a thiol group.

(i)   In the glyoxalase system the thiol of glutathione adds to a carbonyl function to promote a carbon bond rearrangement (methyl glyoxal→D-lactic acid). In the reaction catalysed by maleylacetoacetate and maleylpyruvate isomerases the thiol adds to an olefinic bond promoting enolization, rotation and hence isomerization to produce fumaryl products.

(ii)   A more central role in cells is as an antioxidant. Many processes produce damagingly reactive H_2O_2. In one of the main defense mechanisms, glutathione reduces the active site Se (in selenocysteine) of glutathione peroxidase. This facilitates catalytic reduction of H_2O_2 and of organic peroxides. Two molecules of glutathione participate and are oxidized to the disulfide (see below). The reduced form is regenerated by glutathione reductase using NADPH from the hexose monophosphate shunt.

     In this latter role glutathione is not normally described as a coenzyme, perhaps because it is not vitamin-derived and is present at high concentration (1–5 mM). However, the cycling pool, used by one enzyme and serviced by another, is analogous to the position of, say, the pools of nicotinamide coenzymes.

(iii)   In *Escherichia coli* glutathione and glutaredoxin (a protein) supply one of the routes for supplying reducing equivalents to ribonucleotide reductase, the enzyme that provides deoxyribose units for DNA.

(iv)   Glutathione services the glutathione S-transferases, a family of broad-specificity enzymes, analogous to the cytochrome P_{450} system, responsible for detoxification. Chemically reactive alkenes, epoxides, etc., are removed as glutathione conjugates.

(v) Recently (iv) has led to a biotechnological application in the large-scale production of cloned proteins. The protein is produced as a fusion protein with glutathione S-transferase, facilitating purification on a column of covalently bound glutathione. The unwanted portion of the protein is removed by proteolytic cleavage with thrombin at an engineered site.

Name and abbreviations
Glutathione is conventionally abbreviated GSH in the reduced form, and GSSG in the oxidized disulfide form.

Structure
Glutathione is a tripeptide: γ-glutamyl-cysteinyl-glycine.

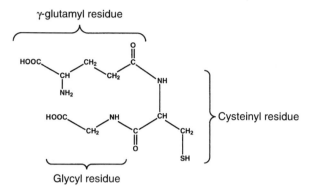

Commercially available forms

Reduced glutathione
  free acid   Formula weight 307.3
  ethyl ester   Formula weight 335.4
Oxidized glutathione
  free acid   Formula weight 612.6
  Na$_2$ salt   Formula weight 656.6
S-lactoyl glutathione,   Formula weight 379.4
  glyoxalase II substrate

Stability
GSH and GSGH are both stable as solids in air, GSH solutions are readily oxidized unless oxygen is excluded.

Analogs
Cysteinyl glycine
Mono (des-Glu) glutathione (disulfide of GSH and Cys-Gly)
Various S-alkyl GSH derivatives
Glutathione sulfonic acid

4.11. Guanine mononucleotide cofactors: GMP, GDP, GTP

Role
Guanine nucleotides are nucleic acid precursors, but in addition play an important cofactor role in energy metabolism and in protein synthesis. GTP, rather than ATP, is formed by substrate-level phosphorylation in the succinic thiokinase reaction of the Krebs cycle:

Succinyl-CoA + GDP + Pi → succinate + GTP + CoASH

GTP is also required to drive gluconeogenesis via the PEP carboxykinase reaction, which converts oxaloacetate (made by pyruvate carboxylase) to phosphoenolpyruvate.

Oxaloacetate + GTP ⟷ PEP + GDP + CO_2

GTP activates the chain elongation factor EF-Tu involved in delivering aminoacyl-tRNAs to the A site of the ribosome for assembly in protein synthesis. GTP also has crucial roles, though not as an ordinary enzyme cofactor, in the widespread signal transduction mechanisms that utilize 'G-proteins', in protein targeting, microtubule assembly, etc. In addition to its formation in the Krebs cycle, GTP can also be supplied by the nucleoside diphosphokinase reaction:

GTP + ADP ⟷ GDP + ATP

Names
Guanosine is the ribose mononucleotide of guanine.

Guanosine 5'-monophosphate ≡ GMP ≡ guanylic acid
Guanosine 5'-diphosphate ≡ GDP
Guanosine 5'-triphosphate ≡ GTP

Structure

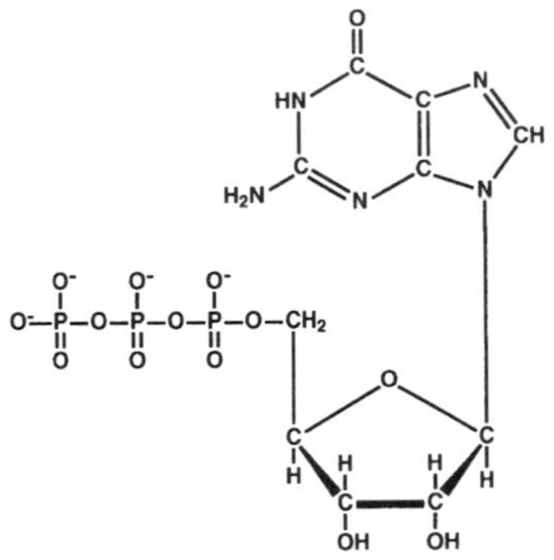

Commercially available forms

GMP
   free acid   Formula weight 363.2
   Na_2 salt   Formula weight 407.2
   Na_2 salt ($7H_2O$)   Formula weight 533.2
GDP
   Li_2 salt   Formula weight 455.0
   Na and Tris salts also available
GTP
   Li_2 salt   Formula weight 535.1
   Na_2 salt   Formula weight 567.1
   Tris salt
   Li salt available in solution

Stability
GMP (crystalline Na_2 salt), reportedly stable at room temperature. GDP, Li_2 salt soluble dry at 4°C. GTP salts, store dry at 4°C; up to 5% decomposition in 6 months.

COFACTORS

Photometric properties
UV absorption spectra of GMP, GDP and GTP essentially identical at pH 7. Maximum at 252 nm (1.37×10^4 mol^{-1} cm^{-1}), deep trough at 228 nm (2.5×10^3 mol^{-1} cm^{-1}), shoulder around 275 nm. The shoulder is more pronounced at low pH.

Analogs
Periodate-oxidized derivatives of GMP, GDP and GTP are available. Also the following derivatives modified in the phosphate ester linkages: GMP PNP, GDPβS, GTPβS and GTPγS (see Section 4.1 for a fuller explanation).

4.12. Heme cofactors

Role
Cells use a range of closely related cofactors, the hemes (haem) (28), for various oxidoreductions, e.g. in the electron transport chains of bacteria, mitochondria and chloroplasts; in oxidative detoxification of organic molecules by the cytochrome P_{450} system; in disposal of H_2O_2 by peroxidases and catalase. They are also used in such oxygen carrier proteins as myoglobin and hemoglobin. Hemes have a central iron atom which, in the catalytic cycle of most hemoproteins, shuttles between the ferrous (Fe^{II}) and ferric (Fe^{III}) state. The oxidation–reduction potential of this conversion varies between different hemes, and, for any one heme, between different hemoproteins. Hemes can be synthesized by mammals and are thus not vitamin-derived.

Structures and names
The *porphin* ring system has four pyrrole rings linked in a circular arrangement by methene bridges. The *porphyrins* are a subgroup of porphins with distinct sets of substituents on the eight possible substitutable carbons in the four rings. Addition of an iron atom coordinated between the four pyrrole N atoms produces *hemes*. The structure shown is that of *protoheme*, in which the parent porphyrin is protoporphyrin IX.

Protoporphyrin IX
Protoheme is the prosthetic group of cytochromes *b* and *o*, the P_{450} cytochromes, catalase and also hemoglobin and myoglobin. A modification is seen in cytochromes *c*, b_4 and *f*, said to contain heme *c*. Heme *c* is in fact protoheme covalently linked to the SH groups of two protein cysteine residues via the two vinyl substituents on positions 2 and 4 of the ring system.

In cytochromes a and a_3, which operate at a higher oxidation–reduction potential than b and c, heme a is found. In this heme the 9-methyl is replaced by -CHO and the 2-vinyl group is replaced by

$$-CH\text{-}CH_2 \left[CH_2CH = C\text{-}CH_2 \right]_3 H$$
$$\quad\; | \qquad\qquad\quad\; |$$
$$\quad OH \qquad\qquad\; CH_3$$

Commercially available forms
Various hemoproteins are readily available (e.g. cytochromes, catalase, peroxidases) but the free hemes are usually not. The more stable porphyrins are more generally available.

Hemin chloride	Formula weight 652.0
Hematin	Formula weight 633.5
Protoporphyrin IX free acid	Formula weight 562.7
Protoporphyrin IX Na$_2$ salt	Formula weight 606.2
[1]Coproporphyrin I	
N-propyl ester	Formula weight 823.0
tetraethyl ester	Formula weight 766.9
tetramethyl ester	Formula weight 710.8
[2]Coproporphyrin III tetramethyl ester	Formula weight 710.8
[3]Uroporphyrin I	
dihydrochloride	Formula weight 1167.4
isopropyl ester	Formula weight 1167.4
octamethyl ester	Formula weight 953.0

[1], 1,3,5,7-tetramethyl 2,4,6,8-tetrapropionic acid.
[2], 1,3,5,8-tetramethyl 2,4,6,7-tetrapropionic acid.
[3], 1,3,5,7-tetraacetic acid 2,4,6,8-tetrapropionic acid.

Hematin is the ferric form of protoheme, and hemin chloride has a chlorine atom coordinated to the central iron atom of hematin.

Photometric properties
The hemes have intense absorbance bands in the visible region of the spectrum and undergo large changes in the position and amplitude of these bands upon reduction.

4.13. Lipoic acid

Role
Lipoate (12 p. 269, 13 p. 129, 29) is the covalently bound prosthetic group of the 'E2' transacetylase component of 2-oxoacid dehydrogenase multienzyme complexes (e.g. pyruvate dehydrogenase) (29). Serves a dual role, first as an oxidoreduction cofactor in the thiamine-dependent oxidative decarboxylation of the oxoacid substrate, and secondly as an acyl-transfer cofactor, receiving an acyl group from thiamine and transferring it to coenzyme A. This leaves the reduced, dithiol form of the lipoate cofactor, which is then reoxidized by NAD$^+$ in a reaction catalysed by E3, lipoamide dehydrogenase. The lipoic acid is covalently attached in amide linkage to the ϵ-NH$_2$ group of a lysine residue in a highly mobile domain of E2 (30), forming a 'swinging arm'. As a substrate for purified enzymes free lipoamide can substitute for E2-bound lipoate.

Synonyms
6,8-thioctic acid, 1,2-dithiolane 3-pentanoic acid.

Structure

Stability
Lipoic acid is available both in the oxidized form (crystalline) and in the reduced form (liquid). The latter is supplied in a sealed ampoule since the dithiol is very susceptible to oxidation. Lipoamide is only available in the crystalline oxidized form which must be chemically reduced as required.

Photometric properties
The dithiolane ring has an absorption band at 335 nm, giving the oxidized cofactor a yellowish color.

Commercially available forms

DL-Lipoic acid, oxidized form ($>99\%$)   Formula weight 206.3
DL-Lipoic acid, reduced form ($\sim 98\%$)   Formula weight 208.3
DL-Lipoamide, oxidized form (99–100%)   Formula weight 205.3

4.14. Nicotinamide cofactors

Role
Oxidoreduction coenzymes (14 p. 147, 23 p. 3, 24 p. 3, 31), receiving or donating a pair of hydrogens as a proton and a hydride equivalent. Dehydrogenases using these coenzymes (e.g. 2, 4) are generally specific either for NAD^+ or for $NADP^+$, although a few enzymes show dual specificity (31). Usually catabolic enzymes (e.g. glyceraldehyde 3-phosphate dehydrogenase in glycolysis) use NAD^+, and biosynthetic reductions (e.g. enoyl CoA reductase in fatty acid biosynthesis) use NADPH. However, a few specialized catabolic dehydrogenases, notably those of the hexose monophosphate shunt (glucose 6-phosphate and 6-phosphogluconate dehydrogenases) exist to generate the NADPH for biosynthesis and accordingly are $NADP^+$-dependent (32, 33).

Vitamin relationship
Derived from the B-vitamin niacin or nicotinamide.

Names, abbreviations and synonyms
Nicotinamide adenine dinucleotide, abbreviated NAD (reduced form $NADH_2$) or NAD^+ (reduced form $NADH + H^+$). The latter is chemically more correct, indicating the positive charge on the nicotinamide ring and the fact that the coenzyme only receives one of the two H atoms transferred. Coenzyme I or cozymase in the early literature, then DPN (diphospho-pyridine nucleotide. Preparation of a range of analogs made this vague name untenable. Replaced by the more correct NAD^+ in an IUB recommendation in the 1960s. 'DPN' was

retained by many US authors and journals through the 1970s. Also, more specifically termed β-NAD$^+$ to distinguish from α-NAD$^+$ in which the nicotinamide's glycosidic linkage is to the α rather than the β anomer of ribose.

Nicotinamide adenine dinucleotide phosphate, abbreviated NADP (reduced form NADPH$_2$) or NADP$^+$ (reduced form NADPH + H$^+$). Coenzyme II in the early literature, then TPN (triphosphopyridine nucleotide). Present designation since 1960s (see above for NAD$^+$).

Structure
These coenzymes, despite the abbreviation NAD(P)$^+$, carry a net negative charge at neutral pH owing to the presence of the phosphodiester linkage and, in NADP$^+$, an extra phosphate group on the 2'-OH of the adenine ribose. NMR shows that in free solution nicotinamide coenzymes adopt a 'stacked' conformation, with the nicotinamide ring overlaying the adenine ring. X-ray crystallography shows that in the active sites of dehydrogenases the coenzyme molecules are extended. In NADH and NADPH C-4 of the nicotinamide ring carries two hydrogen atoms, one derived from the last substrate molecule oxidized. The two positions are stereochemically distinct and dehydrogenases fall into two classes, 'A-specific' and 'B-specific' according to which side of the nicotinamide ring they use. This is explained by the enzyme selecting either the 'syn' or 'anti' conformation in binding the coenzyme.

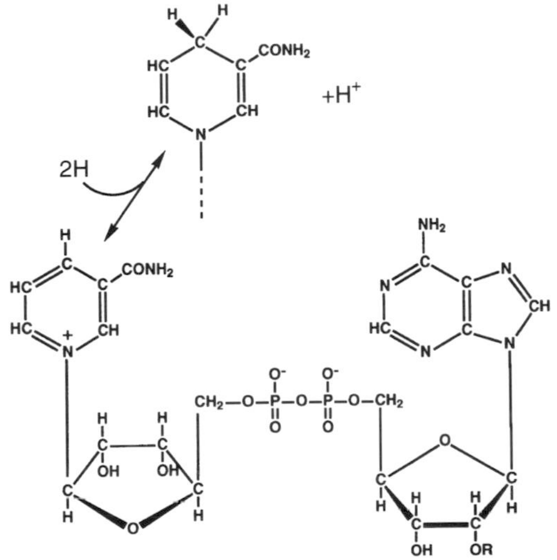

Commercially available forms

NAD$^+$

Free acid	Formula weight 663.4
Free acid, crystallized with 3H$_2$O	Formula weight 717.4
Lithium salt, 2H$_2$O	Formula weight 705.4
Lithium salt	Formula weight 669.4
Sodium salt	Formula weight 685.4

NADH

Disodium salt	Formula weight 709.4
Dipotassium salt	Formula weight 741.6
Di(cyclohexylammonium) salt	Formula weight 863.8
Di Tris salt	Formula weight 907.7
Monosodium salt	Formula weight 765.4

NADP$^+$
  Disodium salt   Formula weight 787.4
  Monopotassium salt (2H$_2$)   Formula weight 817.4
  Monopotassium salt   Formula weight 781.5
  Free acid   Formula weight 743.4
  Tris salt   Formula weight 864.5
NADPH
  Tetrasodium salt   Formula weight 833.4
  Tetrapotassium salt   Formula weight 897.8
  Tetra Tris salt   Formula weight 1230
  Tetra (cyclohexylammonium) salt   Formula weight 1142

Stability
The solids are best stored in a desiccator at 4°C, taking care to warm to room temperature before opening for use, in order to avoid condensation. In this state they may be successfully kept for weeks/months. Less stable when moist. Important to avoid freezing the reduced cofactors as this generates potent inhibitors. For demanding applications (e.g. kinetics, especially fluorimetric) it is best to use fresh samples and/or check purity directly (e.g. by analytical HPLC). In solution, NAD$^+$ and NADP$^+$ are relatively stable at low pH but unstable in alkali. NADH and NADPH are moderately stable at mildly alkaline pH but unstable at acid pH values. For reaction mixtures at pH 7–8 NAD(P)H may safely be mixed with the buffer but NAD(P)$^+$ is best kept at low pH (e.g. by dissolving in water) and mixed with the buffer shortly before use.

Purification
Preparations of 95% purity are much cheaper than the highest grades, and may be purified either by low pressure LC (34) or HPLC (35). The latter has the advantage that the methanol and acetonitrile solvents may be rapidly removed by evaporation. Low pressure chromatography usually entails either a subsequent gel filtration step or lyophilization to remove a volatile salt.

Enzymatic assay
If necessary, the concentration of enzymatically active coenzyme may be checked by quantitative enzyme assay using dehydrogenase reactions driven to equilibrium in one direction or the other.

Photometric properties
The oxidized cofactors have an absorbance peak at 260 nm (1.8×10^4 l mol^{-1} cm^{-1}). Reduction of the nicotinamide ring alters this absorption maximum and also produces a new absorption band maximal at 340 nm (6.22×10^3 l mol^{-1} cm^{-1}). Excitation at 340 nm leads to visible blue fluorescence at 450 nm offering a more sensitive alternative to spectrophotometric monitoring of reactions. In pure samples (i.e. no contaminants other than salt or water) the absorbance at 260 nm [or 340 nm for NAD(P)H] may be used to determine concentration.

Analogs
A number of useful NAD(P)$^+$ analogs are available. These include:

(i)   Acetylpyridine analogs of NAD$^+$, NADH and NADP$^+$. The -CONH$_2$ of nicotinamide is replaced by -COCH$_3$. Their principal importance lies in their higher oxidation–reduction potential relative to NAD(P)$^+$. This allows them to be used in enzymatic metabolite assays (e.g. for lactate, malate) without resorting to excessively high pH or trapping agents to pull the reaction towards complete oxidation. The absorption maximum in the reduced form is at 361 nm with an extinction coefficient of 9.0×10^3 mol^{-1} cm^{-1}.

(ii) Etheno derivatives of NAD^+ and $NADP^+$. An extra carbon atom bridges the two rings of adenine. These derivatives (36) are strongly fluorescent in the oxidized form, emitting maximally at 410 nm when extracted with light in the lowest energy absorbance band (290–310 nm). They are active cofactors for many dehydrogenases and offer a way to study binding of an oxidized cofactor with a spectrophotometric signal.

(iii) Thionicotinamide analogs of NAD^+, $NADP^+$, NADH and NADPH ($-CSNH_2$ replaces $-CONH_2$). Different absorption maximum for reduced forms ($1.19 \times 10^4 \, mol^{-1} \, cm^{-1}$ at 398 nm) allows monitoring of transhydrogenase-type reactions (37).

(iv) Hypoxanthine (Deamino) analogs of NAD^+, $NADP^+$, NADH and NADPH. ($6-NH_2$ of adenine replaced by -OH). Good alternative cofactor in kinetic studies. Also called deamino $NAD(P)^+$.

(v) Fragments. Nicotinamide mononucleotide (both oxidized and reduced) and ADP ribose and 'ATP ribose' (2'-phospho ADP ribose) fragments of NAD^+ and $NADP^+$ are available for kinetics and binding studies.

4.15. Pyridoxal phosphate

Role
Pyridoxal phosphate (6, 12 p. 433, 13 p. 405, 14 p. 97) carries an aldehyde substituent at the 4 position of its pyridine ring. In pyridoxal, the $-CH_2OH$ substituent on the 5 position allows facile cyclization to form a 5-membered internal hemiacetal ring. The 5'-phosphate group in pyridoxal phosphate prevents this. This cofactor serves as prosthetic group for various decarboxylases, aminotransferases and racemases. Normally it is bound in Schiff's base linkage to the ϵ-NH_2 of a lysyl residue in the enzyme's active site. Typically the substrate (usually an amino acid) reacts with the enzyme by transimination so that the cofactor becomes linked to the α-N of the substrate instead of the lysyl ϵ-N. The electron-withdrawing properties of the cofactor then facilitate the next step, for example loss of -COOH as CO_2 in decarboxylases or loss of α-NH_2 to yield the pyridoxamine form of the cofactor and an oxoacid product in the first half of an aminotransferase reaction – an incoming oxoacid reverses the sequence of events to yield an amino acid product in the second half.

Vitamin relationship
The cofactor is derived from vitamin B_6. The term B_6 was first applied to pyridoxine (pyridoxol), the alcohol corresponding to the aldehyde pyridoxal. However, pyridoxal, pyridoxamine and their 5'-phosphates may also be regarded as forms of vitamin B_6.

Names and abbreviations
Pyridoxal phosphate $\equiv$ pyridoxol 5'-phosphate $\equiv$ codecarboxylase $\equiv$ Pal P
Pyridoxine $\equiv$ pyridoxol $\equiv$ vitamin B_6

Structure

Erratum

The structure for pyridoxal phosphate on p. 243 is incorrect. The correct structure is printed below.

Commercially available forms

Pyridoxal HCl	Formula weight 203.6
Pyridoxal 5'-phosphate	Formula weight 247.1
Pyridoxamine dihydrochloride	Formula weight 241.1
Pyridoxamine 5'-phosphate hydrochloride	Formula weight 284.6
Pyridoxine	Formula weight 169.2
Pyridoxine hydrochloride	Formula weight 205.6

Stability
All the above forms of B_6 are stable in solution in cold and dark, but photolabile both as solids and in solution, especially at alkaline pH.

Photometric properties
At neutral pH pyridoxal 5'-phosphate has absorption maxima at 330 nm and 388 nm. Pyridoxamine 5'-phosphate has maxima at 254 nm and 327 nm. Upon reaction to form a Schiff's base, the bound form of pyridoxal 5'-phosphate has a peak at about 425 nm. Reduction with $NaBH_4$, converting imine to amine, gives a spectrum similar to that of free pyridoxamine phosphate.

4.16. Pyrroloquinoline-quinone

Role
The cofactor of quinoproteins (38, 39), a class of bacterial oxidoreductases that do not employ nicotinamide or flavin in cofactors.

Name
Also known as PQQ or methoxatin.

Structure

Commercially available forms

Free acid	Formula weight 330.2
Potassium salt	Formula weight 444.5

Stability
The salt is stable at 4°C if stored dry and in the dark.

4.17. Thiamine pyrophosphate

Role
Thiamine pyrophosphate (12 p. 73, 13 p. 51, 14 p. 15) is the prosthetic group of various enzymes that catalyse the α C-C cleavage of 2-oxoacids and hydroxyketones. These include the

decarboxylation components (E1) of oxoacid dehydrogenase multienzyme complexes, and transketolase and phosphoketolase, both important in sugar metabolism.

Vitamin relationship
Thiamine was the first clearly identified vitamin, the anti-beriberi factor, vitamin B_1.

Names and abbreviations
Thiamine is also sometimes referred to as aneurin. Thiamine phyrophosphate, TPP $\equiv$ carboxylase.

Structure

Commercially available forms

Thiamine chloride hydrochloride	Formula weight 337.3
Thiamine monophosphate chloride	Formula weight 380.8
Thiamine pyrophosphate chloride	Formula weight 460.8

Stability
Dry solids are stable. Solutions of thiamine ClHCl stable at pH 3.5, but unstable at neutral pH and above, especially to heat. TPP is less stable.

4.18. Uracil mononucleotide cofactors: UTP, UDP, UMP

Role
As well as being an RNA precursor, UTP has an important cofactor role in the activation of sugar units in polysaccharide synthesis, for example formation of glycogen in mammalian liver or muscle.

UTP + glucose 1-phosphate⟷UDP-glucose + pyrophosphate
  UDP-glucose pyrophosphorylase

UDP is then displaced by the 4 OH of a terminal glucose unit in glycogen to elongate the polysaccharide.

Likewise, involved in the activation of *N*-acetylglucosamine in the synthesis of cell surface oligosaccharides.

Names
Uridine is the ribose mononucleoside of uracil.
Uridine 5'-monophosphate $\equiv$ UMP $\equiv$ uridylic acid
Uridine 5'-diphosphate $\equiv$ UDP
Uridine 5'-triphosphate $\equiv$ UTP

Structure

Commercially available forms

UMP
   free acid   Formula weight 324.2
   Na$_2$ salt   Formula weight 368.1
UDP   Formula weight 404.2
   available as Na and Tris salts
   also K$_2$ salt (3H$_2$O)   Formula weight 534.4
UTP   Formula weight 484.1
   available as Na$_3$ salt   Formula weight 550.1
   or as Tris salt, or as Li salt in solution

UDP derivatives are available for *N*-acetylglucosamine, galactose, galactronic acid, glucose, glucuronic acid, hexanolamine, mannose and xylose.

Stability
UDP is stable stored dry at 4°C. Under the same conditions UTP may suffer decomposition – a few percent in 6 months. The UTP Li salt in solution is stable at −20°C.

Photometric properties
UMP, UDP and UTP show essentially identical UV absorption spectra at pH 7 with a maximum at 262 nm (10^4 l mol^{-1} cm^{-1}) and a deep trough at 230 nm (2.2×10^3 l mol^{-1} cm^{-1}).

5. REFERENCES

1. Keil, B. (1971) in *The Enzymes*, Vol. 3 (P.D. Boyer, ed.). Academic Press, New York, p. 250.

2. Bränden, C.-I, Jörnvall, H., Eklund, H. and Furugren, B. (1975) in *The Enzymes*, Vol. 11 (P.D. Boyer, ed.). Academic Press, New York, p. 104.

3. Bray, R.C. (1975) in *The Enzymes*, Vol. 12 (P.D. Boyer, ed.). Academic Press, New York, p. 300.

4. Holbrook, J.J., Liljas, A., Steindel, S.G. and Rossmann, M.G. (1975) in *The Enzymes*, Vol. 11 (P.D. Boyer, ed.). Academic Press, New York, p. 191.

5. Engel, P.C. (1992) in *Chemistry and Biochemistry of Flavoenzymes*, Vol. 3 (F. Müller, ed.). CRC Press, Boca Raton, FL, p. 597.

6. Metzler, D.E. (1977) in *The Chemical Reactions of Living Cells*. Academic Press, New York (International Edition), p. 444.

7. Wulff, K. (1983) in *Methods of Enzymatic Analysis*, Vol. 1 (H.-U. Bergmeyer, ed.). Verlag Chemie, Weinheim, p. 340.

8. Williamson, J.R. and Corkey, B.E. (1969)

Methods in Enzymology (J.M. Lowenstein, ed.). Academic Press, New York, Vol. 13, p. 488.

9. Jaworek, D. and Welsch, J. (1985) in *Methods of Enzymatic Analysis* (H.-U. Bergmeyer, ed.). Verlag Chemie, Weinheim, Vol. 7, p. 365.

10. Secrist, J.A. III, Barrio, J.R., Leonard, N.J. and Weber, G. (1972) *Biochemistry* **11**, 3499.

11. Counsell, J.N. and Hornig, D.H. (1981) (eds) *Vitamin C (Ascorbic Acid).* Applied Science Publishers, London.

12. McCormick, D.B. and Wright, L.D. (1970) (eds) *Methods in Enzymology,* Vol. 18A, *Vitamins and Coenzymes* Part A. Academic Press, New York.

13. McCormick, D.B. and Wright, L.D. (1979) (eds) *Methods in Enzymology,* Vol. 62, *Vitamins and Coenzymes* Part D. Academic Press, New York.

14. Chytil, F. and McCormick, D.B. (1986) (eds) *Methods in Enzymology,* Vol. 62, *Vitamins and Coenzymes* Part G. Academic Press, New York.

15. Knowles, J.R. (1989) *Ann. Rev. Biochem.,* **58**, 195.

16. McCormick, C.B. and Wright, L.D. (1971) (eds) *Methods in Enzymology,* Vol. 18C, *Vitamins and Coenzymes* Part C. Academic Press, New York.

17. McCormick, D.B. and Wright, L.D. (1980) (eds) *Methods in Enzymology,* Vol. 67, *Vitamins and Coenzymes* Part F. Academic Press, New York.

18. Chytil F. and McCormick, D.B. (1986) (eds) *Methods in Enzymology,* Vol 123, *Vitamins and Coenzymes* Part H. Academic Press, New York.

19. Stadtman, E.R. (1955) *Methods in Enzymology,* Vol. 1, p. 596 (S.P. Colowick and N.O. Kaplan eds). Academic Press, New York.

20. Lipmann, F. and Tuttle, L.C. (1945) *J. Biol. Chem.,* **158**, 505.

21. Michal, G. and Bergmeyer, H.-U. (1985) in *Methods of Enzymatic Analysis,* Vol. 7 (H.-U. Bergmeyer ed.). Verlag Chemie, Weinheim, p. 169.

22. Lynen, F. and Ochoa, A. (1953) *Biochim. Biophys. Acta,* **12**, 299.

23. McCormick, D.B. and Wright, L.D. (1971) (eds) *Methods in Enzymology,* Vol. 18B, *Vitamins and Coenzymes* Part E. Academic Press, New York.

25. Muller, F. (1992) (ed.) *Chemistry and Biochemistry of Flavoenzymes.* CRC Press, Boca Raton, FL.

26. Ghisla, S. and Massey, V. (1986) *Biochem. J.,* **239**, 1.

27. Dolphin, D., Poulson, R. and Avramovic, O. (1989) *Glutathione Chemistry, Coenzymes and Cofactors,* Vol. 3, Parts A and B. *Biochemistry and Medical Aspects.* John Wiley, New York.

28. Florkin, M. and Stotz, E.H. (1966) (eds) *Comprehensive Biochemistry,* Vol. 14, *Biological Oxidation,* Chapters 4–10. Elsevier, Amsterdam.

29. Yeaman, S.J. (1989) *Biochem. J.,* **257**, 625.

30. Miles, J.S., Guest, J.R., Radford, S.E. and Perham, R.N. (1988) *J. Mol. Biol.,* **202**, 97.

31. Goldin, B.R. and Friedin, C. (1971) *Curr. Top. Cell. Reg.,* **4**, 77.

32. Levy, H.R. (1979) in *Advances in Enzymology* (A. Meister ed.). John Wiley, New York, Vol. **49**, p. 97.

33. Villet, R.H. and Dalziel, K. (1972) *Eur. J. Biochem.,* **27**, 244.

34. Dickinson, F.M. and Engel, P.C. (1977) *Anal. Biochem.,* **82**, 512.

35. Wenz, I., Loesche, W., Till, U., Pettermann, H. and Horn, A. (1976) *J. Chromatog.,* **120**, 187.

36. Barrio, J.R., Secrist, J.A. III and Leonard, N.J. (1972) *Proc. Natl Acad. Sci. USA,* **60**, 2039.

37. Stein, A.M. and Stein, J.H. (1965) *Biochemistry,* **4**, 1491.

38. Duine, J.A. and Jonegejan, J.A. (1989) *Ann. Rev. Biochem,* **58**, 403.

39. McIntire, W.S. (1992) in *Essays in Biochemistry* Vol 27 (K.R. Tipton, ed.). Portland Press, London, p. 119.

Cofactors

CHAPTER 8
EPR SPECTROSCOPY IN ENZYMOLOGY

R. Cammack and J.K. Shergill

1. INTRODUCTION

Electron paramagnetic resonance (EPR) spectroscopy is a technique that has found specialist applications in many aspects of enzymology. Although generally considered to be a chemical method, there have been a number of innovations in the use of the technique for biological studies. EPR is reasonably rapid, sensitive, quantitative and selective in the compounds it detects. It has been applied successfully to detect free radicals and metal centers in the active sites of enzymes. It can also provide structural information, such as the types of atoms around the active site, and the distribution of unpaired electrons. The structural information derived from EPR is often complementary to that obtained from crystallography since it is sensitive to the interactions of enzyme active sites with protons and unpaired electrons, which do not appear in diffraction patterns.

EPR, also known as electron spin resonance (ESR) or electronic magnetic resonance (EMR) spectroscopy, is a magnetic resonance technique used for the study of paramagnetic materials (i.e. those that contain unpaired electrons) (1, 2). It is analogous to nuclear magnetic resonance (NMR) spectroscopy, in that it involves resonant absorption of electromagnetic radiation by the electron spins ($S = 1/2$), in which a splitting of energy levels is induced by an applied magnetic field. Since the majority of molecules contain only paired electrons, EPR is a more selective technique than NMR. The principles of EPR spectroscopy are described in various texts (1, 2), some of which describe biological applications (3, 4).

For some applications in enzymology, the EPR spectrometer may be used like a spectro-photometer. From the characteristics of the EPR spectrum, the nature of a paramagnetic species may be identified. EPR may be applied to complex mixtures such as cell extracts, and used to monitor the formation of free radicals that are the substrates or products of enzyme reactions. In addition, from the spectroscopic parameters, notably the g-factors and hyperfine interactions, EPR can provide detailed information about the chemistry of transition ions and radicals in enzyme active sites. The method has some limitations. For example, measurements of most transition metals must be made at cryogenic temperatures, which precludes normal continuous kinetic measurements. In compensation, the EPR effect allows the determination of information such as molecular motion, which would be difficult or impossible to obtain in other ways (Section 3).

1.1. EPR-detectable species in enzymology

The paramagnetic species found in enzyme systems comprise transition metal ions in certain oxidation states, free radicals and the gases O_2 and NO. Many of the enzymes that are observed by EPR spectroscopy catalyse oxidation–reduction reactions, and contain transition metal ions or cofactors that can generate free radicals. Some other types of enzymes, including transferases, hydratases and mutases, contain transition metal ions and radicals. The different types of metal centers in proteins, with examples, are summarized in *Table 1*. The iron proteins represent the greatest diversity of transition metal enzymes, and, fortunately, the EPR spectra

▶ p. 252

Table 1. Metal centers in enzymes

Type of group	Example	EPR characteristics and comments	Refs
Mn^{II}	Creatine kinase EC 2.7.3.2	Six-line hyperfine pattern due to ^{55}Mn. Superhyperfine splittings due to ^{17}O-labeled ATP allow determination of metal coordination in enzyme	6
Mn dinuclear	Catalase from *Thermus thermophilus* EC 1.11.1.6	Binuclear Mn-complex in the active site. EPR at 80–100 K revealed broad multiline signals (i.e. hyperfine structure due to two $I = 5/2$ Mn nuclei) centered around $g = 2.0$, and assigned to the three species: $[Mn^{II}\text{-}Mn^{II}]$, $[Mn^{II}\text{-}Mn^{III}]$, $[Mn^{II}\text{-}Mn^{IV}]$	7
V^V	Bromoperoxidase from various seaweeds EC 1.11.1.7	Signals with ^{51}V hyperfine splitting: $g_{\|} = 1.948$, $g_\perp = 1.979$, $g_0 = 1.969$ $a_{\|} = 17.6$, $a_\perp = 5.5$, $a_0 = 9.5$ mT	8
High-spin heme Fe^{III}	Catalase from human liver EC 1.11.1.6	Catalase A: $g_{x, y} = 6.45, 5.38, \approx 2.0$ Catalase B (formate bound): $g_{x, y} = 6.80$, 5.07, ≈ 2.0; detected at ≤ 10 K	9
	Cytochrome P_{450} EC 1.14.15.6	Substrate-bound ferric heme ($g = 8.04$) detected at ≤ 6 K, NADPH reducible	10
Low-spin heme Fe^{III}	Cytochrome P_{450} EC 1.14.15.6	Substrate-free ferric heme detected up to ~ 25 K, $g = 2.42, 2.25, 1.92$	10
	Nitric oxide reductase from *Pseudomonas stutzeri* EC 1.7.99.2	Cytochrome bc complex Low-spin c heme: $g = 3.02, 2.29, \sim 1.5$ High-spin b heme: $g = 6.34, 6.03, 5.70$	11
High-spin nonheme Fe^{III}	Lipoxygenase, soybean EC 1.13.11.12	EPR signals detected at ≤ 10 K, at: $g = 7.4, 6.2$ and 5.8 (shoulder). Signals indicate that the symmetry of the iron is tetragonal or near tetragonal	12
	Rubredoxin, *Pseudomonas oleovorans* EC 1.18.1.1	In this case, the high-spin heme Fe^{III} ion is typified by completely rhombic symmetry, with $g = 4.31, 9.4$, and is detected at ≤ 100 K	13
	Phenylalanine hydroxylase, mammalian liver EC 1.14.16.1	(i) $g = 4.3$, in phosphate buffer (ii) $g = 6.7, 5.3, 2.0$ in Tris buffer (iii) $g = 7.0, 5.2, 1.9$, in phosphate buffer + L-norepinephrine Active site iron as detected at ≤ 10 K	14
	4-Methoxybenzoate mono-oxygenase, *Pseudomonas putida* EC 1.14.99.15	Slightly different features detected at $g = 9.4$–4.3, for different substrate–enzyme complexes	15

Table 1. Continued

Type of group	Example	EPR characteristics and comments	Refs
Zn^{II}–Fe^{III} pair	Acid phosphatase, bean EC 3.1.3.2	g = 8.53, 5.55, 2.85, probably from Fe^{III}–Zn^{II} center. The zinc ion is diamagnetic, so the spectra are effectively due to the Fe^{III}	16
Fe–S, FMN	NADH:ubiquinone oxidoreductase, e.g. mitochondria, bacterial respiratory chains EC 1.6.5.3	Fe–S clusters: EPR over the range 8–77 K reveals rhombic signals from reduced (E_m = −20 to −245 mV) [2Fe–2S] and [4Fe–4S] clusters in the g = 2.09–1.88 region Signal at g = 2.00 from FMNH·radical	17, 18
[2Fe–2S], [3Fe–4S], [4Fe–4S], heme-Fe and FAD	Succinate dehydrogenase, e.g. mitochondria EC 1.3.5.1 Fumarate reductase EC 1.3.99.1	Oxidized enzyme: g = 2.01 signal from center 3, a [3Fe–4S] cluster, at ≤20 K Fe^{III} b-type heme, g ~ 3.5, shifting to g ~ 3.1 on purification Reduced enzyme: center 1: [2Fe–2S] cluster; rhombic, g = 2.03–1.90, detected at ≤77 K, and center 2: [4Fe–4S] cluster; broad, weak features at g = 2.5–1.6, detected at ≤20 K	19, 20
Low-spin Co^{II} in B_{12}	Methionine synthase, E. coli EC 2.1.1.13, 4.2.99.10	Demethylated enzyme is EPR-detectable. Signal in g = 2 region displays hyperfine structure due to interaction of electron with the cobalt nucleus (S = 7/2)	21
High-spin Co^{II} substituted for Zn	Cobalt-substituted carboxypeptidase A EC 3.4.17.1	Enzyme is still active Broad signal at g = 6, 2.0	22
Ni^{III}	Nickel–hydrogenase, Desulfovibrio gigas EC 1.18.99.1	Oxidized: $g_{x,y,z}$ = 2.32, 2.23, 2.01 (Ni-A) Hydrogen reduced $g_{x,y,z}$ = 2.19, 2.16, 2.01 (Ni-C)	23
Ni^I	Methyl coenzyme-M reductase (MCR), Methanobacterium thermoautotrophicum	Two signals detected at ≤77 K: MCR_{red1}, $g_{x, y}$ = 2.088, g_z = 2.260, MCR_{red2}, $g_{x,y,z}$ = 2.184, 2.235, 2.285 Signal shifts in inhibitor complexes	24
Cu^{II}	Ceruloplasmin, human blood EC 1.16.3.1	Spectroscopically distinguishable Cu type I, II and III sites Signals in the g = 2 region detected at ~100 K, with ^{65}Cu hyperfine structure: type I: $a_{||}$ < 10 mT type II: tetragonally coordinated Cu type III: EPR-silent, antiferromagnetically coupled binuclear site	25

Table 1. Continued

Type of group	Example	EPR characteristics and comments	Refs
	Laccase, *Rhus vernicifera* (lacquer tree) EC 1.10.3.2	Trinuclear EPR-active copper (type II:III) cluster with $a_\parallel \sim 15.5\text{--}22.5$ mT, $a_\perp \sim 18\text{--}26$ mT. Well-resolved ligand superhyperfine structure from three protein [^{14}N]histidines. The site can accommodate a variety of exogenous ligands, e.g [^{15}N]azide, CN^-	26
Mo^V cofactor with terminal $Mo = S$ ligand	Xanthine oxidase (XO), milk EC 1.1.3.22 Aldehyde oxidase (AO), liver EC 1.2.3.1	Various Mo^V signals observed at ~ 100 K: (i) very rapid: $g = 2.03\text{--}1.92$ (ii) rapid type 1: $g = 2.00\text{--}1.95$ (iii) rapid type 2: $g = 2.13\text{--}1.95$ (iv) slow: $g = 1.99\text{--}1.92$ (v) inhibited: $g = 2.00\text{--}1.92$. Signals (i)–(iii) arise during anaerobic reaction of XO with excess xanthine. The rapid signals exhibit interaction with protons $a = 0.4, 1.0\text{--}1.4$ mT	27
Mo^V cofactor with terminal $Mo = O$ ligand	Respiratory nitrate reductase, *E. coli* EC 1.7.99.4	Rhombic EPR signals at $g = 1.987, 1.976, 1.96$ observed at 120 K from the Mo^V center of oxidized nitrate reductase. ^{1}H splitting at pH 5. Enzyme also contains [3Fe–4S] cluster, detected in the oxidized state, and three [4Fe–4S] clusters detected in the reduced state	28
W^V	Formate dehydrogenase (FDH), from *Clostridium thermoaceticum* EC. 1.2.1.2	Features, $g = 2.101, 1.980, 1.950$, from the reduced enzyme at ~ 200 K attributed to W^V. Enzyme also contains [4Fe–4S] clusters	29

of iron proteins are among the most informative (5). Free radicals in enzymes may be derived from amino acids in the protein (*Table 2*), or from prosthetic groups or reaction intermediates (*Table 3*). Where such paramagnets do not occur naturally in an enzyme it may be possible to introduce one to act as a reporter group; for example, by substituting a paramagnetic metal ion, such as Mn^{2+} for Mg^{2+}, or Co^{2+} for Zn^{2+}, or by attaching a nitroxide spin label.

Relatively few enzyme substrates or products are EPR-detectable. A few enzymes react with transition metal complexes, such as chromate (38). It is possible to examine the production of stable free radicals such as the metronidazole radical (39), and in these cases continuous EPR measurements may be used. More commonly the radicals are unstable, such as the oxygen radicals generated by redox enzymes such as xanthine oxidase; these may be detected by spin trapping (40, 41).

Table 2. Free radicals in enzymes, derived from amino acyl residues

Type of group	Example	EPR characteristics	Refs
Glycine radical	Pyruvate formate-lyase from anaerobic *E. coli* EC 2.3.1.54	A free radical from carbon-2 of glycine-734 detected at $g = 2.0037$, using ^{13}C-labeled (selectively) enzyme at 225 K Hyperfine splitting (1.5 mT) from the solvent-exchangeable α proton. ^{13}C splitting $a_{\parallel} = 4.9$ mT, $a_{\perp} = 0.1$ mT for central nucleus	30
Tyrosine cation radical	Prostaglandin H synthase EC 1.14.99.1	Doublet EPR signal at $g = 2.0025$ observed ≤ 90 K, formed on aerobic addition of arachidonic acid and H_2O_2 or prostaglandin; assigned to a tyrosyl radical. Hypothetical role of radical could be abstraction of a hydrogen at C-13 of arachidonic acid, which is assumed to be the initial step of the cyclo-oxygenase reaction Signal is nearly identical to that of the tyrosyl radical in ribonucleotide reductase	31
Tryptophan cation radical	Cytochrome *c* peroxidase *Saccharomyces cerevisiae* EC 1.11.1.5	Axial radical signal observed at 5 K: $g_{\parallel} = 2.034$, $g_{\perp} = 2.006$, 1.999, with some resolved hyperfine structure. Identified as due to Trp191 which is in close proximity to the proximal heme ligand	32

The other biologically significant paramagnetic species are dioxygen (O_2) and nitric oxide (nitrogen monoxide, NO). O_2 has two unpaired electrons, and is in the so-called triplet state. It may contribute broad, rather poorly defined spectra at low temperatures (42) but tends to create interference when measuring other species. The relaxation effect of O_2 in solution causes broadening of narrow-line radicals. NO has an unpaired electron. It may be detected at very low temperatures (< 10 K) under special circumstances (43). NO is more readily detected in complexes with metal ions, for example Fe^{II}. This effect may be used to detect Fe^{II} states which are normally EPR-silent (5), or to detect NO with Fe-containing spin traps (44).

1.2. Principles of EPR spectroscopy

In an applied magnetic field, an unpaired electron (which possesses both spin and charge, and therefore a magnetic moment) can occupy one of two energy levels. The levels are characterized by a quantum number m_s, which can take values of $-1/2$ or $+1/2$. These correspond, roughly speaking, to having the electron magnetic moments aligned either *with* (parallel state) or *against* (antiparallel state) the field. The phenomenon of EPR occurs when unpaired electrons interact with applied electromagnetic radiation, at a frequency which fulfills the resonance condition; that is the energy of the quantum equals the difference between the two energy levels:

$$\Delta E = h\nu = g\mu_B B_0 \qquad\qquad \text{Equation 1}$$

where:

B_0 is the applied static magnetic field (strictly magnetic flux density). The unit commonly used is the gauss (G); the SI unit is the tesla (T); $1\ \text{T} = 10^4\ \text{G}$;

Table 3. Free radicals in enzymes, derived from prosthetic groups or substrates

Type of radical	Example	EPR characteristics	Refs
Flavin radical – red anionic	Glucose oxidase, pH 9.3 EC 1.1.3.4 Cholesterol oxidase EC 1.1.3.6	Linewidth 1.5 mT Signal at $g \approx 2.004$ Temperature $= 77$–200 K	33, 34
Flavin radical – blue neutral	Glucose oxidase, pH 5.3 EC 1.1.3.4 Xanthine oxidase EC 1.1.3.22 NADPH:cytochrome P_{450} reductase EC 1.6.2.3	Linewidth 2.0 mT Signal narrows to 1.5 mT in ^2H_2O Signal at $g \approx 2.004$ Temperature $= 77$–200 K	33
TPP radical	Pyruvate ferredoxin oxidoreductase EC 1.2.7.1	Radical at $g \approx 2.006$ shows splitting due to substrate protons	35
TOPA quinone	Galactose oxidase EC 1.1.3.9		36
Pyridoxal phosphate – amino acyl	Lysine 2,3-aminomutase from *Clostridium* EC 5.4.3.2	$g \approx 2$ radical signal observed at ~ 77 K with enzyme in the presence of lysine and S-adenosyl methionine. Identified as a π-radical with hyperfine couplings to α, β-^{1}H, and β-^{14}N of lysine	37

TOPA, 6-hydroxydopa quinone; TPP, thiamin diphosphate radical.

ν is the frequency of the applied microwave radiation; units are gigahertz (GHz);
h is Planck's constant, 6.6262×10^{-34} m^2 kg s^{-1};
μ_B is the Bohr magneton for the electron, 9.2741×10^{24} J T^{-1}.

The EPR spectrometer is designed to record an EPR signal when the microwave frequency ν and magnetic field B_0 fulfill the resonance condition for a paramagnetic species. The g-factor feature of an EPR spectrum is calculated from *Equation 1*.

Numerically if ν is in GHz, and B_0 in mT:

$$g = 71.4484 \, \frac{\nu}{B_0} \qquad \qquad \text{Equation 2}$$

or if ν is in GHz, and B_0 in gauss:

$$g = 714.484 \, \frac{\nu}{B_0}. \qquad \qquad \text{Equation 3}$$

The principal differences between EPR and NMR arise from the much greater magnetic moment of the electron compared with those of nuclei (1836 times that of the proton). The spectral range is much greater than in NMR, and spin–spin interactions of electrons with other electrons (the fine-structure interaction) or with nuclei (the hyperfine interaction) are much stronger.

ENZYMOLOGY LABFAX

For the free electron, g is a precisely known constant ($g_e = 2.0023193044...$). However, in EPR of paramagnetic materials the g-factor is treated as a dimensionless spectroscopic variable whose value is a characteristic of the paramagnetic center. It is therefore an aid in the identification of an unknown signal. In most organic radicals the range of g is between 2.000 and 2.006. In paramagnetic metal ions it is larger. Where an ion has more than one unpaired electron, such as high-spin Fe^{III} ($S = 5/2$), values between 1 and 10 are common.

The g-factor is anisotropic; that is, its value varies according to the orientation of the molecule in the applied magnetic field. Since biochemical preparations consist of molecules that are randomly orientated relative to the magnetic field, the spectrum consists of the sum of spectra with different g-factors, and is therefore broadened. The g-factor anisotropy is characterized by three principal values, g_z, g_y and g_x. In the case of axial symmetry, as found for many transition metals, the EPR resonance is characterized by two values, as $g_z \neq g_y = g_x$). By convention, g_z is defined as the g-factor observed when B_0 is parallel to the symmetry axis ($g_{||}$), and g_x and g_y are observed when B_0 is perpendicular to the symmetry axis ($g_\perp$). If $g_z \neq g_y \neq g_x$, the system has rhombic symmetry. For small molecules such as nitroxides in solution, rapid motion averages out this anisotropy, and a narrow line is seen at the average g-factor. However, in the solid state (which, in EPR, includes proteins in solution) the motion is not rapid enough, and the spectrum is spread over the range of all possible g-factors. Hence for transition metal ions the range of the spectrum may be very large.

1.3. Hyperfine interactions

One of the most characteristic features of radicals and transition metal complexes, which assists in their identification, is the hyperfine coupling of the electron spin with nearby nuclear spins. The values of relevant nuclear spins are given in *Table 4*. The hyperfine interaction with a nuclear spin, I, is defined by the number of lines and the magnitude of the hyperfine splitting, A.

Table 4. Some nuclei with magnetic moments

Nucleus	Natural abundance of isotope (%)	Spin	Number of hyperfine lines
^{1}H	99.985	1/2	2
^{2}H	0.015	1	3
^{13}C	1.11	1/2	2
^{14}N	99.63	1	3
^{15}N	0.37	1/2	2
^{17}O	0.037	5/2	6
^{19}F	100	1/2	2
^{31}P	100	1/2	2
^{33}S	0.76	3/2	4
^{51}V	99.76	7/2	8
^{53}Cr	9.55	3/2	4
^{55}Mn	100	5/2	6
^{57}Fe	2.19	1/2	2
^{59}Co	100	7/2	8
^{61}Ni	1.134	3/2	4
^{63}Cu	69.1	3/2	4
^{65}Cu	30.9	3/2	4
^{95}Mo	15.7	5/2	6
^{97}Mo	9.46	5/2	6

If the EPR spectrum is a narrow line, the hyperfine splitting is seen as a splitting into $(2I+1)$ lines; for example, a proton, ^1H, $I = 1/2$, splits the line into 2. If the electron also interacts with a second nucleus, the lines will be further split, and so on. For cases such as radicals where the spectra show so many splittings as to be uninterpretable, or for transition metal spectra where the small hyperfine splittings are concealed by the linewidth of the spectrum, greater resolution is offered by electron–nuclear double resonance (ENDOR) spectroscopy (Section 2.1).

In spectroscopic measurements the splitting is often expressed in magnetic field units, in which case it is given the symbol a. For molecular orbital calculations it is expressed in energy units, such as MHz, and given the symbol A. The relationship between these parameters is

$$A = \frac{g\mu_B}{h \times 10^9} \cdot a \quad (A \text{ in MHz, } a \text{ in mT}) \qquad \text{Equation 4}$$

The magnitude of a or A depends on the magnetic moment of the nucleus, and the extent to which the electron spin interacts with the nucleus. It also decreases proportionately as the electron density is distributed over more atoms. Thus it is possible to deduce the distribution of unpaired electron density over a molecule.

2. INSTRUMENTATION

A typical EPR spectrum is unlike other forms of spectra in that the abscissa is the magnetic field, and the ordinate is the first derivative of the absorption. This is because of the design of the EPR spectrometer, the components of which are illustrated in *Figure 1a*. For a description of the operating principles and construction of spectrometers, see refs 45 and 46. The conventional continuous-wave EPR spectrometer operates in the X-band range of microwave frequency, 9–10 GHz. It employs a microwave cavity, in which the sample sits, which resonates at a fixed microwave frequency, v. This enhances the microwave field at the sample (by a factor known as the Q-factor of the cavity), and hence increases the sensitivity. However, it prevents the acquisition of a spectrum in the conventional way by measuring absorption as a function of frequency. Instead, the applied magnetic field, B_0, is varied or 'swept', by means of an electromagnet (*Figure 1*). As can be seen from *Equation 1*, this will allow signals with different g-factors to be brought into resonance, albeit with certain distortions (47). To further enhance the signal:noise ratio, the magnetic field is rapidly modulated over a small range with additional coils, and the signal is detected in phase with the modulation, resulting in a first-derivative display of the absorption spectrum.

In *Figure 1b*, the sample is shown in a quartz flat cell for measurement of aqueous solutions, in a rectangular cavity. Other designs include cylindrical cavities such as loop–gap resonators (48) which offer a higher filling factor η (the proportion of the cavity volume which is occupied by the sample) . For work at cryogenic temperatures a flow cryostat may be inserted through the cavity, or the cavity may be immersed in liquid helium.

2.1. Specialist EPR techniques *(Table 5)*

Other types of EPR spectrometer exist; in particular, pulsed spectrometers. In a pulsed EPR spectrometer, the sample is irradiated with a sequence of high-power microwave pulses (typically 1 kW power and 10–20 nsec long), and the EPR signal is detected as a spin-echo. This can be measured as a function of magnetic field, as in a conventional spectrometer, but this offers few advantages and gives lower sensitivity. The method is more useful for measuring transient species induced, for example, by a laser flash, or for measuring electron-spin relaxation times.

Figure 1. Diagram of a continuous-wave EPR spectrometer. The microwave source is either a klystron or a Gunn diode. The rest of the microwave circuit, called the microwave bridge, is designed to operate at a 'fixed' frequency, and supplies microwave power to the cavity. Resonant absorption by the sample is detected by a microwave diode detector. The magnetic field B_0 is provided by an electromagnet. (b) Rectangular microwave cavity, with flat cell for aqueous samples.

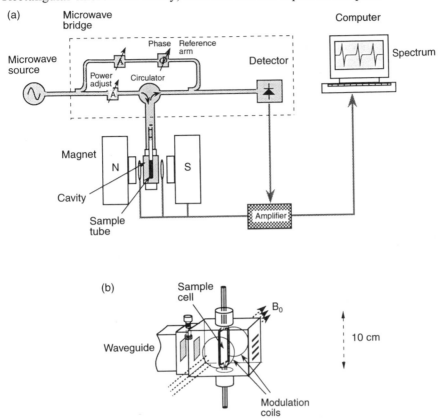

ENDOR is an extension of the EPR technique, specially for detection of electron–nuclear hyperfine interactions. The sample is irradiated with radiofrequency (RF) at NMR frequencies in addition to the microwave radiation. In continuous-wave ENDOR, the EPR signal is partially saturated with microwave power, and the NMR frequencies are swept. In pulsed ENDOR, the sample is irradiated with a sequence of microwave and RF pulses and the signal is detected as an electron spin-echo. In both cases, an enhancement of the EPR signal is observed when a nucleus coupled to the electron spin comes into resonance.

Each nucleus, with spin I, causes a hyperfine splitting of the spectrum into $(2I+1)$ lines. Thus, for example, the spectrum of the nitroxide spin-label is split into three lines by the ^{14}N nucleus ($I=1$). Where there are multiple (n) nuclei, the splittings are multiplicative, that is, each line is split into $(2I_1+1)(2I_2+1)...(2I_n+1)$ lines. In practice, for centers in proteins only the strongest hyperfine splittings are resolved in EPR, because the anisotropy is not averaged out and the spectra are broadened. ENDOR spectroscopy avoids these difficulties. In ENDOR, each additional hyperfine interaction adds just one pair of lines to the spectrum, so the number of

Table 5. Specialized instrumental techniques

Technique	Applications	Equipment needed	Notes	Ref.
Variable-frequency EPR, e.g. S-band (2–4 GHz), Q-band (35 GHz)	Resolution of hyperfine interactions and g-factors, since lines separated by g vary with frequency, but lines separated by A do not	Different microwave bridges and cavities	Needs a different microwave bridge for each frequency	49
Low-frequency EPR 0.25–1 GHz	Measurements in large aqueous samples EPR imaging	Radiofrequency resonator	Lower sensitivity than X-band	50
High-frequency EPR 95–250 GHz	Resolution of g-factors in radicals Zero-field splittings in transition metal ions	Superconducting magnet Special bridge	Difficult to sweep over wide range	51
Parallel-mode EPR	Detection of even-spin systems, e.g. $S = 1, 2$, for which the EPR transitions are forbidden in normal transverse mode	Bimodal cavity with microwave field parallel to the B_0 field		52
ENDOR (continuous wave)	Resolution of hyperfine interactions	Radiofrequency (RF) source; special cavity	Less sensitive than EPR; requires power saturation	53, 54
Pulsed ENDOR	Resolution of hyperfine interactions	Pulsed spectrometer RF source, special cavity	Saturation of the signal is not necessary	58
ESEEM	Hyperfine interactions, especially to ^{14}N, ^2H	Pulsed EPR spectrometer	For weak hyperfine interactions ≤ 10 MHz	56
Saturation-recovery EPR	Measurement of electron-spin relaxation rates: observation of paramagnetic centers at distances ≤ 4 nm	Pulsed EPR spectrometer		55
Flash photolysis	Light-induced EPR signals	Pulsed spectrometer, laser illumination into cavity	Normally requires a reversible reaction as signal must be accumulated	57

lines is $2n$. Also, the spectral resolution in ENDOR is in the MHz range of NMR rather than the GHz range of EPR. The greater resolution of ENDOR spectroscopy is obtained at the expense of sensitivity, and sample concentrations of the order of 100–500 µM are needed.

Electron spin-echo envelope modulation (ESEEM) is a specific pulsed EPR method for measuring anisotropic hyperfine couplings. It is effective for detecting the presence of nuclei with $I > 1/2$, which have a quadrupole moment, and which are more difficult to detect by ENDOR. It is only seen in the solid state, and normally requires cryogenic temperatures. The method involves measuring the intensity of the electron-spin echo resulting from the application of two or more microwave pulses, as a function of the temporal spacing between the pulses. The echo intensity is modulated as a result of interactions with the nuclear spins. The frequency-domain spectrum, derived by Fourier transformation, corresponds to hyperfine transition frequencies. The method is particularly sensitive to ^{14}N nuclei in the vicinity of the paramagnetic center.

2.2. Sample requirements

Samples are held in quartz tubes, since glass contains paramagnetic impurities. It is often an advantage to have as large a sample as possible, to enhance the signal intensity, since paramagnetic centers tend to be dilute in biochemical systems. However, aqueous samples are 'lossy', due to dielectric microwave absorption. The maximum volume permitted depends on the type of cavity; loop–gap and dielectric resonators are smaller but allow larger filling factors. Typically, the volume is limited to < 50 µl. Frozen solutions are less lossy and the size of sample is limited by the dimensions of the cryostat and cavity; the optimum volume required for signal detection is of the order of 100 µl.

The concentration required is dependent upon the properties of the EPR signal detected. As the EPR signal amplitude is inversely proportional to the square of the signal linewidth, free radical signals which typically display narrow linewidths are relatively easy to detect. Generally, free radical signals are detectable down to levels of 0.1 µM, although higher concentrations should be employed for well-resolved EPR spectra. Higher concentrations of up to 50 µM are required for a species with a broad spectrum (e.g. a transition metal ion).

The factors that influence the size of the EPR signal are summarized in *Table 6*. The dependence on microwave power is influenced by the electron-spin relaxation rates, T_1 and T_2, which are analogous to those seen in NMR. The longitudinal, or spin-lattice relaxation rate, T_1, determines the microwave power saturation of the signal. If T_1 is short, the signal will increase as the square root of the microwave power; if it is long (slow relaxation), the signal will not increase and may actually decrease with increasing power. This is because the populations of the $m_S = \pm 1/2$ electron energy levels become equalized and microwave absorption ceases. The transverse, or spin–spin relaxation rate, T_2, affects the linewidth of the EPR spectrum; the longer the T_2, the narrower the line. Generally radicals have longer electron-spin relaxation times (i.e. slow relaxation). Transition metal ions tend to have short relaxation times, often so much so that their spectra are too broad for detection at ambient temperatures. Hence it is necessary to cool to cryogenic temperatures such as liquid nitrogen (77 K) or liquid helium (4.2 K). An advantage of working at lower temperatures is that the Boltzmann factor (i.e. the difference in populations of the $m_S = \pm 1/2$ electron energy levels) varies inversely with temperature.

Kinetic measurements of enzymes may be made by EPR, if the signal of interest is detectable at ambient temperatures. Reactions may be observed continuously by a rapid-mixing apparatus connected to a quartz flat cell, and observing the EPR amplitude at a suitable magnetic field. More commonly, the species are short-lived or require cryogenic temperatures for their detection. In the latter case it is necessary to use rather cumbersome rapid-freezing methods

EPR Spectroscopy

Table 6. Factors that affect the integrated signal intensity of a first-derivative EPR spectrum

Factor	Intensity proportional to
Microwave power	$\sqrt{Power}$ until saturated at high power
Modulation amplitude	Modulation amplitude
g-Factor	g
Field interval between spectral points	$(Interval)^2$
Receiver gain	Gain
Sample tube: small samples	Length × $(diameter)^2$
Long cylindrical samples	$(Diameter\ of\ sample)^2$
Temperature	$1/T$ (Boltzmann factor) until saturated at high power
Cavity Q	Q
Filling factor	η
Linewidth (field units)	$(Width)^2$
Time-constant/scan time	No exact correlation; reduced for excessively short scan time or long time-constant

(59). The reagents are mixed and frozen by squirting into cold isopentane. The shortest reaction time measurable is of the order of 10 msec. Reactions may also be induced by illumination with visible or ultraviolet light, either in solution or in the frozen state (57).

3. INFORMATION FROM EPR SPECTRA OF ENZYMES

The types of information that can be derived about enzymes by EPR spectroscopy are summarized in *Table 7*.

3.1. General

The concentration of a particular species
The integrated intensity of an EPR signal is, in principle, an absolute measure of the number of unpaired electrons in the sample, regardless of its chemical state. In practice, spectra are usually quantified by comparison with a standard sample of known concentration (60).

3.2. Information obtained about transition metal ions

Quantitative measurements
Since not all states of a transition ion are detectable, EPR is not a suitable technique for measuring the *total* amount of a transition ion in a solution. However, it *can* measure the amounts of specific paramagnetic species, in a complex mixture.

Identification of metal
Ions such as Cu^{2+} and Mn^{2+} are readily identified by the hyperfine splittings of the spectra into four and six lines, respectively (*Table 1*). The nuclear spins on the relevant elements are given in *Table 4*. If there is doubt about the identity of the metal ion involved, it may be possible to substitute with another isotope having a nuclear spin (e.g. ^{57}Fe for ^{56}Fe). In order to do this, the enzyme may be extracted from an organism grown on a medium specifically enriched in the isotope. Note that the isotope must be introduced to high abundance, rather than as a tracer as for radioactive isotopes.

Table 7. Types of information obtained by EPR spectroscopy

Information	Method	Notes
General		
Spin quantitation	Double integration	Spectrum is compared with a standard of known concentration. Signals measured under nonsaturating conditions
Kinetics	Freeze-quench	Solutions are mixed and squirted into cold isopentane and packed into EPR tubes
Modifications by site-directed mutagenesis	Conventional EPR	Spectra are often sensitive to subtle changes in protein structure
Monitoring of protein purification	Conventional EPR	EPR can be used in relatively impure systems
Distances between centers	Spin–spin interactions Relaxation rates Spectral simulation	Continuous-wave EPR: power saturation Pulsed EPR: saturation-recovery
Transition ions		
Identity	g-Factors, hyperfine interactions	Often by comparison with known spectra
Types and number of metal ions	Isotope substitution – hyperfine interactions	
Oxidation states	g-Factors	Alternate oxidation states may be EPR-silent or even-spin
Types and number of ligands	Superhyperfine interactions	
Interactions of active-site metals with substrates	Hyperfine interactions, isotope substitution	e.g. Mn^{2+} interaction with ^{17}O-labeled nucleotides
Radicals		
Determination of electron density distribution	Spectral simulation, ENDOR, isotope substitution	Detailed description requires molecular-orbital calculations
Mobility, rate of motion	Spin labels Lineshape analysis	
Mobility, anisotropy of motion		Determined by computer simulation of spectra

EPR Spectroscopy

Where an enzyme contains a metal such as zinc it may be possible to substitute another EPR-detectable metal ion such as Co^{2+}. In order to introduce the metal ion, in some cases it may be possible to extract the native metal ion with chelators and then diffuse in the substitute metal ion (61).

Ligands to the metal ion
These can be detected by hyperfine (sometimes called *superhyperfine*) interactions with the nuclei of ligand atoms, particularly ^{1}H and ^{14}N. Here again other nuclei having spins may be substituted for O, N, S (*Table 4*). The presence of exchangeable water or protons at the active site may be detected by transferring the enzyme into $^{2}H_2O$. It may be possible, using suitable

microbial auxotrophs, to introduce specifically labeled amino acids (308). Splittings from ligand nuclei (such as ^1H and ^{14}N) that are not resolved in conventional EPR spectra may be detected by ENDOR or ESEEM spectroscopy.

An example of the application of ligand hyperfine interactions to enzyme mechanism is the use of stereospecifically labeled ^{17}O-labeled nucleotides for the studies of the mechanisms of Mg^{2+}-dependent enzymes such as creatine kinase. Studies of the EPR spectra of the enzymes with Mn^{2+}-nucleotide complexes have led to detailed models of the transition states (6).

Zero-field splittings

For ions with $S > 1/2$, the electrons on a transition metal become coupled to give $(2S + 1)$ spin states. For example, for high-spin Fe^{3+} or Mn^{2+} there are three pairs of states, $m_s = \pm 1/2, \pm 3/2, \pm 5/2$. EPR transitions can occur between these states. Depending on the coordination geometry of the metal site, the energy levels may be shifted by electrostatic interactions, giving rise to so-called zero-field splittings. These splittings give rise to large variations in the g-factor of the ions. In principle, the zero-field splittings can provide information about the coordination geometry (5, 47). For example, an octahedral iron site with axial symmetry, as in a heme protein, will have $g_x = g_y = 6$, $g_z = 2$. Rhombic distortions of the heme plane, as in catalase, cause g_x and g_y to deviate from 6.

Coordination geometry

The spectra are an indication of the symmetry properties of the coordination site. Predictions of geometry based on simple inorganic complexes may be unhelpful due to the distorted geometries found in proteins. However the g-factors are often very sensitive to perturbations of the ligands around the metal ion, such as the addition of formate to catalase (*Table 1*). Hence EPR is a good probe of *changes* in coordination induced, for example, by the binding of substrates or by site-directed mutagenesis.

The oxidation state of the metal ions

The different oxidation states are usually easily distinguished. The one-electron oxidation or reduction of the sample may change a paramagnetic state into a state which is 'EPR-silent' (undetectable by EPR), because the spin state is either zero or an integral number. The so-called 'even-spin' states ($S = 1, 2...$), such as high-spin Fe^{2+} ($S = 2$), are usually undetectable by conventional EPR, though they may be observed by special techniques such as high-field EPR or by the use of a parallel-mode cavity (*Table 5*). The redox potential of the center can also be determined by EPR, by poising the sample at various potentials and measuring their spectra (63).

The spin state of a metal center

The spin state of a metal center, such as high- or low-spin Fe^{3+}, is also usually obvious from the spectrum (5). The spin state can affect the reactivity of a metal ion, for example in cytochrome P_{450}, where the redox potential, and hence reactivity, of the high-spin substrate-bound form is different from the low-spin substrate-free form (*Table 1*).

Spin–spin interactions

Often redox enzymes contain more than one paramagnetic center, for example an internal redox chain of metal centers or flavin. Spin–spin interactions between such paramagnetic centers are detected in EPR (*Table 8*). The effects depend on the magnitude of the interaction, giving an indication of distance between the paramagnetic centers.

If two paramagnets are covalently linked, their spins are generally strongly coupled, by a process known as the exchange interaction. If the resulting exchange-coupled system has a ground state with a spin that is the sum of the spins of the two paramagnets, this is termed *ferromagnetic*

Table 8. Spin–spin interactions

Interacting centers	Example	Effect on EPR spectrum	Refs
Tyrosyl radical + Fe^{III}	Ribonucleotide reductase	Relaxation enhancement	64
B_{12} Co^{II} + radical	Methylmalonyl-CoA mutase	Strong antiferromagnetic interaction	65
Copper + radical	Amine oxidase	Strong antiferromagnetic interaction	66
Thiamine diphosphate radical + [4Fe–4S]	Pyruvate ferredoxin oxidoreductase	Ferromagnetic interaction; triplet spectrum	35
Cu + TOPA quinone	Galactose oxidase	Splitting of radical spectrum	67
Low-spin hemes	Hexaheme nitrite reductase	Weak exchange coupling	68
Mo^{V} + [2Fe–2S]	Xanthine oxidase	Relaxation enhancement and splitting	69
Siroheme + [4Fe–4S]	Sulfite reductase	Strong antiferromagnetic interaction	70
[4Fe–4S] + flavin radical	Trimethylamine dehydrogenase	Ferromagnetic interaction; triplet spectrum	71
[2Fe–2S] + [3Fe–4S] + [4Fe–4S] clusters + FADH· radical	Succinate dehydrogenase, fumarate reductase	Relaxation enhancement	20, 72

coupling; for example two $S = 1/2$ spins couple to give $S = 1$, called a triplet state. If the spin of the ground state is the difference between those of the paramagnets, this is *antiferromagnetic coupling*. For example in the Fe^{3+}–Fe^{3+} dimer of ribonucleotide reductase (73) a strong antiferromagnetic exchange coupling between the $S = 5/2$ ions leads to an $S = 0$ ground state. No EPR signal is detected from this, but it betrays its presence by the spin–spin relaxation effect on the nearby tyrosine radical. The Fe–Fe dimer only becomes EPR-detectable if it is reduced to the mixed-valence (Fe^{3+}–Fe^{2+}) state, as occurs in reduced [2Fe–2S] iron–sulfur clusters (5) or methane mono-oxygenase (74). Here the exchange coupling produces an $S = 1/2$ state, with an EPR signal at $g_{av} < 2$ (5).

Over greater distances, the interactions between the spins are more dependent on magnetic dipole–dipole interactions, which cause splittings of the spectra of the two centers. The electron-spin relaxation rate of the slower-relaxing center is enhanced. These effects may be detected at distances up to about 4 nm.

3.3. Information obtained about radicals
EPR spectra of radicals in solution consist of numerous sharp lines. The lines are narrow because the anisotropy of g and A is averaged by rapid motion, and the multiple lines are due to hyperfine splittings of the electron energy levels by protons and other nuclear spins. However, in proteins the spectra tend to be broadened by anisotropy.

The number and types of nuclei in the radical
Strong hyperfine couplings to protein radicals may be detected by EPR; weaker couplings are determined by ENDOR and ESEEM.

EPR Spectroscopy

The unpaired electron distribution on the radical

The magnitude of each hyperfine splitting A is dependent on the extent to which electron density is delocalized on to the nucleus. A detailed description of the electronic distribution on a radical molecule may be derived by molecular orbital calculations. This is relevant, for example, to mechanisms of electron transfer.

3.4. Information from spin labels and spin probes

The properties of the EPR spectrum may be exploited by introducing stable free radicals into biochemical systems (75). *Spin labels* are radicals that are introduced specifically into particular sites, such as amino acid residues of a protein, to measure properties of the molecule, such as rates of motion. *Spin probes* are radicals that are introduced into the system to measure general properties of their environment. The most commonly used radicals for this purpose are nitroxides, for which the theory of line-broadening is well developed. Specific spin-label derivatives may be obtained, which attach covalently to amino acid side-chains (75). An interesting recent development is the use of genetic engineering to produce specific spin-label binding sites (76). Some applications of spin labels are given in *Table 9*.

The EPR spectra of paramagnets in solution, particularly radicals, are sensitive to rotational and translational motion, in the time regime 10^7–10^{10} sec^{-1} (77). The spectra become broader for slower motion. By the use of special operating conditions, known as *saturation-transfer EPR*, it is possible to extend the time range of molecular motion to 10^4 sec^{-1}, where comparisons can be made with NMR. The spectra may also be simulated to determine if the motion is isotropic or anisotropic (restricted in one direction) (78).

3.5. Spin traps

Ideally, a spin trap is a diamagnetic compound, which is introduced into a biological system *without perturbing* it, and which reacts *rapidly* and *specifically* with a transient free radical, to form a *stable* radical which can be *identified* from its EPR signal. These requirements are somewhat contradictory. The two most commonly used spin-traps for organic radicals are nitroso compounds and nitrones (*Table 10*). These react with radicals to form nitroxides.

Reaction scheme

Nitroso

$$X\cdot + R-N=O \rightarrow R-\underset{\underset{X}{|}}{N}-O\cdot$$

Nitrone

$$X\cdot + R-CH=\underset{\underset{O}{\downarrow}}{N}-R \rightarrow R-\underset{\underset{X}{|}}{C}H-\underset{\underset{O\cdot}{|}}{N}-R$$

Fe-DETC$_2$

$$\cdot NO + \,\diagdown\!\!Fe\!\!\diagup \rightarrow \,\diagdown\!\!\underset{Fe}{\overset{\overset{O}{\|}}{N}}\!\!\diagup$$

Radicals detected by spin traps may be centered on oxygen, carbon or sulfur. The nitroso compounds give an adduct in which the radical added is bonded directly to the nitroxide nitrogen, so that the EPR signal is more diagnostic of the original radical; however, the adducts are less stable and may decay within minutes. The nitrones tend to give more stable adducts which are more difficult to identify.

Table 9. Applications of spin labels to studies of enzymes

Application	Example	Notes	Refs
Motion of protein side-chains	Myosin headgroups	Anisotropy of motion can be calculated	79
Saturation transfer EPR – slow motion of protein	Protein motion in membranes	Spectra are interpreted by computer simulation	77, 80
Spin-label oximetry	Activity of cytochrome oxidase	Linewidth of radical is increased by O_2 Other radicals with narrow lines are also used	81
Production of spin-labeled enzyme products	Acetylcholinesterase; hydrolysis of acetyl thiocholine	Not affected by opaque sample	82
Rates of metabolic reduction of spin label	Enzyme-catalysed reduction of nitroxides in cells	Nitroxide is reduced to a hydroxylamine	83

Table 10. Types of spin traps

Type/spin trap	Structure	Suitable for
2-Methyl-2-nitrosopropane (MNP)		C, S-centered radicals
5,5-Dimethyl-1-pyrroline-N-oxide (DMPO)		C,S,O-centered radicals
N-tert-butyl-α-phenylnitrone (PBN)		C,S,O-centered radicals
Fe^{II}-(diethyl-dithiocarbamate)$_2$		Nitric oxide

The majority of redox enzymes do not release free radicals because of their damaging effects in cells. The most commonly investigated radicals derived from enzymes are oxygen radicals, which are of great interest because of their cytotoxic effects and involvement in processes such as inflammation. These include superoxide ($\cdot O_2^-$), hydroxyl ($\cdot OH$) and peroxyl ($\cdot OOH$, $ROO\cdot$)

radicals. Carbon-centered radicals include $\cdot CCl_3$ radicals formed, for example, by the reaction of carbon tetrachloride with cytochrome P_{450} (84).

Another type of spin trap which is specific for nitric oxide is the iron complex of diethyldithiocarbamate, as introduced by Vanin's group (44). This forms a stable Fe^{2+}–NO complex which is detectable at micromolar concentrations at cryogenic temperatures.

4. REFERENCES

1. Weil, J.A., Bolton, J.R., and Wertz, J.E. (1994) *Electron Paramagnetic Resonance: Elementary Theory and Practical Applications.* John Wiley & Sons, New York.

2. Atherton, N.M. (1993) *Principles of Electron Spin Resonance.* Ellis Horwood, Chichester.

3. Knowles, P.F., Marsh, D. and Rattle, H.W.E (1976) *Magnetic Resonance of Biomolecules.* John Wiley & Sons, London.

4. Swartz, H.M., Bolton, J.R. and Borg, D. C. (1972) *Biological Applications of Electron Spin Resonance.* Wiley-Interscience, New York.

5. Cammack, R. and Cooper, C.E. (1993) *Methods Enzymol.,* **227,** 353.

6. Reed, G.H. and Markham, G.D. (1984) in *Biological Magnetic Resonance* (L.J. Berliner and J. Reuben, eds). Plenum Press, New York, Vol. 6, p. 73.

7. Khangulov, S.V., Barynin, V.V., Voevodskaya, N.V. and Grebenko, A.I. (1990) *Biochim. Biophys. Acta,* **1020,** 305.

8. Krenn, B.E., Izumi, Y., Yamada, H. and Wever, R. (1989) *Biochim. Biophys. Acta,* **998,** 63.

9. Williams-Smith, D.L. and Morrison, P.J. (1975) *Biochim. Biophys. Acta,* **405,** 253.

10. Tsai, R., Yu, C.A., Gunsalus, I.C., Peisach, J., Blumberg, W., Orme-Johnson, W.H. and Beinert, H. (1970) *Proc. Natl Acad. Sci. USA,* **66,** 1157.

11. Kastrau, D.H.W., Heiss, B., Kroneck, P.M.H. and Zumft, W.G. (1994) *Eur. J. Biochem.,* **222,** 293.

12. de Groot, J.J.M.C., Veldink, G.A., Vliegenthart, J.F.G., Boldingh, J., Wever, R. and van Gelder, B.F. (1975) *Biochim. Biophys. Acta,* **377,** 71.

13. Peisach, J., Blumberg, W.E., Lode, E.T. and Coon, M.J. (1971) *J. Biol. Chem.,* **246,** 5877.

14. Martinez, A., Andersson, K. K., Haavik, J. and Flatmark, T. (1991) *Eur. J. Biochem.,* **198,** 675.

15. Twilfer, H., Bernhardt, F.-H. and Gersonde, K. (1981) *Eur. J. Biochem.,* **119,** 595.

16. Beck, J.L., de Jersey, J., Zerner, B., Hendrich, M.P. and Debrunner, P.G. (1988) *J. Am. Chem. Soc.,* **110,** 3317.

17. Beinert, H. (1978) *Methods Enzymol.,* **54,** 133.

18. Ohnishi, T. (1993) *J. Bioenerget. Biomem.,* **25,** 325.

19. Ackrell, B.A.C., Johnson, M.K., Gunsalus, R.P. and Cecchini, G. (1992) in *Chemistry and Biochemistry of Flavoenzymes* (F. Müller, ed.). CRC Press, London, Vol. 3, p. 230.

20. Cammack, R., Patil, D.S. and Weiner, J.H. (1986) *Biochim. Biophys. Acta,* **870,** 545.

21. Frasca, V., Banerjee, R.V., Dunham, W.R., Sands, R.H. and Matthews, R.G. (1988) *Biochemistry,* **27,** 8458.

22. Martinelli, R.A., Hanson, G.R., Thompson, J.S., Holmquist, B., Pilbrow, J.R., Auld, D.S. and Vallee, B.L. (1989) *Biochemistry,* **28,** 2251.

23. Cammack, R., Fernandez Lopez, V.M. and Hatchikian, E.C. (1994) *Methods Enzymol.,* **243,** 43.

24. Rospert, S., Voges, M., Berkessel, A., Albracht, S.P.J. and Thauer, R.K. (1992) *Eur. J. Biochem.,* **210,** 101.

25. Musci, G., di Patti, M.C.B. and Calabrese, L. (1993) *Arch. Biochem. Biophys.,* **306,** 111.

26. Solomon, E.I., Baldwin, M.J. and Lowery, M.D. (1992) *Chem. Rev.,* **92,** 521.

27. Hille, R. and Massey, V. (1985) in *Molybdenum Enzymes* (T. Spiro, ed.). John Wiley & Sons, Chichester, p. 443.

28. Guigliarelli, B., Asso, M., More, C., Augier, V., Blasco, F., Pommier, J., Giordano, G. and Bertrand, P. (1992) *Eur. J. Biochem.,* **207,** 61.

29. Deaton, J.C., Solomon, E.I., Watt, G.D., Wetherbee, P.J. and Durfor, C.N. (1987) *Biochem. Biophys. Res. Commun.,* **149,** 424.

30. Wagner, A.F.V., Frey, M., Neugebauer, F.A., Schäfer, W. and Knappe, J. (1992) *Proc. Natl Acad. Sci. USA,* **89,** 996.

31. Karthein, K., Dietz, R., Nastainczyk, W. and Ruf, H. (1988) *Eur. J. Biochem.,* **171,** 313.

32. Hori, H. and Yonetani, T. (1985) *J. Biol. Chem.,* **260,** 349.

33. Müller, F. (1990) *Chemistry and Biochemistry of Flavoproteins.* CRC Press, Boca Raton, FL.

34. Medina, M.M., Vrielink, A. and Cammack, R. (1994) *Eur. J. Biochem.*, **222**, 941.

35. Cammack, R., Kerscher, L. and Oesterhelt, D. (1980) *FEBS Lett.*, **118**, 271.

36. Baron, A.J., Stevens, C., Wilmot, C., Senevir-atne, K.D., Blakeley, V., Dooley, D.M., Phillips, S.E.V., Knowles, P.F. and McPherson, M.J. (1994) *J. Biol. Chem.*, **269**, 25095.

37. Ballinger, M.D., Frey, P.A. and Reed, G.H. (1992) *Biochemistry*, **31**, 10782.

38. Shi, X.G., Dalal, N.S. and Vallyathan, V. (1991) *Arch. Biochem. Biophys.*, **290**, 381.

39. Moreno, S.N.J., Mason, R.P. and Docampo, R. (1984) *J. Biol. Chem.*, **259**, 8252.

40. Chamulitrat, W., Takahashi, N. and Mason, R.P. (1989) *J. Biol. Chem.*, **264**, 7889.

41. Rosen, G.M., Pou, S., Britigan, B.E. and Cohen, M.S. (1994) *Methods Enzymol.*, **233**, 105.

42. Kon, H. (1973) *J. Am. Chem. Soc.*, **95**, 1045.

43. Brudrig, G.W., Stevens, T.H. and Chan, S.I. (1980) *Biochemistry*, **19**, 5275.

44. Kubrina, L.N., Caldwell, W.S., Mordvintcev, P.I., Malenkova, I.V. and Vanin, A.F. (1992) *Biochim. Biophys. Acta*, **1099**, 233.

45. Poole, C.P., Jr (1983) *Electron Spin Resonance (2nd edn)*. John Wiley & Sons, New York.

46. Czoch, R. and Francik, A. (1989) *Instrumental Effects in Homodyne Electron Paramagnetic Resonance Spectrometers*. Ellis Horwood, Chichester.

47. Pilbrow, J.R. (1990) *Transition Ion Electron Paramagnetic Resonance*. Oxford University Press, Oxford.

48. Hyde, J.S., Froncisz, W. and Oles, T. (1989) *J. Magn. Reson.*, **82**, 223.

49. Basosi, R., Antholine, W.E. and Hyde, J.S. (1993) in *EMR of Paramagnetic Molecules* (L.J. Berliner and J. Reuben, eds). Plenum Press, New York, Vol. 13, p. 103.

50. Eaton, G.R., Eaton, S.S. and Ohno, K. (1991) *EPR Imaging and In Vivo EPR*. CRC Press, Boca Raton, FL.

51. Möbius, K. (1993) in *EMR of Paramagnetic Molecules* (L.J. Berliner and J. Reuben, eds). Plenum Press, New York, Vol. 13, p. 253.

52. Hendrich, M.P. and Debrunner, P.G. (1989) *Biophys. J.*, **56**, 489.

53. Lowe, D.J. (1992) *Prog. Biophys. Mol. Biol.*, **57**, 1.

54. Hoffman, B.M., DeRose, V., Doan, P.E., Gurbiel, R.J., Houseman, A.L.P. and Telser, J. (1993) in *EMR of Paramagnetic Molecules* (L.J. Berliner and J. Reuben, eds). Plenum Press, New York, Vol. 13, p. 151.

55. Hirsh, D.J., Beck, W.F., Lynch, J.B., Que, L. and Brudvig, G.W. (1992) *J. Am. Chem. Soc.*, **114**, 7475.

56. Mims, W.B. and Peisach, J. (1989) in *Advanced EPR: Applications Biology and Biochemistry* (A.J. Hoff, ed.). Elsevier, Amsterdam, p. 1.

57. Snyder, S.W., Morris, A.L., Bondeson, S.R., Norris, J.R. and Thurnauer, M.C. (1993) *J. Am. Chem. Soc.*, **115**, 3774.

58. Thomann, H. and Bernardo, M. (1993) in *EMR of Paramagnetic Molecules* (L.J. Berliner and J. Reuben, eds). Plenum Press, New York, Vol. 13, p. 275.

59. Ballou, D.P. (1978) *Methods Enzymol.*, **54**, 85.

60. Randolph, M.L. (1972) in *Biological Applications of Electron Spin Resonance* (H.M. Swartz, J.R. Bolton and D.C. Borg, eds). Wiley-Interscience, New York, p. 119.

61. Maret, W. and Zeppezauer, M. (1988) *Methods Enzymol.*, **158**, 79.

62. Gurbiel, R.J., Batie, C.J., Sivaraja, M., True, A.E., Fee, J.A., Hoffman, B.M. and Ballou, D.P. (1989) *Biochemistry*, **28**, 4861.

63. Cammack, R. (1995) in *Bioenergetics: a Practical Approach* (G.C. Brown and C.E. Cooper, eds). IRL Press, Oxford, in press.

64. Sahlin, M., Petersson, L., Gräslund, A., Ehrenberg, A., Sjöberg, B.-M. and Thelander, L. (1987) *Biochemistry*, **26**, 5541.

65. Zhao, Y., Abend, A., Kunz, M., Such, P. and Retey, J. (1994) *Eur. J. Biochem.*, **225**, 891.

66. Collison, D., Knowles, P. F., Mabbs, F.E., Rius, F.X., Singh, I., Dooley, D.M., Cote, C.E. and McGuirl, M. (1989) *Biochem. J.*, **264**, 663.

67. Dooley, D.M., McGuirl, M.A., Brown, D.E., Turowski, P.N., McIntire, W.S. and Knowles, P.F. (1991) *Nature*, **349**, 262.

68. Blackmore, R.S., Brittain, T., Gadsby, P.M.A., Greenwood, C. and Thomson, A.J. (1987) *FEBS Lett.*, **219**, 244.

69. Bertrand, P., More, C., Guigliarelli, B., Fournel, A., Bennett, B. and Howes, B. (1994) *J. Am. Chem. Soc.*, **116**, 3078.

70. Krueger, R.J. and Siegel, L.M. (1982) *Biochemistry*, **21**, 2892.

71. Stevenson, R.C., Dunham, W.R., Sands, R.H., Singer, T.P. and Beinert, H. (1986) *Biochim. Biophys. Acta*, **869**, 81.

72. Maguire, J.J., Johnson, M.K., Morningstar, J., Ackrell, B.A.C. and Kearney, E.B. (1985) *J. Biol. Chem.*, **260**, 10909.

73. Atta, M., Andersson, K.K., Ingemarson, R., Thelander, L. and Graslund, A. (1994) *J. Am. Chem. Soc.*, **116**, 6429.

74. Woodland, M.P., Patil, D.S., Cammack, R. and Dalton, H. (1986) *Biochim. Biophys. Acta,* **873,** 237.

75. Berliner, L.J. (1978) *Methods Enzymol.,* **49,** 418.

76. Farahbakhsh, Z.T., Hideg, K. and Hubbell, W.L. (1993) *Science,* **262,** 1416.

77. Fajer, P., Knowles, P.F. and Marsh, D. (1989) *Biochemistry,* **28,** 5634.

78. Howard, E.C., Lindahl, K. M., Polnaszek, C.F. and Thomas, D.D. (1993) *Biophys. J.,* **64,** 581.

79. Barnett, V.A. and Thomas, D.D. (1989) *Biophys. J.,* **56,** 517.

80. Hyde, J.S. (1978) *Methods Enzymol.,* **49,** 480.

81. Jiang, J.J., Bank, J.F., Zhao, W.W. and Scholes, C.P. (1992) *Biochemistry,* **31,** 1331.

82. Khramtsov, V.V., Gorunova, T.E. and Weiner, L.M. (1991) *Biochem. Biophys. Res. Commun.,* **179,** 520.

83. Belkin, S., Mehlhorn, R.J., Hideg, K., Hankovsky, O. and Packer, L. (1987) *Arch. Biochem. Biophys.,* **256,** 232.

84. Poyer, J.L., Floyd, R.A., McCay, P.B., Janzen, E.G. and Davis, E.R. (1978) *Biochim. Biophys. Acta,* **539,** 402.

CHAPTER 9
SOLUTIONS USED IN ENZYMOLOGY

A. BUFFERS

L. Stevens

1. INTRODUCTION

The concept of a buffer as an agent that resists changes in hydrogen ion concentration, is how the term is most frequently used in a biochemical context. However, it can be broadened to include agents that resist changes in ion concentration, generally metal ions, and also agents used to maintain a constant redox potential (1). Only buffers used to resist changes in hydrogen ion concentration are considered here.

In almost all experimental procedures in which enzymes are used, and in which their catalytic activity has to be maintained, buffers are used to maintain a stable pH. The choice of buffer depends very much on the experimental procedure being carried out, and on the particular properties of the enzyme in question. In this section the factors determining the choice of buffer are first considered, the salient features of particular buffers are summarized in *Table 1*, and finally suggestions and hints are given on how best to make up buffers.

Buffer solutions are solutions that resist changes in pH, following the addition of acids or bases. They generally comprise either weak acids together with their conjugate base, or weak bases together with their conjugate acid. The reader is referred to several articles that consider the theory of buffer action for further information (1–6). Buffer solutions are most effective at pHs close to their pK, and as a general rule are best used within the range p$K \pm 1$. Outside this range they are less effective. The buffering capacity of a solution is directly related to its concentration. Within a 10-fold range of concentration the pH of a buffer will show little change, although its buffering capacity will be diminished 10-fold.

2. FACTORS DETERMINING THE CHOICE OF BUFFER

The choice of buffer will be determined by the following general considerations:

(i)   the type of experimental procedures being performed;
(ii) the characteristics of the enzyme under study; and
(iii) the range of buffers available.

3. EXPERIMENTAL PROCEDURES

These may involve purification, structural analysis or studies of the kinetic behavior and reaction mechanism. During enzyme purification the buffered media in which the enzyme extracts are suspended are required to give the enzyme maximum stability. Although this is often close to the pH optimum for the enzyme, it need not necessarily be so. Provided the enzyme is not irreversibly inactivated by the buffered medium, it will probably be satisfactory. In the early stages of purification when cells or tissue are disrupted, it is often necessary to stabilize an enzyme against acid released from lysosomes. Often a medium which provides osmotic stabilization for subcellular organelles is used. Chromatographic procedures are amongst the most frequently used methods for enzyme purification. For ion-exchange chromatography and affinity chromatography, the pH chosen is selectively to bind or release the enzyme from

▶ p. 274

Table 1. Properties of selected buffers

pK_a [a]	Compound [b]	$d(pK_a)/dT$	Saturated solution at 0°C (M)	Interaction with metal ions [d] w, m or s	Price [c]	Comments
4.64	Acetic acid	0.0002	>10		L	
5.28	Succinic acid	0	0.36	Cu^{2+}, m	L	Interferes with Lowry
5.80	Citric acid	0	>2	Mg^{2+}, Ca^{2+}, Mn^{2+} and Cu^{2+}, m	L	Interferes with Lowry
6.02	Mes	−0.011	0.65	Negligible	M	
6.32	Bis-Tris	−0.017			H	
6.32	Pyrophosphate	−0.01	0.1	Mg^{2+}, Ca^{2+}, m; Cu^{2+}, s	L	Product of ligases and may inhibit
6.62	Ada	−0.011	V. sol.	Mg^{2+}, w; Ca^{2+}, Mn^{2+}, m; Cu^{2+}, s	H	Interferes with Lowry, absorbs <260 nm
6.67	Aces	−0.020	0.22	Cu^{2+}, m	H	Interferes with Lowry, absorbs <260 nm
6.77	Mopso	−0.015	0.75	Negligible	M	Interferes with Lowry
6.84	Phosphate	−0.0028	0.2	Mg^{2+}, Ca^{2+}, Mn^{2+} and Cu^{2+}, m	L	Inhibits kinases, dehydrogenases, carboxypeptidase, fumarase, urease, aryl sulfatase, adenosine deaminase phosphoglucomutase and enzymes involving phosphate esters. Stabilizes phospho-ribosepyrophosphate synthase
6.86	Pipes	−0.0085	V. sol	Negligible	H	Interferes with Lowry
6.97	Imidazole	−0.020	V. sol	Mn^{2+}, Cu^{2+}, m	L	Reactive, unstable, Mops or Bes generally better alternative

pKa	Buffer		Solubility	Metal interactions		Comments
6.98	Bes	−0.016	3.2	Cu^{2+}, m	M	Interferes with Lowry
7.02	Mops	−0.015	3.09	Negligible	M	Interferes with Lowry
7.27	Tes	−0.020	2.6	Cu^{2+}, m	H	Interferes with Lowry
7.39	Hepes	−0.014	2.25	Negligible	M	Interferes with Lowry
7.42	Dipso	−0.015	0.24		H	
7.49	Tapso	−0.018	1.0		H	
7.77	Heppso	−0.010	2.2		H	Interferes with Lowry
7.78	Triethanolamine	−0.020	V. sol		L	
7.82	Popso	−0.013	V. sol		H	
7.85	Hepps	−0.015	1.58	Negligible	M	Interferes with Lowry
7.92	Tricine	−0.021	0.8	Mg^{2+}, w; Ca^{2+}, Mn^{2+}, m	M	Interferes with Lowry
8.00	Tris	−0.031	2.4	Cu^{2+}, Zn^{2+}, m	L	Interferes with Lowry and Bradford, reacts with carbonyl groups
8.09	Glycylglycine	−0.028	1.1	Ca^{2+}, Mg^{2+}, w; Mn^{2+}, Cu^{2+}, m	M	Not suitable for peptidases
8.17	Bicine	−0.018	1.1	Mg^{2+}, Ca^{2+}, m; Mn^{2+}, m; Cu^{2+}, s	L	Interferes with Lowry
8.19	Taps	−0.018			H	
8.88	Diethanolamine	−0.024	V. sol		L	
9.08	Borate	−0.008	0.05		L	Binds vic diols, not recommended with carbohydrates or ribonucleotides
9.23	Ches	−0.029	1.14		M	Interferes with Lowry
9.47	Ethanolamine	−0.029	V. sol		L	Reacts with carbonyl groups

Table 1. Continued

pK_a [a]	Compound [b]	$d(pK_a)/dT$	Saturated solution at 0°C (M)	Interaction with metal ions [d] w, m or s	Price [c]	Comments
9.55	Glycine	−0.025	4.0	Mg^{2+}, Ca^{2+}, w; Mn^{2+}, m; Cu^{2+}, s	L	Interferes with Lowry and Bradford
9.96	Carbonate	−0.009	0.8		L	Requires closed system
10.05	Caps	−0.032	0.47		M	Interferes with Lowry

[a] pK_a, the practical dissociation constant defined as:

$$pH = pK^*a + \log[(A^-)/(HA)] - (2z + 1)[\{0.5I^{0.5}/(1 + I^{0.5})\} - 0.1I], \text{ where } I = \text{ionic strength and } z = \text{charge on the conjugate ion.}$$

[b] The structures of compounds with abbreviated names are given in *Figure 1*.
[c] The price categories are based on 1993 manufacturers' catalogs. L, less than £20 l[−1] of 1 M buffer solution; M, between £20 and £50 l[−1] of 1 M buffer solution; and H, greater than £50 l[−1] of 1 M buffer solution.
[d] Interaction with metal ions; $\log K_M = 1$–2 for weak binding (w), 2–5 for medium binding (m) and > 5 for strong binding (s), where K_M is the metal–buffer binding constant.

Figure 1. Structures of buffer components referred to in *Table 1*.

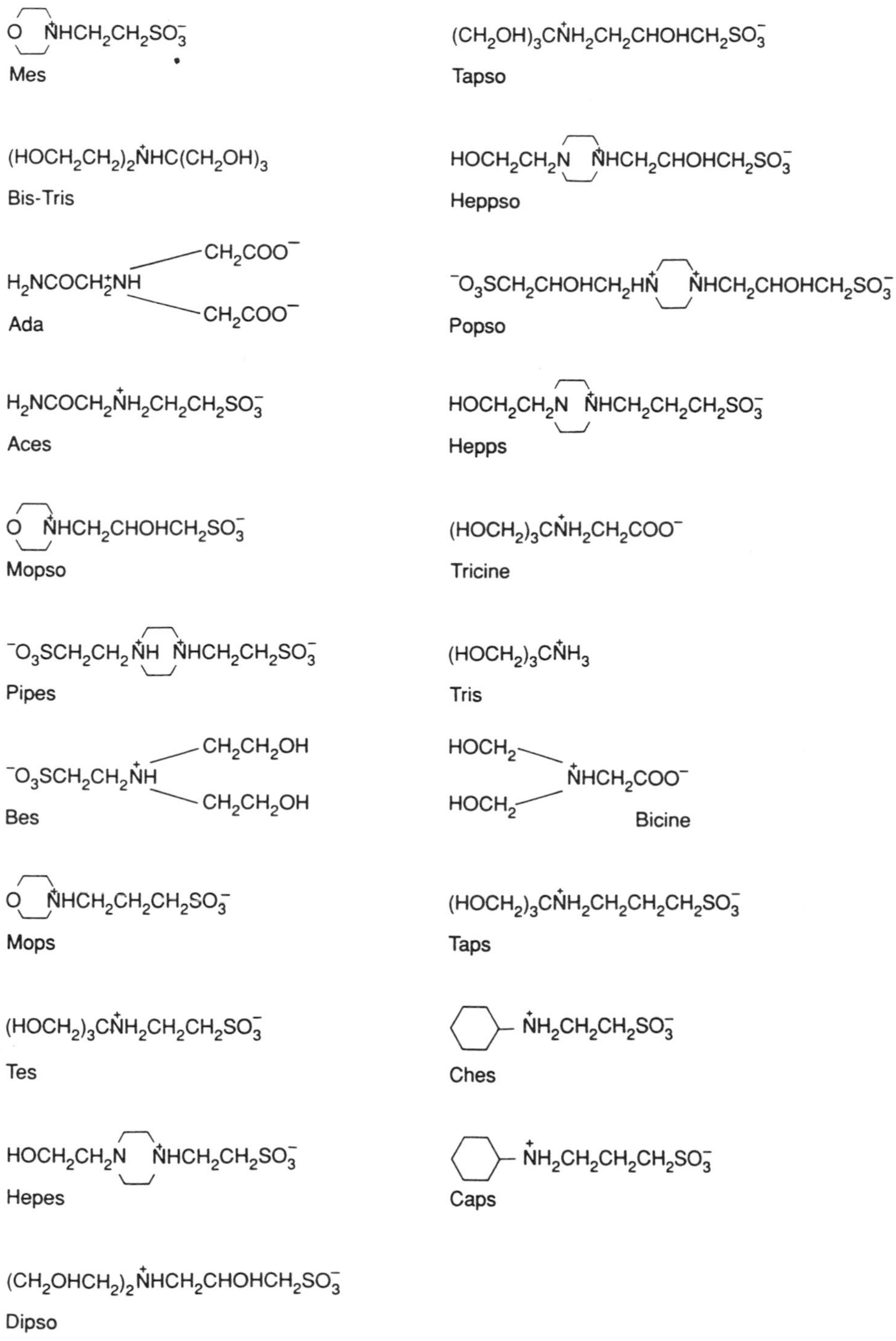

$\overset{\frown}{O}\quad \overset{+}{N}HCH_2CH_2SO_3^-$

Mes

$(CH_2OH)_3C\overset{+}{N}H_2CH_2CHOHCH_2SO_3^-$

Tapso

$(HOCH_2CH_2)_2\overset{+}{N}HC(CH_2OH)_3$

Bis-Tris

$HOCH_2CH_2\overset{\frown}{N}\quad \overset{+}{N}HCH_2CHOHCH_2SO_3^-$

Heppso

$H_2NCOCH_2^+\overset{+}{N}H\overset{\diagup CH_2COO^-}{\underset{\diagdown CH_2COO^-}{}}$

Ada

$^-O_3SCH_2CHOHCH_2H\overset{+}{N}\quad \overset{+}{N}HCH_2CHOHCH_2SO_3^-$

Popso

$H_2NCOCH_2\overset{+}{N}H_2CH_2CH_2SO_3^-$

Aces

$HOCH_2CH_2\overset{\frown}{N}\quad \overset{+}{N}HCH_2CH_2CH_2SO_3^-$

Hepps

$\overset{\frown}{O}\quad \overset{+}{N}HCH_2CHOHCH_2SO_3^-$

Mopso

$(HOCH_2)_3C\overset{+}{N}H_2CH_2COO^-$

Tricine

$^-O_3SCH_2CH_2\overset{+}{N}H\quad \overset{+}{N}HCH_2CH_2SO_3^-$

Pipes

$(HOCH_2)_3C\overset{+}{N}H_3$

Tris

$^-O_3SCH_2CH_2\overset{+}{N}H\overset{\diagup CH_2CH_2OH}{\underset{\diagdown CH_2CH_2OH}{}}$

Bes

$HOCH_2\diagdown \atop HOCH_2\diagup \overset{+}{N}HCH_2COO^-$

Bicine

$\overset{\frown}{O}\quad \overset{+}{N}HCH_2CH_2CH_2SO_3^-$

Mops

$(HOCH_2)_3C\overset{+}{N}H_2CH_2CH_2CH_2SO_3^-$

Taps

$(HOCH_2)_3C\overset{+}{N}H_2CH_2CH_2SO_3^-$

Tes

$\bigcirc - \overset{+}{N}H_2CH_2CH_2SO_3^-$

Ches

$HOCH_2CH_2\overset{\frown}{N}\quad \overset{+}{N}HCH_2CH_2SO_3^-$

Hepes

$\bigcirc - \overset{+}{N}H_2CH_2CH_2CH_2SO_3^-$

Caps

$(CH_2OHCH_2)_2\overset{+}{N}HCH_2CHOHCH_2SO_3^-$

Dipso

the column. For anion exchange chromatography the buffer chosen is generally such that it provides the cation (e.g. Tris–HCl for DEAE-cellulose), and for cation exchange chromatography the buffer provides the anion (e.g. MES or acetate for CM-cellulose). For gel filtration, conditions are generally chosen so that the proteins being separated have least tendency to adsorb to the column. This is generally achieved by ensuring that the eluent has a minimum ionic strength of about 0.1, but the choice of buffer is not critical. For hydroxyapatite ($Ca_{10}(PO_4)_6(OH)_2$) and calcium phosphate gel chromatography, phosphate buffers are most frequently used for elution. They probably act by desorbing proteins off Ca^{2+} binding sites on the gel.

Electrophoretic separation is sometimes used in the later stages of purification, most frequently with polyacrylamide as support. The pH chosen is generally that which gives best resolution from other proteins. An important consideration with electrophoresis is to have the maximum buffering capacity for a given ionic strength, and at the same time to minimize the concentrations of small highly mobile anions and cations, such as Na^+, K^+, Mg^{2+}, Cl^-, SO_4^{2-}, HPO_4^{2-}. These small mobile ions carry most of the current, and their presence will lead to heating effects, and also decrease the mobility of the proteins being separated. Minimizing the concentrations of small mobile ions is generally achieved by using two buffers differing by about 1 pH unit, one of which provides the anion and the other the cation (e.g. for pH 4.5, β-alanine$^+$–acetate$^-$; pH 6.8–7.5, imidazole$^+$–MOPS$^-$; pH 8–9, either Tris$^+$–borate$^-$ or Tris$^+$–glycine$^-$).

Certain stages in enzyme purification often require large volumes of buffered medium, either for elution, or for dialysis. The large volumes favor the choice of the less expensive buffers.

During purification procedures enzyme activity is assayed to monitor the success or otherwise of the process. Assays are generally carried out near the pH optimum of the enzyme to give maximum sensitivity, and this may not always coincide with the pHs used in the purification steps. Many spectrophotometric assays, especially continuous assays, depend on absorbance measurements in the UV spectrum, usually in the range 260–400 nm. Most of the N-substituted glycine and taurine buffers listed in *Table 1* have low absorbance in this range. Other factors affecting the choice of buffer depend on the characteristics of the particular enzyme and are considered in Section 4.

Enzymes are often concentrated at various stages during purification procedures, either by ultrafiltration, dialysis against polyethylene glycol, precipitation with ammonium sulfate or by freeze drying. For the last of these procedures the protein has first to be dialysed to remove the solute, although this can be avoided if a volatile buffer is used. A list of volatile buffers is given in ref. 2.

The success of individual steps in an enzyme purification procedure is assessed on the basis of yield and increase in specific activity. For the latter the amount of protein has to be determined, and the methods used are given in Chapter 2, Part C. Two of the most frequently used are the Lowry method and the Coomassie blue binding method, most frequently that of Bradford. Both methods are subject to interference by certain buffers (see *Table 1*), but the range of interfering buffers is much greater with the Lowry method. Whether or not this is a serious problem will depend on the relative concentrations of the proteins and buffer. For concentrated protein solutions, which have to be diluted before assay, little interference may result, and this can be corrected satisfactorily by a buffer blank.

For studies on the catalytic activity of enzymes, using steady-state kinetics or transient kinetics, aimed at understanding enzyme mechanisms, both the pH and the ionic concentration of the buffer are important. If the effect of pH on the rate of the enzyme-catalysed reaction is being studied, then it is important to ensure that the same ionic strength is being maintained at each pH. With the normal buffers of fixed molarity, the extent to which the ionic strength will vary over its useful pH range will depend on the charge on the conjugate base. The theory and practice of buffers for studying pH-dependent processes are reviewed in ref. 7.

4. THE CHARACTERISTICS OF THE ENZYME

There is a wide range of stabilities of different enzymes. Some will only tolerate relatively narrow ranges of pH and environmental conditions, whereas others are much more robust. In general, extracellular enzymes will tolerate a wider range of conditions than intracellular enzymes.

The nature of the reaction catalysed can affect the nature of the buffer used. $NAD(P)^+$-dependent dehydrogenases make up at least 15% of the enzymes listed by the Enzyme Commission (1992) (8). All of these generate protons during the reduction of $NAD(P)^+$, and so have to be adequately buffered so that there is no significant change in pH during the measurement of activity. Thus if assayed in the direction of $NAD(P)^+$ reduction, where protons are generated, the preferred pK of the buffer would be slightly below the pH of assay. Enzymes for which orthophosphate or pyrophosphate is either a substrate or product make up nearly 10% of EC listed enzymes (8). For these, phosphate or pyrophosphate buffers may affect the measured rate of catalysis, or interfere with the method of assay. There are many different kinases, and these may also be affected by phosphate. However, phosphate is one of the principal intracellular anions, and also stabilizes a number of enzymes.

Some buffers react with either the substrates or products. Included in this category are buffers containing primary amino-groups (e.g. Tris, glycine, glycylglycine and ethanolamine), which are likely to react with carbonyl groups to form Schiff bases. Borate complexes with *vic*-diols, which occur in a number of pentoses and hexoses, including ribonucleotides. A number of enzymes have a requirement for metal ions for catalysis, and for these the metal ion binding properties of the buffers should be taken into account (see *Table 1*).

5. PREPARATION OF BUFFERS

When making up buffers for enzyme work, the pH of the buffer is generally more important than its concentration. Concentrations are generally chosen so that they are in the correct range to buffer adequately under the conditions being used (e.g. 0.01 M, 0.1 M, 0.2 M, etc.) but the exact concentration is generally not critical. The most dilute buffer solution that will suffice is usually used. The amounts of the two components can be calculated from the pK, using the Henderson–Hasselbalch equation:

$$pH = pK_a + \log[A^-]/[HA]$$

and then weighed out and dissolved. However, if this is done, it is necessary to check the final pH using a pH meter since the components may contain moisture and this will affect concentration, or in some cases the free base may be liquid which is hygroscopic. More frequently, acidic buffers are often made up by taking the required amount of free acid to give the required molarity and then adding concentrated base until the required pH is reached (e.g. acetic acid and NaOH). In the case of an alkaline buffer the required amount of the base to give the required molarity is taken and acid is added until the required pH is reached (e.g. Tris and HCl). Sometimes the starting solution is the salt (e.g. triethanolamine HCl), especially if this is more readily obtained in purified crystalline form. In this case alkali is added to obtain the required pH, and the buffer will then also contain NaCl. If a low concentration of metal ions is required, tetramethylammonium hydroxide may be used as an alternative strong alkali.

Many manipulations with enzymes, unless they are strongly hydrophobic, are carried out at 4°C in order to increase the stability of the enzyme preparation, but buffers are rarely made up in the

Solutions

cold room. At the other extreme, enzymes from thermophilic organisms may be assayed at 50°C or higher. The $d(pK_a)/dT$ for carboxylic acids is generally very low (*Table 1*), but for many other buffers it is sufficiently high to make a significant change. The primary standard buffers used to calibrate pH meters have very low $d(pK_a)/dT$; for example, potassium hydrogen phthalate, which is most commonly used as the pH 4 standard, has a pH value of 4.000 at 15°C, 4.011 at 0°C and 4.061 at 50°C. To adjust the pH of a buffer for a particular temperature (e.g. 4°C), first equilibrate the pH standard to 4°C. For most purposes it will be sufficiently accurate to assume that the pH of the standard is unchanged. Set the temperature compensation control on the meter to the temperature required (e.g. 4°C). This usually compensates for the change in the relationship of emf to pH with temperature. Calibrate the pH reading on the meter to the pH of the standard. Measure and adjust the pH of the buffer solution that is being made up only after it has also been equilibrated at the temperature required.

B. CHELATING AGENTS AND DENATURANTS

P.C. Engel

6. INTRODUCTION

Enzymology makes use of a wide range of accessory compounds in addition to the obvious components of a reaction mixture (i.e. a buffer, the reactants and the enzyme itself). Some of these are specific to individual systems or kinds of reaction and are too numerous to list here. A few, however, are of such widespread use that it seems appropriate to include some information on their properties in this volume. They include metal-chelating agents, denaturing agents, detergents and thiol compounds.

7. CHELATING AGENTS

Chelating agents are used to control metal ion concentrations in reactions or to remove metal ions essential for structural or functional integrity. A very frequent use in enzymology is as a scavenger for stray heavy metal ions present as impurities in reagents, in water, or on the surface of labware. Such ions often have adverse effects on enzyme activity and may catalyse the oxidation of essential -SH groups. Thus EDTA is often a constant addition in an enzyme buffer, at a concentration of 10^{-4} or even 10^{-3} M. It has to be borne in mind, however, that occasionally a chelating agent used in this way may have the opposite effect! If the enzyme in question has a site with a very high metal affinity, greater than that of the chelating agent, then the latter may simply serve as a vehicle for conveying metal ions to the enzyme. In such a situation solutions must be pretreated with chelating agents or chelating columns and reactions should be carried out in plasticware rather than glass, since glass has a much higher capacity for binding metal ions.

The affinity of a chelating agent for a metal is indicated by the stability constant. This is the $\log_{10}$ of the dissociation constant for the complex. Although it is generally thought that there is a simple binary reaction:

$$\text{metal} + \text{chelating agent} \rightleftarrows \text{complex}$$

involved in the formation of a complex, in practice a series of minor complexes may also be formed. In such cases it is the stability constant for the most stable complex that is reported. For some ligands (e.g. EDTA) the complex contains one chelating agent molecule and one metal ion, but for others there may be more than one chelating agent molecule with a series

Table 2. Chelating agents: solubility and pK_a values

	EDTA [60-00-4]	EGTA [67-42-5]	1,10-Phenanthroline [66-71-7]	2,2'-Bipyridine [366-18-7]	Citric acid [77-92-9]	Phosphoric acid [7664-38-2]
Formula	$C_{10}H_{16}N_2O_8$	$C_{14}H_{24}N_2O_{10}$	$C_{12}H_8N_2$	$C_{10}H_8N_2$	$C_6H_8O_7$	H_3PO_4
Mol. wt	373.2 (2Na·2H$_2$O)	380.4				
Solubility (20°C, g/100 ml H$_2$O)	11.1	v. sol.				
pK_a (25°C)						
pK_a1	10.17	9.40	4.93	4.42	5.69	11.74
pK_a2	6.11	8.78	1.9	–	4.35	6.57
pK_a3	2.68	2.66[a]	–	–	2.87	1.72
pK_a4	2.0	2.0[a]	–	–	–	–

[a] At 20°C.

Abbreviations: EDTA, ethylenediaminetetraacetic acid; EGTA, ethylene glycol-bis(β-aminoethyl ether)N,N,N',N'-tetraacetic acid; v. sol., very soluble.

of intermediates; in these cases the stability constant for the most stable form is indicated in *Table 3* and the presence of intermediates is marked.

The ratio of stability constants for a chelating agent with a pair of metal ions shows how effectively the two would be bound in a mixture. For EDTA, the stability constants for Ca^{2+} and Mg^{2+} are 10.61 and 8.83. The difference between these two is ~ 1.8, which is $\log_{10}$ of ~ 60. Accordingly, the ratio of complexed Ca^{2+} to Mg^{2+} in an equimolar mixture is about $60:1$. For EGTA, with a lower stability constant for Mg^{2+}, this ratio is about 4×10^5.

Tables 2 and *3* include the two most widely used chelating agents, EDTA and EGTA, and two others that have some specialized uses. Citrate is used as a weak chelating agent in some systems and phosphate is included for comparison.

8. DENATURING AGENTS

8.1. Mechanisms of denaturing agents
Protein denaturing agents are chaotropic agents that disrupt the higher-order structure (secondary and higher) of proteins. Most often this is by disrupting the hydrogen bonds and

Table 3. Chelating agents: stability constants

	EDTA	EGTA	1,10-Phenan-throline	2,2'-Bipyridyl	Citric acid	Phosphoric acid
Common physiological cations						
Na^+	1.64	–	–	–	0.7	0.60
K^+	0.8	–	–	–	0.59	0.49
Ca^{2+}	10.61	10.86	0.7^a	–	3.5	3.4^a
Mg^{2+}	8.83	5.28	1.2	–	3.37	1.5
Other physiological cations						
Co^{2+}	16.26	12.35	19.8^b	15.9^b	5.0	2.18
Co^{3+}	41.4	–	–	–	–	–
Cu^+	–	–	15.8^a	12.95	–	–
Cu^{2+}	18.7	17.57	10.69	6.33	5.9	3.2
Fe^{2+}	14.27	11.8	21.0^b	17.2^b	4.4	3.6
Fe^{3+}	25.0	20.5	14.1^b	16.29	11.5	8.3
Mn^{2+}	13.81	12.18	10.3^b	2.62	4.15	–
Ni^{2+}	18.52	13.5	24.3	20.16^b	5.4	2.08
Zn^{2+}	16.44	12.6	–	13.2^b	5.9^b	2.4
Other cations						
Ag^+	7.32^a	6.88^a	12.6^b	6.67^b	–	–
Al^{3+}	16.5	13.9	–	–	–	–
Cd^{2+}	16.36	16.5	14.6^b	10.3^b	4.54^b	–
Hg^{2+}	21.5	22.9	23.35^b	19.5^b	10.9	–
Pb^{2+}	17.88	14.5	4.65	2.9	6.1	–

[a]At 20°C.
[b]Most stable complex.
Abbreviations: EDTA, ethylenediaminetetraacetic acid; EGTA, ethylene glycol-bis(β-aminoethyl ether)*N,N,N',N'*-tetraacetic acid.

other interactions that maintain these higher-order structures. Denaturing detergents act by a slightly different mechanism.

Denaturants can be either solubilizing (*Table 4*) or precipitating (*Table 5*). Urea, sodium dodecyl (or lauryl) sulfate (SDS or SLS) and guanidinium salts are effective solubilizing denaturants, whereas trichloroacetic acid (TCA), ethanol, methanol and chloroform/isoamyl alcohol are effective precipitants.

In enzymology, precipitating denaturants may be used:

(i)   in purification – ethanol fractionation is common in some of the older protocols;
(ii) to release tightly bound ligands in a protein-free state (e.g. to estimate the content of a prosthetic group);
(iii) to remove enzyme at the end of a reaction mixture if protein is likely to interfere with the extraction or analysis of products.

Table 4. Solubilizing denaturants (see also Section 10)

Name	Formula	Mol. wt	Solubility (g/100 ml 20°C)	Working concn	Comments
Sodium dodecyl sulfate [151-21-2]	$C_{12}H_{26}O_4S \cdot Na$	273.3	25[a]	0.1–2%	Cationic detergent, salt-sensitive, cold-sensitive
Sodium dodecyl sarcosinate [137-16-6]	$C_{12}H_{29}NO_3 \cdot Na$	294.4	30[b]	0.5–2.5%	Cationic detergent, less sensitive to salt and cold than SDS but also less effective
Lithium dodecyl sulfate [2044-56-6]	$C_{12}H_{26}O_4S \cdot Li$	257.3	30	0.1–2%	Less sensitive to salt and cold than SDS or Sarkosyl
Urea [57-13-6]	CH_4N_2O	60.06	48–60	4–10 M	Nonionic, chaotropic agent; does not disrupt ionic bonds
Guanidinium chloride [50-01-1]	CH_6N_3Cl	95.53	76	4–6 M	Ionic chaotropic agent; disrupts ionic bonds
Guanidinium thiocyanate [593-84-0]	$C_2H_6N_4S$	118.1	47	3–4 M	Similar to the chloride but more effective

[a]At 25% SDS is very sensitive to temperature and salt particles, 20% is a useful working concentration. It is important to involve K^+-containing buffers when using SDS because potassiuim dodecyl sulfate precipitates.
[b]Sold as a 30% solution.

Table 5. Precipitating denaturants

Name	Formula	Mol. wt	Solubility	Working concn	Comments
Acetic acid	$C_2H_4O_2$	60.05	Freely miscible	5–10 vol. %	Also ppts nucleic acids
Acetone	C_3H_6O	58.08	Freely miscible	2–6 vols	Can be used for selective pptn; also ppts nucleic acids
Ethanol	C_2H_6O	46.1	Freely miscible[a]	1–2 vols	Also ppts nucleic acids
Methanol	CH_4O	32.04	Freely miscible with water[a]	1–2 vols	

[a] A salt (preferably chloride or acetate) must be present at concentrations greater than 0.1 M for effective precipitation.
Abbreviations: aq. solns, aqueous solutions; pptn, precipitation; ppts, precipitates.

Solubilizing denaturants, on the other hand, apart from their major use in electrophoresis, are essential tools in the study of membrane-bound enzymes. They are also often used in separating the components of a multi-enzyme complex or the different subunits of a hetero-oligomeric enzyme.

Generally, denaturants are used in a large excess but care should be taken to ensure that this is the case. Agents that disrupt hydrogen bonding should be used at a ratio of at least two moles per mole of amino acids. SDS binds proteins in a well-defined complex of 1.4 g SDS/g protein. Many denaturing solutions, especially solubilizing ones, contain a high concentration of a thiol reagent. This helps to reduce disulfide bonds to thiols to complete the loss of tertiary structure.

8.2. Mixing denaturants
Quantitative studies show that the binding of two denaturants to proteins when presented as a mixture, results in reduced binding of both denaturants; that is, they act antagonistically (similar to detergents). The best choice of denaturant is the most powerful single component system that is compatible with your requirements.

8.3. Characteristics of denaturants and precipitants

SDS, Sarkosyl, lithium dodecyl sulfate
SDS is soluble up to 25% w/v but is then very sensitive to reductions in temperature and dust particles in solution. A stock of 20% w/v is most convenient. Sarkosyl is usually available as a 30% stock solution. LiDS is more effective at cold-room temperatures than either SDS or Sarkosyl. Potassium dodecyl sulfate is insoluble so it is important to avoid introducing potassium salts into solutions containing SDS.

Urea
Urea is soluble up to 10 M but above 8 M care must be taken with temperature to prevent precipitation. Urea slowly decomposes in water, with the formation of ammonia and the highly reactive cyanate ion. When using urea solutions, pretreat them with a mixed-bed ion-exchange resin or include a low concentration of an amine-containing compound (e.g. Tris buffer, lysine) to react with the cyanate as it is formed and avoid high temperatures.

Guanidinium salts
The chloride is readily available at high purity at a reasonable price. It is generally useful for all except some of the most recalcitrant proteins, in which case the thiocyanate is to be preferred. Even the purest grades of the thiocyanate contain significant insoluble material and solutions should be filtered before use. The thiocyanate reacts with Cleland's reagent, so 2-mercapto-ethanol is the preferred thiol reagent.

Trichloroacetic acid (TCA) and perchloric acid (PCA)
These are both extremely effective precipitants and will precipitate both nucleic acids and proteins. TCA is more widely used because of the reactivity of perchloric acid as an oxidizing agent. TCA is opaque in the ultraviolet, whereas PCA is transparent.

Acetone
The addition of 6 volumes of acetone will effectively precipitate most proteins, but it can be used as a selective precipitant for some classes of proteins at low temperatures.

Ethanol (EtOH) and methanol (MeOH)
Proteins are denatured and precipitated by the addition of two volumes of EtOH. At higher ratios the proteins may resolubilize. Methanol appears to be more effective. Either may be used in combination with about 10% acetic acid to ensure complete and irreversible precipitation.

9. THIOL REAGENTS

Thiol reagents have two uses in biochemistry: to prevent the uncontrolled oxidation of thiol groups in proteins or elsewhere (e.g. that of coenzyme A) to disulfides, and to maintain a reducing environment for some enzyme reactions. In all cases these reagents work by mass action. Several reagents have been used and some still are preferred for specific reactions. The most widely used are the dithiol butanols (e.g. dithioerythritol and dithiothreitol, DTT) and 2-mercaptoethanol (*Table 6*). Glutathione is frequently used because it is a physiological reducing agent and can be regenerated *in situ* with glutathione reductase. In general, the dithiol butanols are the reagents of choice, not only because their low volatility minimizes the potent and offensive thiol odor, but also because of their low oxidation–reduction potential (-0.33 V at pH 7). This arises because these are 'designer compounds' chosen because of their tendency to form a thermodynamically favored internal disulfide. This means that they can bring about essentially complete reduction of other disulfides at much lower concentrations than required with 2-mercaptoethanol (also very smelly!). There may, however, be contraindications on grounds of chemical reactivity or stereochemistry in individual cases (for example, DTT appears to be incompatible with guanidine thiocyanate).

10. DETERGENTS

The primary uses of detergents in biochemistry are:

(i)   solubilization of membranes and membrane lipids;
(ii) solubilization and stabilization of proteins, especially membrane proteins and others with significant hydrophobic surfaces;
(iii) as protein denaturants;
(iv) as wetting agents; and
(v)   as emulsifying agents.

Structurally, detergents all have a hydrophilic 'head' region and a long hydrophobic 'tail'. This causes them to form aggregates (micelles) with a hydrophobic core into which a hydrophobic

Table 6. Thiol reagents

Name	Formula	Mol. wt	Solubility
Cysteine [52-90-4]	$C_3H_7O_2NS$	121.16	V. sol.
Dithiothreitol [3843-12-3]	$C_4H_{10}O_2S_2$	154.25	Sol.[a]
Dithioerythritol [6892-68-8]	$C_4H_{10}O_2S_2$	154.25	Sol.[a]
2-Mercaptoethanol [60-24-2]	C_2H_6OS	78.1	Liquid freely miscible with water[b]
Glutathione [70-18-8]	$C_{10}H_{17}N_3O_6S$	307.3	Sol.
2,3-Dimercaptopropanol [59-52-9]	$C_3H_8OS_2$	124.2	8.7 g/100 ml H_2O[c]

[a]Typical working concentrations for these compounds are 1–5 mM and they can be prepared as stocks of > 100 mM.
[b]The typical working concentration is 0.1–2% by volume. Molarity of the pure liquid is 14.3 M.
[c]Much more soluble in vegetable oils. Commercially sold as a stock solution in 10% peanut oil.

molecule may partition from an aqueous medium, with the resulting stabilization of the hydrophilic molecule in the aqueous system. Many chemical classes of detergent are recognized, but generally they are discussed in terms of their ionic properties; there are four classes in this respect: nonionic, zwitterionic, cationic and anionic (*Table 7*). Generally speaking, nonionic and zwitterionic detergents are nondenaturing and are useful in the solubilization of proteins. Ionic detergents are often denaturing, especially SDS and Sarkosyl; CTAB is widely used as a germicide. Detergents may be quite effective in solubilizing a specific protein without loss of biological activity but without necessarily being effective in membrane solubilization, and vice versa.

Table 7. Detergents and surfactants

Name[a]	Mol. wt	HLB	CMC (μM)	Aggregation no. (20–25°C, 0–0.1 M Na^+)
Anionic detergents				
Sarkosyl[a] [137-16-6]	295.4			
SDS (SLS) [151-2-3]	288.4	≈ 40	8200	620
Cationic detergents				
CPB [140-72-7]	402.5			
CTAB [57-09-0]	364.5		1000	170

Table 7. Continued

Name[a]	Mol. wt	HLB	CMC (μM)	Aggregation no. (20–25°C, 0–0.1 M Na$^+$)
Zwitterionic detergents				
CHAPS [75621-03-3]	614.9		6–10×10^3	10
CHAPSO [82473-24-3]	630.9		8000	11
Zwittergent 3.08 [15178-76-4]	280		33×10^4	
Zwittergent 3.10 [15163-36-7]	308		25–40×10^3	41
Zwittergent 3.12 [14933-09-6]	336		2–4×10^3	55
Zwittergent 3.14 [14933-09-6]	364		100–400	83
Zwittergent 3.16 [2281-11-0]	392		10–60	155
Non-ionic detergents				
Brij 35 [3055-98-9]	≈ 1200	16.9	75	40
Brij 56 [9004-95-9]	682	12.9	2	40
Brij 58 [9004-95-9]	1122	15.7	77	40
Octyl-β-D-glucopyranoside [29836-26-8]	292.4		25×10^3	84
Triton X-100 [9002-93-1]	625	13.5	250	100–155
Triton X-114 [9036-19-5]	537	12.4		
Nonidet P40 [9036-19-5]	625	13.1	250	100–155
Tween 20 [9005-64-5]	1228	10.7	59	
Tween 60 [9005-67-8]		14.9		
Tween 80 [9005-65-6]	1310	15.0	12	58

[a]Names: Sarkosyl, *N*-laurylsarcosine, sodium salt; SDS, sodium dodecyl (lauryl) sulfate; CPB, cetylpyridinium bromide; CTAB, cetyltrimethylammonium bromide; CHAPS, 3-[(3-cholamidopropyl)-dimethylammonio]-1-propanesulfonate; CHAPSO, 3-[(3-cholamidopropyl)-dimethylammonio]-2-hydroxy-1-propanesulfonate; Zwittergent 3.08, *N*-octylsulfobetaine; Zwittergent 3.10, *N*-decylsulfobetaine; Zwittergent 3.12, *N*-dodecylsulfobetaine; Zwittergent 3.14, *N*-tetradecylsulfobetaine; Zwittergent 3.16, *N*-hexadecylsulfobetaine; Brij 35, polyoxyethylene (23) lauryl ether. Brij 56, polyoxyethylene (10) cetyl ether; Brij 58, polyoxyethylene (20) cetyl ether; Triton X-100, nonaethylene glycol octylphenol ether; Triton X-114, heptaethylene glycol octylphenyl ether; Nonidet P40, polyoxyethylene (9)-*p*-*t*-octylphenol; Tween 20, polyoxyethylene sorbitan monolaurate; Tween 60, polyoxyethylene sorbitan monostearate; Tween 80, polyoxyethylene sorbitan monooleate.

By definition, all detergents are soluble in water to some extent, and a stock solution of 5–10% detergent can usually be prepared. When the solution is saturated with detergent it separates into two phases. The phase behavior of detergent solutions is extremely complex. Solubility is salt-dependent and so care should be taken when preparing a concentrated stock as the presence of salts may lead to phase separation. In cases where this is likely to happen, unless the detergent is a waxy solid that dissolves only slowly in water (e.g. the Brij series), it is best to add the detergent directly to the diluted reagent immediately before use. Solubility is also temperature-dependent. Although detergents dissolve more readily upon heating, there is an upper limit called the cloud point. At this temperature the detergent reversibly separates from the aqueous phase to form a separate phase as a cloudy suspension. For most detergents it acts as an upper temperature limit for its use and usually lies at temperatures >60°C but Triton X-114 has a cloud point of 22°C. This property of Triton X-114 has been found to be useful in the purification of membrane proteins.

10.1. Terms used in the tables

Aggregation number
The average number of molecules per micelle.

CMC: critical micellar concentration
The minimal concentration at which the detergent forms micelles. This is an important guideline in solubilization because the formation of micelles is necessary for protein stabilization and lipid solubilization. A high CMC also indicates that the detergent is more readily dialysable. Compare octyl pyranoside, which is dialysable, with Triton X-100, which is not.

HLB: hydrophile–lipophile balance
A quantitative measure of the hydrophilicity and lipophilicity of a detergent, and therefore of its solubilization properties. There is a wide range of HLB values covering many applications. For solubilization of biological membranes and lipids an HLB value of ≈15 is most effective. An HLB value between 12 and 14.5 is recommended for solubilization of membranes, and detergents with an HLB value betwen 15 and 20 are useful for solubilization of extrinsic proteins. Detergents with HLB values outside this range are useful as wetting agents and permeabilizing agents.

11. REFERENCES

1. Perrin, D.D. and Dempsey, B. (1974) *Buffers for pH and Metal Ion Control.* Chapman & Hall, London.

2. Stoll, V.S. and Blanchard, J.S. (1990) *Methods Enzymol.,* **182,** 243.

3. Blanchard, J.S. (1984) *Methods Enzymol.,* **87,** 405.

4. Good, N.E. and Izawa, S. (1972) *Methods Enzymol.,* **24,** 53.

5. Scopes, R.K. (1982) *Protein Purification: Principles and Practice,* chapter 6. Springer-Verlag, New York.

6. Dawson, R.M.C., Elliot, D.C., Elliot, W.M. and Jones, K.M. (1986) *Data for Biochemical Research (3rd edn).* Oxford University Press, Oxford, p. 417.

7. Ellis, K.J. and Morrison, J.F. (1982) *Methods Enzymol.,* **87,** 405.

8. *Enzyme Nomenclature: Recomendation (1992) of the Nomenclature Committee of the International Union of Biochemistry.* Academic Press, New York.

INDEX

acyl–enzyme intermediate   61
assay   78
classification   6
inhibition   122, 163
rate equations   61
temperature dependence   195–196
Cibacron dyes   11, 13, 16, 17
circular dichroism   34
Cleland notation, for enzyme mechanisms   87
CMP   231–232
cobalamin cofactors   228
cobalt ions
chelation stability constants   278
EPR   251
substitution for Zn^{2+} 252
coenzyme A   229–230
metabolic role   224
coenzyme Q   230–231
cofactors   223–247
compulsory-order mechanisms
initial-rate parameters   88
product inhibition patterns   129–130
continuous assays   80
Coomassie Blue
gel staining   47, 50
protein estimation   35, 39–40
cooperativity
negative   107, 210
positive   107, 206
copper ions
chelation stability constants   278
EPR   251, 260
Cornish-Bowden plots, for inhibition   127, 128
corrin ring   228
coupled assays   80
critical micellar concentration   284
CTP   2, 231–232
curvature, in rate assays   82–83
cytochromes   238–239
cytosine mononucleotide cofactors   231–232

Dalziel initial-rate parameters   87
Dalziel maximum rate relationships   87, 90, 92
deamidation   31
denaturants   278–281
precipitating   280
solubilizing   279
detergents   281–284
anionic   282
cationic   282
nonionic   283
zwitterionic   283
diffusion coefficient   27
dihydrofolate reductase, dye chromatography   17
dihydrofolic acid   234–235
dye mimicry   17
direct linear plot   98, 102, 116, 121
dissociation constants   94, 200
determination   201–206
disulfide bridges   30, 31
dithiothreitol   10, 281, 282

Dixon plot
inhibition   126, 127
pH dependence   110, 112
DNA
–protein interactions   216–217
sequence, inference of amino acid sequence and composition   30
DTNB, thiol reagent, pH dependence   63
DTT   10, 281, 282
dye–ligand chromatography   9–18
column regeneration   10, 11
column storage   11
dye immobilization   12–16
dye leakage   14
metal ions   9, 10, 12
pH effects   9, 10
temperature effects   10

E-64, proteinase titrant   65
Eadie–Hofstee plot   98
EDTA
in dye–ligand chromatography   10, 11
properties   276, 278
EGTA, chelating properties   277–278
electrode assay methods   78
electron paramagnetic resonance
pulsed   257–259
spectroscopy   249–268
electron spin relaxation rates   259
electrophoresis   41–59
capillary   23–25
for assessing purity   19, 22
marker proteins   46, 48–50
nondenaturing   19, 22, 41, 42
problems in   23, 55–59
SDS–PAGE   19, 22, 41, 43
with denaturant   19, 22, 41, 43
ELISA
in active site labeling   65
to detect dye leakage   14
ENDOR   256, 257, 262, 263
endosmotic flow, in capillary electrophoresis   24
enolase
activity staining   52
gel marker   49
purification   18
enthalpy of reaction   191, 194
entropy of reaction   191, 194
enzyme activity
constant   19
in estimating purity   21
measurement   7, 77–83
units   7, 34
enzyme
classification   1–6
cofactors   223–247
heterogeneity   30
instability   81–83, 111
kinetics   84–96
nomenclature   1–6
purification   9–17, 25–26, 41–48
purity, assessment   18–24, 41–55

commercial, purity 21
dye–ligand purification 17
gel marker 49
in ATP-coupled assay 68
isotope exchange 91
polarimetric assay 79
tight-binding inhibition 154
turnover 167
ligand binding 199–221
ligases 4
Lineweaver–Burk plot 84, 98
lipoic acid 239–240
Lowry protein assay 35, 38
low-spin Fe^{III} 250, 262
lyases 4

magnesium ions, requirement for kinases and
ligases 223
manganese ions
chelation stability constant 278
in EPR 250, 260
substitution for Mg^{2+} 252, 262
manometry 78
mass spectrometry 19, 28–30
electrospray ionization 28
plasma desorption 28
time-of-flight 28
maximum-rate relationships 87, 90, 92
mechanism-based inhibitors 64, 164–165
2-mercaptoethanol 10, 28, 282
metal ion dependence 149–150, 223
methanol, as precipitant 280
methotrexate 235
4-methylumbelliferone substrates 62
Michaelis–Menten equation 85, 97, 115–116, 191
integrated form 97, 99, 100
microcalorimetry 79
molybdenum ions in EPR 252
Monod–Wyman–Changeux model 206
monoamine oxidase inhibitors 162, 164–165, 167,
169
myosin
active enzyme centrifugation 68
gel marker 50
motion detected by spin labeling 265

NAD^+ 240–243
dye mimicry 17
$NADP^+$ 240–243
nicotinamide cofactors 240–243
nickel ions
chelation stability constants 278
in EPR 251
nitric oxide, EPR detection 249, 253, 265
nitrocellulose membranes
for blotting 47, 53
in filter binding assays 216

ping-pong mechanism 88
for enzymes with prosthetic groups 224
pK_a 177–180, 186–190
plotting methods 96–112

polarimetric assay 79
polyvinyldifluoride (PVDF), membranes for
blotting 54
porphin ring 238
porphyrins 238
post-translational modification, detection 19, 30
Procion dyes 10, 11, 12, 16
product inhibition 87, 92, 93, 134–136
patterns for two-substrate reactions 129
prosthetic groups 224
protein concentration determination 34–41
A_{205} 37
A_{280} 36
amino acid analysis 37–38
BCA 39
biuret 36
Bradford method 39
buffer interference 174
gravimetric 36
Lowry (Folin) 38
OPA (o-phthalaldehyde) 40
protein fluorescence, in binding studies 217–220
proteolysis, detection 19, 30, 31
protoporphyrin IX 238–239
pyridoxal phosphate 243–244
-dependent enzymes 165
prosthetic group 224
radical 254
2-pyridyl disulfides, active-site titrants for thiol
proteinases 63
pyrroloquinolinequinone (PQQ) 244
pyruvate kinase
activity staining 53
classification 6
coupled assay with LDH 80
gel marker 50
in ATPase detection 68
naming 1
turnover 167

quinoproteins 244

radiochemical assay 79, 80
rapid equilibrium random order mechanism 89
product inhibition patterns 132
rapid reaction techniques 94, 176, 200
in EPR 259–260
rate constraints
direct measurement 94
temperature effects 192
resonance energy transfer 217–218
restriction endonucleases 3, 199
riboflavin 233

Scatchard plot 106, 201, 213
for cooperative proteins 209
SDS–PAGE — see Electrophoresis
secondary plots 84–85, 108
sedimentation
coefficients 66
equilibrium 27–28
velocity 27

serine hydroxymethyltransferase, dye chromatography   17
silver stain   22, 47
sodium dodecyl sulfate (SDS)
    as a denaturant   279, 280
    in dye–ligand chromatography
    in gel electrophoresis   41–48
sodium thiocyanate, for dye column regeneration   11
specific activity   21, 34, 77
spin labels   264
spin probes   264
spin trapping   252, 253, 264
staining procedures
    activity   50
    Alcian blue   50
    blots   54
    Coomassie blue   47, 50, 51, 54
    copper   50
    Fast Green FCF   50
    immunoprecipitation   50
    nigrosin   51
    silver   22, 47, 50
    Sudan Black   50
standard assay   77
steady-state rate equations   84, 86–92
stopped assays   80
stopped flow   90, 94, 220–221 — see also Rapid reaction
substrate binding measurements   90–91, 93, 94, 199–221
substrate competition   123–125
substrate contamination by inhibitor   148, 151
substrate inhibition   144, 146–147
succinate dehydrogenase
    classification   6
    competitive inhibition by malonate   122
    metal center, EPR properties   251
suicide substrate   64, 166–169
superoxide dismutase
    activity staining   53
    classification   6
    naming   1
synthases and synthetases   5

temperature effects   190–196
ternary complex mechanism   87
tetrahydrofolic acid   234 235
Theorell–Chance mechanism   88
    Haldane relationship   91, 93, 94
    product inhibition patterns   129–130
thiamine pyrophosphate   244–245
    radical   254
thiol reagents   28
three-substrate enzymes   86–87, 94–96

tight-binding inhibitors   149–150, 151–152
transferases   4
transition metal ions, EPR detection   249, 259, 260–261
transition state   192–193
triazine dyes   11–12, 17
trichloroacetic acid
    as precipitant   279, 281
    to remove cofactors   223
triplet state, of O_2 253
Triton X-100
    as detergent   283
    to regenerate or elute dye columns   11
trypsin
    active-site titration   61, 62
    assay   78
    BAPNA as artificial substrate   80
    calcium dependence   223
    classification   3, 5, 6
    inhibition   17
Tsou plot   65, 110
tungsten ions, in EPR   252
turnover number   34
Tween detergents   283
two-substrate enzymes   84, 87, 91, 103
    product inhibition patterns   129
tyrosinase   12

UDP   245–246
ultracentrifugation   19, 27–28, 66
ultrafiltration, in binding studies   214
UMP   245–246
unpaired electron   253
uracil mononucleotide cofactors   245–246
urea
    as denaturant   279, 280
    dye column elution, regeneration, washing   10, 11, 16
UTP   2, 245–246

vanadium ions, in EPR   250
van't Hoff equation   191
vitamin B1   245
vitamin B6   243
vitamin B12   228
V_{max} 85, 97, 98, 99, 102

Warburg and Christian, method for estimating nucleic acid contamination of proteins   18, 20

xanthine oxidase
    classification   6
    EPR   252, 264
X-gal, assay for β-galactosidase   80

Index